Bandeira, Tylkowski (Eds.)
Chemical Synergies

Also of interest

Mathematical Chemistry and Chemoinformatics.
Structure Generation, Elucidation and Quantitative Structure-Property Relationships
Kerber, Laue, Meringer, Rücker, Schymanski, 2014
ISBN 978-3-11-030007-9, e-ISBN 978-3-11-025407-5

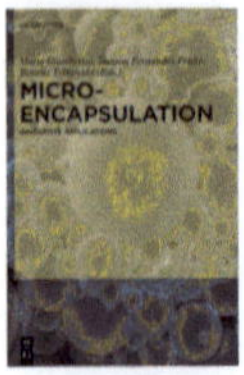

Microencapsulation.
Innovative Applications
Giamberini, Fernandez Prieto, Tylkowski (Eds.), 2015
ISBN 978-3-11-033187-5, e-ISBN 978-3-11-033199-8

Computational Sciences.
Ramasami (Ed.), 2017
ISBN 978-3-11-046536-5, e-ISBN 978-3-11-046721-5

Polymer Engineering.
Tylkowski, Wieszczycka, Jastrzab (Eds.), 2017
ISBN 978-3-11-046828-1, e-ISBN 978-3-11-046974-5

Physical Sciences Reviews.
e-ISSN 2365-659X

Chemical Synergies

From the Lab to In Silico Modelling

Edited by
Nuno A.G. Bandeira and Bartosz Tylkowski

DE GRUYTER

Editors
Dr. Nuno A.G. Bandeira
Centre for Chemistry and Biochemistry
Faculty of Science
University of Lisbon
Campo Grande Ed. C8
Lisbon, 1749–016
Portugal

Centro de Química Estrutural
– Instituto Superior Técnico
Universidade de Lisboa
Av. Rovisco Pais
1049–001 Lisbon
Portugal

Institute of Chemical Research of Catalonia (ICIQ)
Avda. Països Catalans, 16
43007 Tarragona
Spain
nbandeira@iciq.es

Dr. Bartosz Tylkowski
Centre Tecnològic de la Química de Catalunya,
Carrer Marcelli Domingo s/n
Tarragona, 43007
Spain

Department of Chemical Engineering
Rovira i Virgili University
Av. Paisos Catalans 26
Tarragona, 43007
Spain
bartosz.tylkowski@urv.cat

ISBN 978-3-11-048135-8
e-ISBN (PDF) 978-3-11-048206-5
e-ISBN (EPUB) 978-3-11-048136-5

Library of Congress Control Number: 2018936216

Bibliographic information published by the Deutsche Nationalbibliothek
The Deutsche Nationalbibliothek lists this publication in the Deutsche Nationalbibliografie; detailed bibliographic data are available on the Internet at http://dnb.dnb.de.

Typesetting: Integra Software Services Pvt. Ltd.
Printing and binding: CPI books GmbH, Leck
Cover image: JK1991/iStock/Getty Images Plus

www.degruyter.com

Preface

The idea for this book project came up during one of our many evening coffee meetings in downtown Tarragona. The goal was to produce an overview of the current state of the art in computational molecular science with a broad interdisciplinary appeal. While experts may still gain valuable insight from this book, it is mostly addressed to those working outside of the field, the focus being kept on the collaborative synergies between the lab bench and the computer screen.

Computational Chemistry has come a long way since the days of Heisenberg and Schrödinger. The Nobel prize given to the late Walter Kohn and John A. Pople in 1998 and most recently in 2013 to Martin Karplus, Michael Levitt and Arieh Warshel are major hallmarks of a mature branch of chemistry.

Admittedly even before the advent of wave mechanics which revolutionized physics, chemists like Richard Abegg and Gilbert Newton Lewis were already grappling with concepts like valency and electron pair bonding. This is what should provide the *raison d'être* to computational practitioners rather than just commit to pointless exercises in number crunching. The extraction of trends and formulation of theoretical concepts with a firm anchor on reality has oftenmost been lost in the current literature.

A nice overview of this historical background of chemistry has been given in Chapter 1 by Mauro Fiancchini and this sets the tone of the book in the guise of several case studies in different fields where theoretical tools help explain a given phenomenon at the molecular level.

The understanding of dynamic phenomena such as aggregation and nucleation is of prime importance to explain the formation of biominerals at the atomistic level. The role that bio-polymers (i.e. DNA) play in acting as templating agents for the formation of nuclei that will result in the formation of hydroxyapatite is presented in the Chapter 2 by Guillem Revilla-Lopez and his colleagues where the deployment of classical force field simulations allows the gathering of statistically representative information of these phenomena.

The nascent field of halogen bonding has recently received attention from the community. An introduction to the σ-hole concept is reviewed along with the latest findings by Paulo J. Costa in Chapter 3. This field of enquiry interfaces with the long-established areas such as non-covalent interactions, molecular recognition and crystal engineering and will surely provide essential know-how to the current state of the art in the field.

Bulk metal oxides have traditionally been the purview of solid state physicists due to their electronic properties, but their molecular equivalents otherwise known as polyoxometalates have been known by chemists for nearly two centuries since the days of Berzelius from his account of molybdate and phosphate mixtures. The interplay between their use as electron sponges *vis-à-vis* their chemical constitution, degree of protonation and structural features is discussed in a collection of case

https://doi.org/10.1515/9783110482065-201

studies in Chapter 4 with a didactic approach assuming no prior knowledge of the terminology used in the field.

Azobenzene is by far the most studied photochromic molecule and its applications range from optical storage to bio-engineering. To exploit the great potential of azobenzene (i.e. its application in molecular machines or in photoactivation of biomolecules), one must achieve deep understanding of its photochemistry as single molecule in solution as well as in-chain moiety and pendent group in macromolecular structures. Chapter 5 provides an overview on the modelling of azobenzene-containing molecules in different conditions.

High-resolution infrared spectroscopy is essential to the analysis of molecular rotation-vibration spectra. The high-resolution spectra deliver much information about structure and dynamic of molecules, but often they are very complex. Iwona Gulaczyk in Chapter 6 performs a deep investigation on methylamine as a classic example of a nonrigid molecule in which two large amplitude motions, inversion and torsion, occur simultaneously.

The determination of equilibrium constants is an important process for many branches of chemistry. Chapter 7 provides the readers with a discussion on computer methods which have been applied for elaboration of potentiometric experimental data generated during complexes formation in solution. The chapter describes both: general basis of modelling tools and examples of the use of calculated stability constants.

State of knowledge about a metal transport in two-phase system is discussed 8. Karolina Wieszczycka not only provides the fundamental knowledge on the distribution law and main factors determination in classical solvent extraction, but she also discusses modelling studies that enable to understand the extraction process. The topics detailed in Chapters 9-11 cover the traditional concepts of reactive distillation or membrane and membrane bioreactor processes, to emphasize the modelling studies and the prospective needed for quantum calculations.

We would like to express we gratitude to the contributing authors in making this project a success, as well as to Lena Stoll and Esther Markus from De Gruyter for their assistance and encouragement in this venture.

Nuno Bandeira and Bartosz Tylkowski

Contents

List of contributing authors

Carlos Alemán
Universitat Politècnica de Catalunya
School of Industrial Engineering
Av. Diagonal 647
Campus Sud
08028 Barcelona
Spain

Veronica Ambrogi
University of Naples
Department of Chemical, Materials and Production Engineering
Piazzale Tecchio 80
80125 Naples
Italy
ambrogi@unina.it

Oscar Bertran
Universitat Politècnica de Catalunya
Department of Physics
Avinguda Pla de la Massa 8
08700 Igualada
Spain

Pierfrancesco Cerruti
Institute for Polymers, Composites and Biomaterials (IPCB-CNR)
Via Campi Flegrei 34
80078 Pozzuoli
Italy
cerruti@unina.it

Paulo J. Costa
Faculty of Sciences of the University of Lisbon
Campo Grande
1749-016 Lisboa
Portugal
pjcosta@fc.ul.pt

Mauro Fianchini
Institut Català d'Investigació Quimica
Av. Paisos Catalan 16
43007 Tarragona
Spain
mfianchini@iciq.es

Marta Giamberini
Universitat Rovira i Virgili
Department of Chemical Engineering
Av. Països Catalans 26
43007 Tarragona
Spain
marta.giamberini@urv.cat

Iwona Gulaczyk
Adam Mickiewicz University
Faculty of Chemistry
Umultowska 89b
61-614 Poznań
Poland
iwona.gulaczyk@amu.edu.pl

Renata Jastrzab
Adam Mickiewicz University
Faculty of Chemistry
Umultowska 89b
61-614 Poznań
Poland
renatad@aqmu.edu.pl

Małgorzata T. Kaczmarek
Adam Mickiewicz University
Faculty of Chemistry
Umultowska 89b
61-614 Poznań
Poland

Xavier López
Universitat Rovira i Virgili
Department of Physical and Inorganic Chemistry Marcel·lí Domingo 1, Campus Sescelades
43007 Tarragona
javier.lopez@urv.cat

Valentina Marturano
Institute for Polymers, Composites and Biomaterials (IPCB-CNR)
Via Campi Flegrei 34
80078 Pozzuoli
Italy
valentina.marturano@unina.it

Akira Odani
Kanazawa University
Dept. of Clinical & Analytical Sciences
Kakuma-machi
Kanazawa
920-1192
Japan
odani@p.kanazawa-u.ac.jp

Jordi Puiggalí
Universitat Politècnica de Catalunya
School of Industrial Engineering
Av. Diagonal 647
Campus Sud
08028 Barcelona
Spain

Guillem Revilla-López
Universität Regensburg
Institut für Organische Chemie
Universitätsstr. 31
93053 Regensburg
Germany
guillermo.revilla-lopez@chemie.uni-regens-burg.de

Katarzyna Staszak
Poznán University of Technology
Institute of Chemical Technology & Engineering
Berdychowo 4
60-965 Poznań
Poland
katarzyna.staszak@put.poznan.pl

Maciej Staszak
Poznán University of Technology
Institute of Chemical Technology & Engineering
Berdychowo 4
60-965 Poznań
Poland
maciej.staszak@put.poznan.pl

Juan Torras
Universitat Politècnica de Catalunya
Department of Chemical Engineering
C/ Eduard Maristany 10-14
08019 Barcelona
Spain

Irene Tsibranska
Bulgarian Academy of Sciences
Institute of Chemical Engineering
akad.G.Bonchev str. bl.103
1113 Sofia
Bulgaria
tsibranska@yahoo.com

Pau Turón
B. Braun Surgical
Department of Research and Development, Regulatory Affairs and Quality Management
S.A. Ctra. de Terrassa 121
08191 Rubí, Barcelona
Spain

Serafim Vlaev
Bulgarian Academy of Sciences
Institute of Chemical Engineering
akad.G.Bonchev str. bl.103
1113 Sofia
Bulgaria

Karolina Wieszczycka
Poznán University of Technology
Institute of Chemical Technology & Engineering
Berdychowo 4
60-965 Poznań
Poland
karolina.wieszczycka@put.poznan.pl

David Zanuy
Universitat Politècnica de Catalunya
Department of Chemical Engineering
C/ Eduard Maristany 10-14
08019 Barcelona
Spain

Mauro Fianchini

1 Synthesis meets theory: Past, present and future of rational chemistry

Abstract: Chemical synthesis has its roots in the empirical approach of alchemy. Nonetheless, the birth of the scientific method, the technical and technological advances (exploiting revolutionary discoveries in physics) and the improved management and sharing of growing databases greatly contributed to the evolution of chemistry from an esoteric ground into a mature scientific discipline during these last 400 years. Furthermore, thanks to the evolution of computational resources, platforms and media in the last 40 years, theoretical chemistry has added to the puzzle the final missing tile in the process of "rationalizing" chemistry. The use of mathematical models of chemical properties, behaviors and reactivities is nowadays ubiquitous in literature. Theoretical chemistry has been successful in the difficult task of complementing and explaining synthetic results and providing rigorous insights when these are otherwise unattainable by experiment. The first part of this review walks the reader through a concise historical overview on the evolution of the "model" in chemistry. Salient milestones have been highlighted and briefly discussed. The second part focuses more on the general description of recent state-of-the-art computational techniques currently used worldwide by chemists to produce synergistic models between theory and experiment. Each section is complemented by key-examples taken from the literature that illustrate the application of the technique discussed therein.

Keywords: theory, model, rational synthesis, first principles, computational chemistry

1.1 Introduction

Since the days of Galileo Galilei, science has been approached using the scientific method. The routine of "curiosity sparking observation, formulation and testing of new hypotheses and birth of new theories" represents the first rudimentary iterative block-diagram algorithm to approach "scientifically" any natural phenomenon. Both the predictive and explanatory power of science, and so its evolution, reside in the

---

This article has previously been published in the journal *Physical Sciences Reviews*. Please cite as: Fianchini, M. Synthesis meets theory: Past, present and future of rational chemistry. *Physical Sciences Reviews* [Online] **2017**, *2* (12). DOI: 10.1515/psr-2017-0134

https://doi.org/10.1515/9783110482065-001

use of models. Models are rescaled and simplified visions of reality that scientists create to afford an explanation and prediction of phenomena within certain levels of confidence. Leonhard Euler himself commented on the power of the model with his famous maxim: "Although to penetrate into the intimate mysteries of nature and thence to learn the true causes of phenomena is not allowed to us, nevertheless it can happen that a certain fictive hypothesis may suffice for explaining many phenomena". Euler's "fictive hypothesis" is an approximated, yet powerful tool to account for and, ultimately, influence or even control phenomena to the benefit of humanity. Nowadays, mathematical models are routinely employed to predict events in physics, engineering and natural sciences [1, 2].

What could be inferred about chemistry, then? Any dedicated scholar sooner or later stumbled upon the uncomfortable question: *Can complex chemical reactions be predicted by rational models*? Chemical sciences are intrinsically more bound to their empirical nature than other scientific disciplines, given the fact that the ultimate goal of chemistry is to make new molecules for market, technology, society or simply for scientific interest [3]. The construction of *a priori* models of complete reactions is currently regarded over-ambitious, given the extreme difficulty to characterize accurately Avogadro's numbers of solvated molecules, predict their reactive events in such environment and simulate the course of *n*-parallel and competitive channels during a desired interval of time. While such search for rationality poses unparalleled challenges, it would also represent, if successful, the pinnacle of chemical manipulation and establish an evolved concept of understanding and doing chemistry. This evolution will lead to a new level of rational design of target compounds and transform chemistry into a discipline scientists can fully control, rather than merely observe and improve through trial-and-error strategy. The structure of the chapter has been organized in a way that the reader will receive a general, yet detailed, wide-angle history of chemistry, followed by a more specific treatment of theoretical disciplines. Each section will be integrated with interesting examples singled out from a massive multidisciplinary pool of excellent scientific works. The selection criteria were not only based on relevance, impact factor and citations of the works but also on the diversity of the field and author's interests and familiarity with the works. The common trait d'union of these works is represented, however, by the synergistic interdisciplinary approach between theory, spectroscopy and experiment. The use of theoretical assets in solving chemical problems, whether *a priori* or *a posteriori*, serves as a proof of concept of rational chemistry. The author wishes to convey the idea that chemistry is a multifaceted scientific ground where experimental and theoretical approaches are far from being rivals; on the contrary, they imparted together an evolutionary rational momentum to the discipline.

This momentum allowed chemistry to blossom into a modern science during the last 400 years, from nothing more than a collection of esoteric notions. We could say that chemistry is a relatively young discipline since a scientific, systematic and non-serendipitous approach to this field has been introduced not earlier than the sixteenth century. Technological advances in synthetic methodologies, spectroscopy and data sharing allowed chemistry to reach unprecedented goals and successes. The development of computers carved an important role for theory into the full landscape of chemical disciplines. This chapter reviews the synergistic integration between practical and theoretical fields for the betterment of rational principle underlying chemistry. In line with the guiding philosophy of this project, there is not a preferred target audience for this chapter. The author simply hopes that it will be descriptive enough to interest theoreticians of any level and useful enough to show spectroscopists and experimentalists "what they can actually do" with theoretical methods, should they decide to use them as a viable first term of comparison in their researches.

1.2 The past: Birth of a discipline

In ancient Egypt the idea of a rational principle behind natural transformations was associated to the existence of Thoth, the god of wisdom, magic and writing. Back then, and for the rest of the ancient history, the use of simple chemical transformations was mostly relegated to common life activities, like metallurgy [4]. At the turn of the first millennium, the first sparkle of "scientific" approach made its entrance in history, with the work of the Persian mathematician and astronomer Ibn al-Haytham on light and optic [5]. Some decades later on, a new discipline grew strong in continental Europe, drenched in all the esoteric mysticism and symbolism that Medieval Age could spawn: alchemy [4]. The word from Greek-influenced Arabic language, *al-kīmiyā*, means the art of transmuting metals. Alchemic cult of the four natural elements (water, air, fire and earth) and the two philosophical elements (sulfur and mercury) rooted back to Hellenistic [4] and Arabic beliefs, like those of the Persian alchemist Jābir ibn Hayyān [6]. The alchemic period (Figure 1.1) lasted more or less until the Seventeen Century and was characterized by bizarre individuals like Paracelsus or N. Flamel, whose mysterious vaunted powers failed to achieve the envisioned goals, given the fact that neither the philosopher's stone nor the elixir of eternal life has ever become solid realities. Alchemy incurred often in the wrath of the religious authorities, since it was regarded as a practice inspired by the Devil itself; ironically enough, it is exactly the art of combining substances that provided the Church with the most beautiful heritage for posterity, like the exceptional stained glasses we can still admire in Chartres cathedral.

Modern chemistry, as we intend it, was born in the late sixteenth century. The recognized fathers of modern chemistry are R. Boyle and A.L. Lavoisier [4]. They introduced a rational and scientific approach to chemistry that led them to an unprecedented understanding in the chemistry of gases, air and combustions and to the definitive destruction of the J.J. Becher's phlogiston theory [4]. Lavoisier has been also credited for the first attempt in dissemination of chemistry by writing the

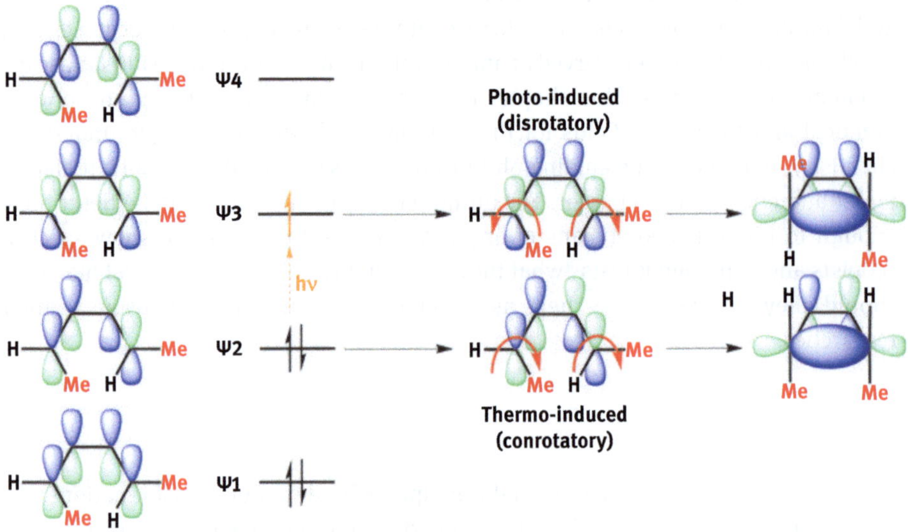

Figure 1.1: Electrocyclic reactions can be predicted through simple Huckel diagrams. (E,Z)-2,4-hexadiene can cyclize giving either cis- or trans-dimethylcyclobutene (on the right) depending on orbital control (and vice-versa for the ring-opening reaction).

first chemical book, *Traité Elémentaire de Chimie* (1789) [4]. As the years went by, J.L. Proust, J. Dalton and A. Avogadro postulated three famous laws that still carry their names and are taught to our freshmen within the first week of a general chemistry course: the law of constant composition, the law of partial pressures and Avogadro's principle, respectively [4]. The interaction of electrical current with chemicals and solutions was pioneered by H. Davy, J.J. Berzelius and M. Faraday [4], who started the field of modern electrochemistry. The first statistical ordering of the properties of the elements by their atomic weight and valence was attempted independently by J.L. Meyer and D.I. Mendeleev through the construction of the first periodic tables. The power of such embryonic models was suddenly clear. A few elements and related properties could be predicted by Mendeleev in 1869–1871 (Table 1.1): for instance, Mendeleev's eka-boron, eka-aluminium and eka-silicon were discovered in 1879, 1875 and 1886, respectively, and renamed scandium (1879), gallium (1875) and germanium (1886) [7]. It is finally with F. Wöhler and H. Kolbe that the word "chemical synthesis" starts to assume the meaning we ascribe to it today. Wöhler and Kolbe contributed to destroy vitalism demonstrating that organic molecules can be made from "anorganisch" compounds [8, 9]. The nineteenth century was a golden century for chemistry. F.A. Kekule's studies on the structure of hydrocarbons and benzene opened the way for a systematic pursuit of organic chemistry. H.L. Le Chatelier, J.H. Van'T Hoff, W. Ostwald, J.W. Gibbs, S.A. Arrhenius and W. Nernst set the bases for the modern interpretation of thermodynamics, kinetics and

Table 1.1: Mendeleev's early periodic table shows the power of prediction of a simple model.

Chemical properties	Mendeleev's prediction in 1871: Eka-silicon	Element properties: Germanium
Atomic weight	72	72.61(2)
Density/g•cm^{-3}	5.5	5.323
Molar volume/cm^{3}•mol^{-1}	13.1	13.64
MP/°C	high	945
Specific heat/J•g^{-1}•K^{-1}	0.305	0.309
Valence	4	4
Color	Dark grey	Greyish-white

electrochemical processes. L. Boltzmann recognized the strong connection between probability and thermodynamics and founded statistical thermodynamics.

H.E.L. Fischer initiated the interest in biological chemistry with his studies on sugars, their reactivity and structural properties (i. e. stereochemistry); he also proposed the "lock and key" model to explain enzyme-substrate interactions. J.C. Maxwell formulated the classical theory of electromagnetism that holds together electricity, magnetism and light [5] and inspired the successive two greatest theories, relativity and quantum mechanics. At the end of nineteenth century and beginning of twentieth century new discoveries shed light on the structure of the atom. The fundamental particles that compose atoms were discovered by J.J. Thomson (electron in 1897), E. Rutherford (nucleus in 1911 and proton in 1919) [4], J. Chadwick (neutron, in 1932). In 1913, H.G.J. Moseley proved by x-ray spectra that Z, the number of protons in an element, defines its position in the periodic table [4]. Another meaningful example of simple models laid out to rationalize more complex phenomena is provided by Blomstrand-Jørgensen-Werner diatribe which arose in the late nineteenth century hoping to explain the peculiarity of the bond in metal-containing compounds. Some of these compounds, like Prussian blue, $KFe[Fe(CN)_6]$, and aureolin, $K_3[Co(NO_2)_6]0.6H_2O$, have been known since ancient times and used as pigments for their intense colors. The developing analytic techniques provided molecular formulas for these compounds; hence the need for a clear explanation about the nature of the bonding in these compounds, since they formally exceeded their valence or the concept of allowed valence developed until that point. Two models responded the call: Werner's and Blomstrand-Jørgensen's. A. Werner can be considered the father of modern coordination chemistry. His model consisted of metal centers able to expand their valence to bind extra "ligands". In the case of the compounds that cobalt(III) trichloride forms with ammonia he hypothesized that the cobalt was in an octahedral environment surrounded by ammonia molecules (Table 1.2). According to different stoichiometries, the chloride ions could have been tightly bound (those primarily connected to the cobalt) or loosely bound (highlighted in red in Table 1.2). C.W. Blomstrand and later S.M. Jørgensen tried to explain the bonding

Table 1.2: The first hypotheses on the structure of cobalt coordination compounds: Werner versus Blomstrand.

Werner formula	# Ions predicted	Blomstrand chain formula	# Ions predicted
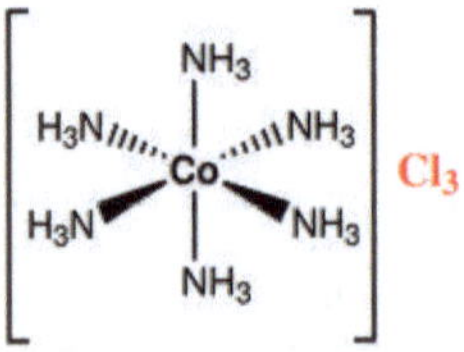	4	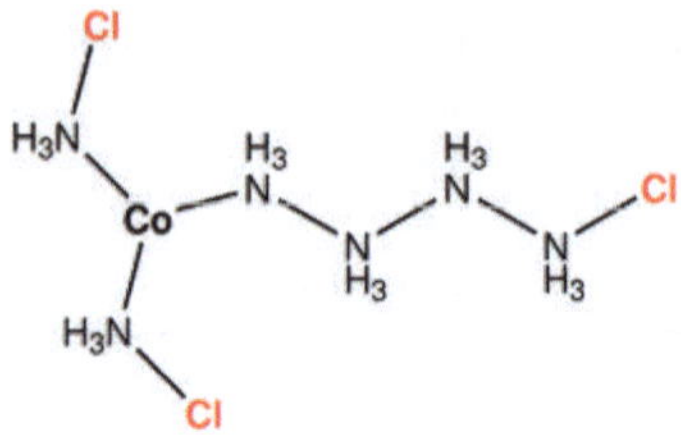	4
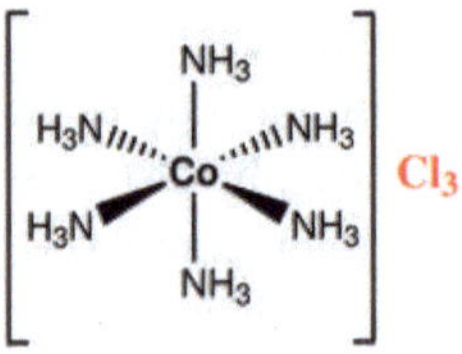	3	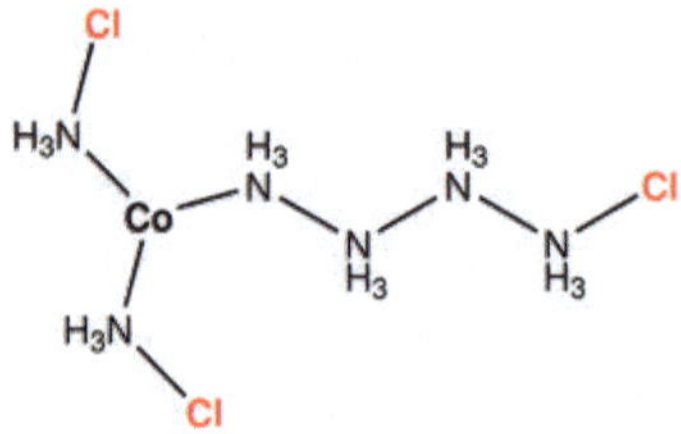	3
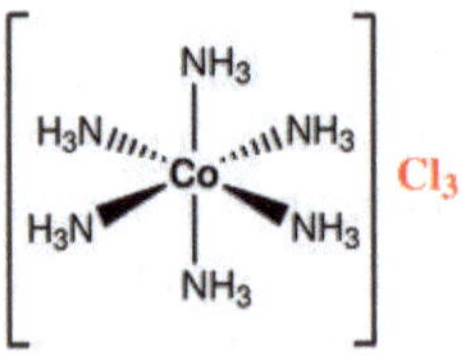 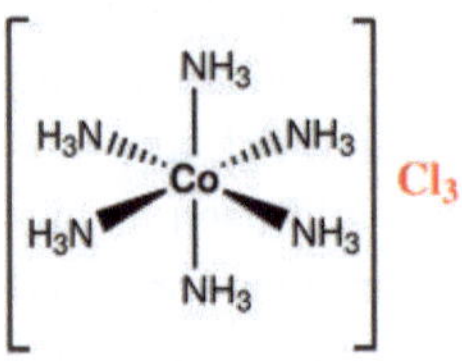	2	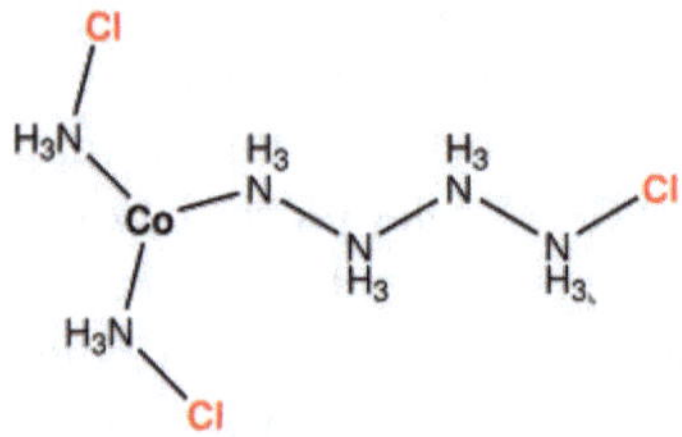	2

mode of such cobalt compounds using a conventional model, the chain theory, more akin to the developing tetravalent carbon chemistry. They supposed that the loosely bound chlorides were those connected to NH_3 moieties. Interestingly both models afforded the same answer in the rationalization of the ionic nature of some chloride ligands present in these complexes, as seen in Table 1.2. No evidence could put the final word to Werner-Jørgensen's long-lasting rivalry until Blomstrand-Jørgensen's model was finally disproved in 1907 when Werner synthesized two isomeric forms of $[Co(NH_3)_4Cl_2]^+$, one called (later on) *cis*, showing a violet color and one called *trans*, showing an intense green color. The chain theory was unable to explain this simple differentiation within the same molecular formula.

The twentieth century will be remembered as the century that saw the birth of quantum mechanics and relativity, two theories that challenged classical mechanics. N. Bohr [10], A. Sommerfeld, M. Planck and L. De Broglie fathered this early quantum theory. It rose to explain theoretically the position of line spectra in the Balmer series of hydrogen [4]. The first fundamental notion embedded in this new theory was that the energy state of any system is discreetly expressed in integer multiples of a constant, h (Planck's constant). This managed to satisfactorily explain hitherto unexplained phenomena like black body radiation, the photoelectric effect and the Compton effect [11]. The second assumption was that particles have both matter and wave-like properties, having associated wave-lengths inversely proportional to their masses ($\lambda \infty m^{-1}$, de Broglie) [4]. The new quantum theory acquired the foundation of the early theory and expanded it with the new notion of wave-mechanics. E. Schrodinger formulated the famous eigenvalue equation that still bears his name and is at the foundation of modern quantum theory. Among several contributions, M. Born proposed the interpretation of the square of the wavefunction as the probability amplitude to find an electron in the r space and P. Erhenfest proposed his theorem on the behavior of expectations values for momentum and position operators. These theoretical insights brought about the famous three atomic quantum numbers (s, l, m_l). A fourth quantum number, m_s, was proposed by W. Pauli in 1924, after his famous formulation on the exclusion principle of two electron in the same quantum state [4]. The existence of the electronic spin was discovered by the Stern-Gerlach experiment in 1921 [4]. In 1927 W. Heisenberg formulated his uncertainty principle ($\lambda_p \bullet \lambda_x \geq \hbar/2$) on the impossibility to determine with absolute precision both momentum and position of a particle. These principles proposed by quantum mechanics struck a hard blow at the heart of the classical theories, whose principal aspect was the determinism of all the properties in a system. This was difficult to accept for many physicists. A. Einstein, for instance, regarded quantum mechanics as an "incomplete" theory in his famous Einstein-Podolsky-Rosen principle [12]. Supporting proofs of the consistency of the new theory, however, came with the discovery of particle tunneling through potential energy barriers and particle diffraction, typical aspects of wave-like behavior. The first phenomenon is known as radioactivity and deals with unstable nuclei expelling α-particles and internal β-electrons (both discovered by E. Rutherford in 1899, albeit the latter were erroneously believed to be rays). Radioactivity was discovered by the pioneering studies of H. Becquerel, W. Crookes, W. Rontgen, P. and M. Curie. The second phenomenon is electron diffraction, proved independently by G.P. Thomson and C.J. Davisson with L.H. Germer, which is the base of modern transmission electron microscopy (TEM) [13] and scanning electron microscopy (SEM) [14]. In the same years, P. Dirac, the founder of quantum electrodynamics, formulated his relativistic version of the Schrödinger equation [11, 15]. One of the final supporting proof on the non-locality nature of quantum entanglement, in violation of Bell inequality and EPR principle, came from experiments performed by R. Hanson's group in 2015 [16].

All these discoveries sealed a definitive connection between physics and chemistry: the common ground allowed the development of a new theoretical body of investigation that will be treated at length in the next section. The Physics-chemistry partnership also lead to the development of new spectroscopic techniques that, in turn, provided structural insights as well as new chemical synthetic strategies, analysis and planning; nowadays, chemists can use powerful spectroscopic tools to characterize their products or test ongoing reactions in situ in a fast and efficient manner. Together with the improvement of laboratory technology and equipment, detection techniques like ionization mass spectrometry [17] and purification/separation techniques like high-performance liquid chromatography [18] or gas-chromatography [19] contributed to non-trivial betterments in the field of synthesis. In the Sixties, the process of rationalization of chemistry took a further step ahead with E. Corey's retrosynthesis, a new way to look at and plan organic total syntheses [20]. Another fundamental improvement in the scientific community, often disregarded, came from the creation of the scientific literature, a platform for data-mining, dissemination and exchange of scientific peer-reviewed results. H. Oldenburg is the father of peer-reviewing mechanism and founding editor of the oldest and still active scientific journal *The Philosophical Transactions of The Royal Society* (1665) [21]. Thanks to the improvement of social media and the invention of cyberspace (i. e. Internet), scientific literature is a globally-accessible tool, essential for any starting and developing research project. Despite all this development, modern synthetic research still relies heavily upon serendipitous discoveries and empirical "trial-and-error" methodology to obtain target compounds. These combined approaches have brought chemistry to sensational discoveries throughout the years, but also contributed to generate chemical waste like volatile organic compounds (VOCs) [22, 23]. In the Nineties, Green Chemistry concept was proposed mainly in reply to the US Pollution Prevention Act and was codified into twelve rules by P. Anastas and J. Warnet [24]. Green chemistry provides a list of strategies to mitigate the environmental impact of large preparations and manufactures. In two centuries, von Liebig's bulky glassware has been more and more replaced by fine glassworks, microreactors, pumps and robotics; new ways of doing chemistry have been developed, like flow chemistry [25] and combinatorial chemistry [26].

The ninth statement of Green Chemistry suggests that catalytic, rather than stoichiometric, protocols should be implemented [24], igniting the spark of an unprecedented "catalyst" rush; ever since, more and more efficient catalysts have been designed experimentally and trial-and-errored to assess their activity or lack of it. Nature exerts the same trial-and-error strategy in designing biologically active molecules to achieve efficiency, improved selectivity and durability. This endless process is called evolution and, unlike our methodologies so far, it is always economic and clean. The fourth statement suggests that chemicals should be designed to improve their desired function, while minimizing toxicity [24]. This is precisely where theoretical disciplines supported by computational methods make their grand entrance.

1.3 The present: Rise of a rational discipline

Early models based on the newly discovered atomic theories have been effectively used to solve experimental issues during the last century. Few very famous examples include:

- **Lewis Structures** proposed by G.N. Lewis [27] to rationalize the nature of the chemical bond.
- **Valence Shell Electron Pair Repulsion (VSEPR)** theory by R.J. Gillespie and R. Nyholm [28] to rationalize the nature of chemical bond and the shape of molecules.
- **Huckel Molecular Orbital (HMO) method** proposed by E. Huckel [29, 30] in 1930 to explain the properties of conjugated polyenes and **Woodward-Hoffman rules** proposed by R. Woodward and R. Hoffman [31] to explain the stereoselective outcome of electrocyclic reactions using Huckel method (Figure 1.1).
- **Crystal Field Theory (CFT)** proposed by H. Bethe and J.H. Van Vleck [32] to rationalize the bond in complexes between metal ions and simple ligands. The interactions were treated as simple fields of electrostatic repulsions between the ligands and the *d* orbitals of the metal pointing in the same direction (Figure 1.2). Crystal field theory and Racah parameters [33] gave birth to Tanabe-Sugano diagrams [34–36], powerful tools to explain and predict spectroscopic and magnetic properties of coordination complexes.

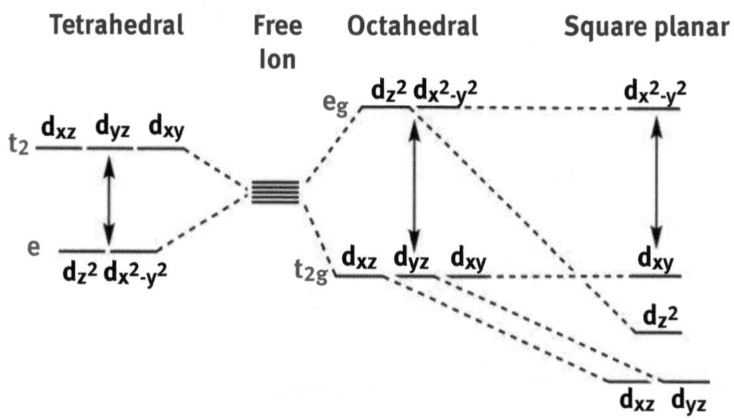

Figure 1.2: The destabilization of the orbitals of the metal, hence the crystal field splitting energy, depends upon the geometric arrangements of the ligands around it. The main classes of crystal fields are represented in this figure.

- **Molecular Orbital Theory (MO)**, also called Hund-Mulliken theory, originally proposed by F. Hund, R. Mulliken, J.C. Slater and J. Lennard-Jones [37] in the late Twenties to rationalize the nature of the chemical bond (localized and delocalized), in light of the new quantomechanical findings. It successfully predicted in 1929 the paramagnetism of molecular O_2 in its ground state [37]. Mulliken predicted the existence of two excited states, above the ground state, $^3\Sigma_g^-$ [38]. $^1\Delta_g$ is commonly referred as singlet oxygen and it is 94.7 kJ/mol higher in energy than $^3\Sigma_g^-$, while $^1\Sigma_g^+$ is 157.8 kJ/mol higher than $^3\Sigma_g^-$ [39]. Molecular orbitals are expanded as linear combination of atomic orbitals (LCAO) to generate bonding, non-bonding and anti-bonding orbitals (Figure 1.3).
- **Valence bond theory (VB)** proposed by L. Pauling [40] in 1927 to rationalize the nature of the chemical bond. It also successfully explained in 1931 the paramagnetism of molecular O_2 in its ground state [41].
- **Ligand Field Theory** by J.S. Griffith and L.E. Orgel [42] is an evolution of CFT as a combination between CFT, molecular orbital theory and hybridization [32].

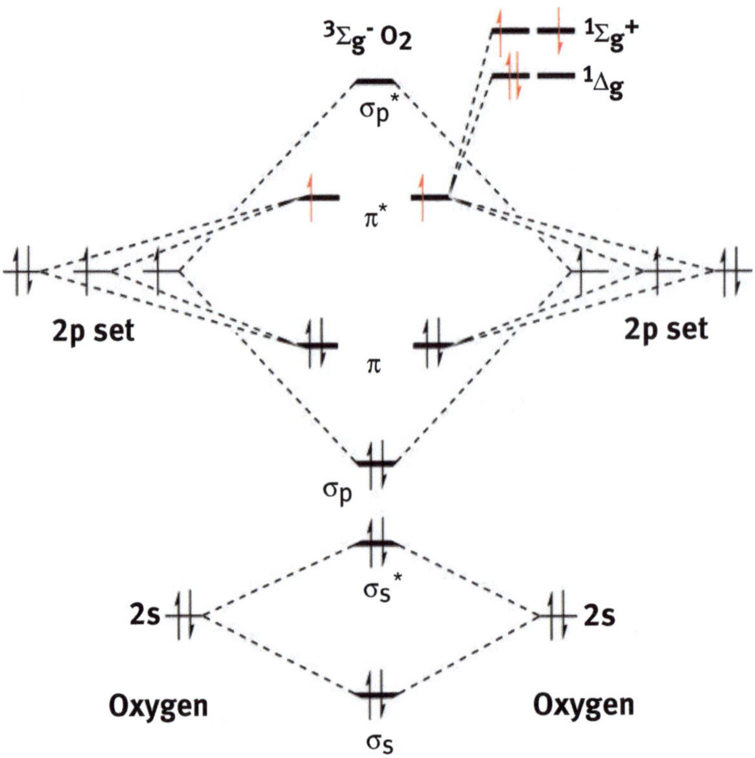

Figure 1.3: Molecular orbital sketch of O_2. The simplest prediction extrapolated from it is that the triplet state is the most stable since it fully complies with Hund maximum multiplicity rule.

Though some of these models are still invoked by experimentalists to justify phenomena, chemistry has evolved and so did the challenges, goals and problems connected to it. Luckily, advanced quantum mechanics coupled to increased computational power (Figure 1.2), made possible by the development of hardware (e. g., J. Bardeen's transistors in 1940–1950) and software (e. g., Gaussian 70 in the Seventies), triggered the construction of more powerful and descriptive *in silico* models in this last two decades. The next section covers the state-of-the-art in modern theoretical computation.

1.3.1 Electronic structure

Hartree–Fock (HF) theory uses a single Slater determinant in its wavefunction definition to comply with the anti-symmetric nature of electron spin. Initial trial molecular spinorbitals are expressed as linear combinations of atomic orbitals and optimized through the Fock operator (Roothaan-Hall equations) and an iterative process known as self-consistent field [43]. The limit of HF resides in its inexistent treatment of electron correlation. Atomic orbitals can be constructed using Slater functions (STO), $\phi_{\xi,n,1,m}(r,\Theta,\Phi) = N \cdot r^{n-1} \cdot e^{-\xi r} \cdot Y_{1,m}(\Theta,\Phi)$, Pople Gaussians (GTO), $\phi_{\xi,n,l,m}(r,\Theta,\Phi) = N \cdot r^{(2n-2-1)} \cdot e^{-\xi r2} \cdot Y_{l,m}(\Theta,\Phi)$, or plane waves, $\phi(r) = A \cdot e^{-iK \cdot r}$. Gaussians are computationally advantageous since they decay with a squared dependence on the radius, but a linear combination of them might be necessary to achieve accurate reproduction of the atomic orbital near and far off the nucleus. Basis sets like 6–311++G(d,p) or aug-ccpvTZ are set of functions using a linear combination of three Gaussian functions to reproduce the valence of the atom; they are complemented by polarization and diffuse functions [44] for accuracy. Post-HF methods are improvement to the HF theory and they usually provide high quality electronic structures, when coupled to saturated basis sets, but they are also time-consuming and resources-intensive [45]. Most of these methods feature multi-determinant expansions of the wavefunction and respect size-consistency [15, 43]. The principal methods [45] are configuration interaction (CI), Moller-Plesset perturbation theory (MP2, MP3, MP4 and MP5, following the expansion of the perturbation term), coupled cluster (CCD, CCSD and CCSD(T)), quadratic configuration interaction (QCIS, QCISD, QCISD(T), QCISD(TQ)), Brueckner Doubles (BD, BD(T), BD(TQ)), compound methods (G2, G3 or CBS).

The early alternative to HF-based methods for large systems was represented by semi-empirical methods: they feature minimal quantum-mechanical treatment of the atomic valences through small basis sets, combined with zero differential overlap (ZDO) approximation [15] and parameterization with terms derived from experimental reference data [15]. Today neglect of diatomic differential overlap (NDDO) methods like PM7 [46] are excellent and inexpensive way to obtain semi-quantitative information on a system, particularly useful when conformational analysis is required. The second alternative is represented by density functional theory (DFT) methods. DFT theory has its roots in the work of W. Kohn, P. Hohenberg and L.J. Sham [47, 48]. Kohn–Hohenberg theorems bind the ground state electron-density of a molecular system, $\rho(r) = \sum \eta_i |\psi_i(r)|^2$, with the ground state energy, $E_{KS}[\rho(r)]$, through an exact and unique functional $F[\rho(r)]$. Thus, a simplified version of these theorems could be written as $E_{KS}[\rho(r)] = T_S[\rho(r)] + J[\rho(r)] + E_{EN}[\rho(r)] + E_{XC}[\rho(r)]$, where $T_s[\rho(r)]$ term is the Hamiltonian of the system, $J[\rho(r)]$ term is the Coulomb electron-electron repulsion, $E_{en}[\rho(r)]$ term is the electrons-nuclei interactions and $E_{XC}[\rho(r)]$ is the exchange-correlation energy, which includes several corrections including quanto-mechanical treatments of Coulomb and Fermi holes [49]. Once $E_{XC}[\rho(r)]$ term is approximated, the computation of the other potential energy terms is generally fast, thus DFT methods are usually faster than post-HF counterpart and scale very favorably with number of functions (~10^1 to 10^3, versus 10^4 of MP2 and 10^7 of CCSD(T) methods). This makes DFT methods very suitable for investigations on large non-truncated systems, where accuracy and speed must go hand in hand [45]. It is precisely the form in which $E_x[\rho]$ and $E_c[\rho]$ are "approximated" and combined together that defines the name of the DFT method (Table 1.3). J. Perdew envisioned the development of DFT accuracy and performances using the Biblical metaphor of Jacob's ladder [50]: each rung of the ladder, with its associated strengths and weaknesses, leads to a global improvement of the calculated DFT results over the previous one. The "heaven" is represented in this case by a density functional method using a fully non-local formalism. **LDA**, or local density approximation, is the first and most rudimentary DFT (Rung 1 on Jacob's ladder). Proposed by P. Dirac [51], the first local density approximation was used in conjunction with Thomas-Fermi gas model [52]. It is a fully local approach, since the exchange-correlation energy depends upon the local electron density at a certain point, ρ(r). Despite its limitations, it can afford

Table 1.3: Summary of DFT methods, definitions, dependencies and place on the Jacob's ladder. PT2 stands for perturbational theory truncated at the 2nd term.

Functionals combination	Type (# rung in Jacob's ladder)	Dependencies
SVWN, PZ81, CP, PW92	**LDA (#1)**	ρ(r)
BLYP, BP86, PBE, B97D	**GGA (#2)**	ρ(r), $\|\nabla\rho(r)\|$
t-HCTH, TPSS, VS98, VSXC, M06L	**Meta-GGA (#3)**	ρ(r), $\|\nabla\rho(r)\|$, $\nabla^2\rho(r)$ (τ)
B3LYP, B3PW91, B3P86, PBE0, mPW1PW91, BMK	**Hybrid GGA (#4)**	ρ(r), $\|\nabla\rho(r)\|$, %HF exchange
B1B95, BB1K, MPW1B95, MPW1KCIS, PBE1KCIS, TPSS1KCIS, TPSSh, M06, M06-2X, M11, MN12-SX	**Hybrid Meta-GGA (#4)**	ρ(r), $\|\nabla\rho(r)\|$, $\nabla^2\rho(r)$ (τ), %HF exchange
MC3BB, MC3MPW, B2PLYP, B2KPLYP, B2TPLYP, mPW2PLYP, XYG3	**Double hybrid (#5?)**	ρ(r), $\|\nabla\rho(r)\|$, %HF exchange, PT2 corrections to the correlation functional

good predictions, mostly due to cancellation of errors when estimating exchange and correlation. **GGA**, or generalized gradient approximation, introduces a higher degree of space inhomogeneity (rung 2, Table 1.3). Exchange-correlation term depends upon local electron density and local gradient of the electron density, $|\nabla\rho(r)|$, at a certain point. The results obtained by GGA are usually better than those obtained by LDA, thought this is more true for small molecules than for solids. **Meta-GGA** is an improvement over GGA since the dependencies have been expanded including the Laplacian of the electron density, $\nabla^2\rho(r)$, and the kinetic energy density, τ (rung 3, Table 1.3). The next rung in the ladder introduces a degree of non-locality by replacing some local exchange energy density with exact HF exchange energy density; these methods are called **hybrid GGAs** (rung 4, Table 1.3). The degree of HF exchange cannot be estimated a priori, but needs to be fitted empirically. The improvement in performances of this class of functionals over the previous ones has been quite drastic, as the hybrid GGAs statistically outperform any previous functional in calculating a wide range of chemical properties. When the kinetic energy density is also taken into account, the result is called **hybrid meta-GGA** functional (rung 4, Table 1.3). B3LYP combination (B3 exchange functional by Becke and LYP correlation functional by Lee-Yang-Parr) is a hybrid GGA functional containing 20 % HF-exchange and it is regarded by most as the standard in organic chemistry, although Minnesota functionals like M06 (27 % HF-exchange) and M06-2X (54 % HF-exchange) have received particular attention, since they show superior performances in thermochemistry, non-covalent interactions and kinetics [53, 54]. DFT description of dispersive interactions (i. e., hydrogen bonds, London dispersions), however, is generally regarded as unsatisfactory [49, 55]. **Double hybrid** functionals are supposed to deliver superior performances for weak interactions by including corrections at perturbation theory level, but they generally show similar computational costs to MP2 [56–61]. The performances of XYG3 functional in calculating thermochemistry, kinetics, weak interactions are remarkable; it is claimed to fill the 5th position of the ladder [62]. The reproduction of long-range dispersive forces can also be achieved using molecular mechanics corrections to DFT; empirical dispersion formulas have been implemented at no extra computational cost for almost any functional [63–66]. Functionals like B97D [64, 66] or ωB97XD [67] already contain dispersion corrections. Figure 1.4 has been imported from R. Peverati and D. Truhlar's recent work [68] and represents a superb way to capture 30 years of improvements in DFT performances at first glance. Summaries of several aspects of DFT and DFT-related performances and implementations can be found in a few manuscripts that combine in-depth expertise with excellent readability (particularly for non-experts) [68, 69].

Heavy metals of the third row, especially Au [70], display electrons moving at nearly-relativistic speeds due to a high Z number. The moving mass of these fermions

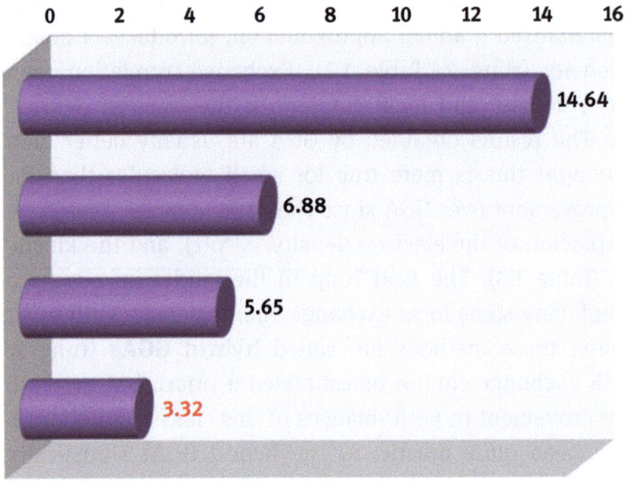

Figure 1.4: Average of mean unsigned errors (in kcal/mol) for all DFT functionals tested in Truhlar's database CE345 and grouped according to their year of publication. This database includes 15 subgroups of different chemical properties.

undergoes relativistic increment given by the equation $m = m_0\sqrt{1-\left(\frac{v}{c}\right)^2}$. This phenomenon leads to important scalar effects, like the contraction of the *s* and *p* shells and the expansion of the *d* and *f* shells, and vectorial effects, like spin-orbit coupling [71]. These effects, in turn, influence the chemistry of heavy transition metals in terms of reactivity, bonding preferences and strengths, coordination numbers and spectroscopy [71]. Relativistic calculations can be introduced via:

- Effective core potential, ECP, like LAN [72], SDD [73, 74] and CEP [75]. Basis functions are replaced by potential functions at the core and the extent of the replacement defines large-core (frozen core and valence), medium-core or small-core (frozen core + valence treated by basis set) pseudopotentials. Accurate results are more likely to be obtained using small-core ECPs [76]. The basis sets complementing medium and small-core ECPs are designed to fit the pseudopotential used, generally through appropriate primitive contractions [77].
- Relativistic Hamiltonians coupled with full-electron basis sets. Relativity can be brought into the Schrodinger equation through the Dirac equation. Douglas–Kroll–Hess (DKH) transformation is commonly used for its accuracy, efficiency and effectiveness [78]. Appropriate basis sets are built by contracting primitives [77].

The goal of electronic structure calculations is to generate potential energy hypersurfaces, alias "mapped" polydimensional surfaces that correlate the potential energy in function of the nuclear coordinates of the atoms. Wells or minima on these surfaces represent stable structures. Calculated metric parameters for these minima provide useful information to synthetic chemists because they can be compared with metric parameters derived from phase scattering of x-ray or neutron single-crystal diffraction experiments [79]. Vice versa, this is also an approach used by theoreticians to validate bona fide the goodness of their calculations. This approach is, however, a risky procedure and must be done carefully for two reasons: the maxima in scattering amplitudes may not coincide with nuclear positions in light atoms and structures in solid state are affected by crystal packing and dielectric fields, not present in the isolated calculated counterparts [80].

Electronic structures predicted the existence of "impossible" compounds like $AuXe^+$ and $XeAuXe^+$ in 1995 [81]. The calculations were carried out by P. Pyykkö et al. using methods like MP2, MP4 and CCSD(T): the bond distance Au-Xe calculated at MP2 level was found to be 2.691 Å for $AuXe^+$ and 2.66 Å for $XeAuXe^+$. The stability of these compounds was attributed to the high electronegativity of gold, due to relativistic effects; the bond dissociation energy for $AuXe^+$ decreases from 0.910 to 0.376 eV when relativistic effects are omitted at CCSD(T) level [81]. The existence of AuXe and $XeAuXe^+$ was confirmed experimentally in 1998 by Schröder, Pyykkö et al. [82]. Finally, Seidel and Seppelt isolated the first stable gold compound with xenon, $[AuXe_4][Sb_2F_{11}]_2$ (Figure 1.5), in 2000 [83]. The compound confirmed the stability postulated for this class of compounds being stable up to 40°C. Metric parameters for Au-Xe bond distance, 2.728(1)-2.750(1) Å, extracted from X-ray diffraction agrees well with distances calculated at MP2 level, 2.787 Å [83]. Electronic structures and enthalpy of formation predicted also the existence of another "impossible" compound in the solid state, Na_2He. The predicting algorithm in the USPEX code [84] suggested the existence of a cubic-phase structure of Na_2He, stable at pressures above 160 GPa. The structure was effectively isolated as a cubic-phase, stable from 113GPa up to 1000 GPa, and characterized in matrix by A.R. Oganov et al. [85]. The electronic structure of the solid was studied by density of states, ELF and Bader's analysis [85].

Electronic structures can be directly linked to magnetic spectroscopies based on the Zeeman effect [86]. In nuclear magnetic resonance (NMR) [87] and electronic paramagnetic resonance (EPR) [88], low frequencies waves (radiofrequencies for

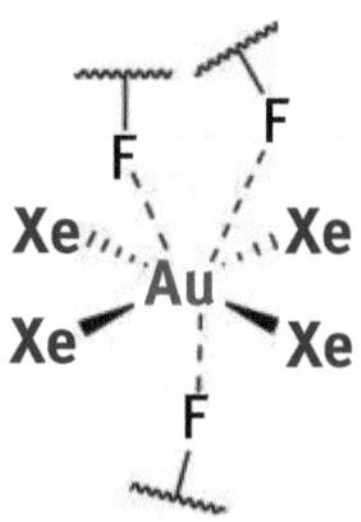

Figure 1.5: Seidel-Seppelt unique compound. Au•••F contacts with the counterions are shown.

NMR and microwaves for EPR) are pulsed to make net nuclear (NMR) or electronic (EPR) spin magnetization process about an axis in a magnetic splitting field B_0 [89]. This precession is sensitive upon the chemical environment of each nuclear or electronic spin. NMR spectra (shift and spin-spin coupling) can be efficiently simulated in gas or in solvent phase by calculating 1D-NMR shielding tensor and susceptibilities using Gauge-Independent Atomic Orbital (GIAO) [90–92] or Continuous Set Of Gauge Transformations (CSGT) [92–94] methods. Accurate results versus experiment can be achieved by using high level of theory, usually DFT or MP2 [92, 95, 96], in conjunction with large, polarized and diffused basis sets. Similarly, hyperfine coupling constants for EPR spectra can be calculated using EPR-II and EPR-III basis sets [97]. GIAO calculations using a pure GGA functional, BP86 (implemented in G03 [98]), in conjunction with SBK basis [99] for metals and 6–311++G(d,p) for carbons and hydrogens have been employed by H.V.R. Dias and T.R. Cundari et al. to provide a comparison to experimental ^1H and $^{13}C\{^1H\}$ NMR spectra in exceedingly rare adducts of ethane with coinage metals [100, 101], Table 1.4. A further model [102] was generated using a hybrid GGA, B3PW91, in conjunction with SDD pseudopotential and K.A. Peterson's correlation consistent basis sets [77]. Theoretical values shown in Table 1.4 for both models along the homoleptic series $[M(\eta^2\text{-}C_2H_4)_3][SbF_6]$ (for M = Cu, Ag, Au) show amazing structural and spectroscopic reproduction of the studied species. Similar computational studies were carried out to explain the NMR and bonding patterns of a peculiar complexation between trans,trans,trans-1,5,9-cyclododecatriene, $CuSbF_6$ and carbon monoxide [102, 103].

Electronic structures can clarify the dynamic behavior of molecules whose potential energy surfaces change upon the absorption of photons of visible or ultra-violet light. Organic chromophores fall under this category, since light can cause transitions from bonding to antibonding orbitals [104]. Heavy metal complexes show also spin-allowed, Laporte-allowed transitions, ligand-to-metal (LMCT) and metal-to-ligand (MLCT) charge transfers [32]. The transition momentum from ψ_1 to ψ_2 eigenstate, $M_{2,1} = \langle\psi_2|\mu|\psi_1\rangle$, can be calculated for vertical excitations (complying with Born-Oppenheimer approximation) to provide

Table 1.4: Theoretical computation can be used to validate experimental results, as in the case of reactive coinage wheels. Reference temperature for NMR data is 298 K.

Compound	Data	^1H NMR, δ	$^{13}C\{^1H\}$ NMR, δ	M-C, C=C bond distances (Å)
$[Cu(\eta^2\text{-}C_2H_4)_3][SbF_6]$	Exp.	**5.44** (CD_2Cl_2)	**109.6** (CD_2Cl_2)	2.193(6),1.359(8) average
	Comp.	**5.2** (gas)	**100.0** (gas)	**2.172, 1.374** (D_{3h})
$[Ag(\eta^2\text{-}C_2H_4)_3][SbF_6]$	Exp.	**5.83** (CD_2Cl_2)	**116.9** (CD_2Cl_2)	2.410(9), 1.323(14) average
	Comp.	**5.6** (gas)	**110.9** (gas)	**2.408, 1.367** (D_{3h})
$[Au(\eta^2\text{-}C_2H_4)_3][SbF_6]$	Exp.	**4.94** (CD_2Cl_2)	**92.7** (CD_2Cl_2)	2.268(5), 1.364(7) average
	Comp.	**5.0** (gas)	**92.6** (gas)	**2.308, 1.388** (D_{3h})

transition energies, multiplicities, symmetry and oscillator strengths, thus simulate UV-Vis and circular dichroism spectra. UV-Vis spectroscopy [89] is used by experimental chemists to identify and quantify an active substance through absorbance and Beer-Lambert law; the description of this approach and its limits are well known [105]. Time dependent density functional theory (TDDFT) coupled with medium-to-large basis sets generally provides reasonably accurate results (i. e., within 0.4 eV from experimental transitions) at a very affordable cost, but it has limitations like the absence of double excitations [106]. Most used DFT functionals are B3LYP and PBE0, although the popularity of functional including long-range corrections to the Coulomb term (e. g., cam-B3LYP, LC-ωPBE and ωB97XD) is rising. More refined and resource-consuming alternatives to TDDFT feature CASSCF and CASPT2, techniques based on the complete active orbital space (including PT2 corrections in the latter) [107]. Modeling excited states encompasses the study of the reorganization energy of the molecule and the solvation layer after the absorption of light; bathochromic shifts can be obtained this way to simulate experimental fluorescence spectra. Study of intersystem crossing phenomena (i. e., conical intersections between different spin potential surfaces) can be valuable to study phosphorescence [108]. Intersystem crossings are more common for heavy metal complexes with strong spin-orbit coupling. Diffuse functions and long-range corrected functionals might be mandatory when studying excited states of higher energy (i. e. other than the first one). Photo-induced single transfer redox reactions via outer sphere mechanism involve bimolecular events between oxidants (electron acceptor) and reducing species (electron donor); Marcus theory was designed to calculate the barrier of activation in terms of free energy difference for such processes, $\Delta G^{\ddagger} = (\Delta G + \lambda)^2/4\lambda$ (λ is the reorganization energy of all the atoms involved in the reaction) [109, 110]. The thermodynamic function that expresses quantitatively the propensity for the transfer of electron(s) is the standard reduction potential; it can be measured experimentally by cyclic voltammetry and calculated using a variety of methods involving implicit, explicit or mixed solvation schemes [111]. Mean unsigned errors of computed values versus experimental ones are reported in the range of 64 mV (~6 $kJ \cdot mol^{-1}$) and as low as 50 mV for selected classes of compounds [111]. Electrochemical methodologies are fundamental in processes like artificial photosynthesis and water splitting [112, 113].

1.3.2 Solvation schemes

The solvent influences kinetic and thermodynamic aspects of a chemical reaction, thus the accurate evaluation of solvation effects should be brought into the calculations. Several different solvation schemes are available: implicit, explicit or mixed implicit-explicit. Implicit methods allow for quanto-mechanical

treatment of the solute and for bulk treatment of solvents as continuous envelopes around solutes, characterized by macroscopic properties like dielectric constants and microscopic properties like solvent radii. Usually computationally very affordable for spectator solvents and reasonably accurate, implicit solvation models are listed below:
- Polarizable Continuum Models include D-PCM [114], C-PCM [115] and IEF-PCM [116].
- Solvation Model Based On Density or SMD [117].
- Models using the GB approximation for the electrostatics include SM8 [118], SM8AD [119], SM12 [120].
- Conductor Like Screening Model or (COSMO) [121] and its real solvent variant (COSMO-RS) [122].
- Others [123].

Explicit methods allow for quantum mechanical treatment of the solute and treat solvents as discrete entities around the solute. They are strongly suggested when interactions between solvent and solute need to be characterized, as in the case of strong hydrogen bonds or coordination to heavy metals, but demanding in terms of resources. The nuclear motion of the solute can be simulated using molecular dynamics or multi-scale models [111]. Mixed methods are probably the best chemical option, although they should be chosen and set-up wisely to provide accurate results at affordable costs.

1.3.3 Ensemble properties

Physical observables represent macroscopic bulk properties of a chemical sample. Simulating those properties at microscopic level requires the average of all the possible states of a chosen ensemble. The taxonomy of statistical thermodynamics encompasses some notable ensembles for N-particle systems: microcanonical or NVE (constant N, volume and energy), canonical or NVT (constant N, volume and temperature), grand canonical or μVT (constant chemical potential, volume and temperature), isoenthalpic-isobaric or NPH (constant number of particles, pressure and enthalpy), isothermal-isobaric or NPT (constant N, pressure and temperature of the system). Calculating all the possible states within an ensemble is impossible, thus few clever simplifications are normally introduced, in order to combine accuracy with computational feasibility. Dynamics methods listed below are excellent examples of methods working on ensemble properties of many-particle systems. Molecular Dynamics, Langevin Dynamics and Quantum Dynamics are particularly suitable for heterogeneous systems, since they are composed of different phases, thus tend to be more affected than homogeneous systems by the multi-scale complexity of the system: the kinetics of diffusion-controlled reaction, for instance, is

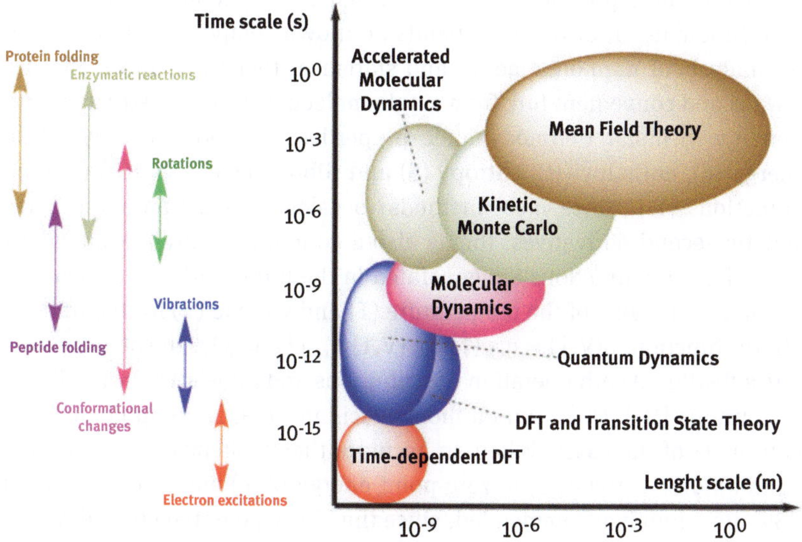

Figure 1.6: Multi-scale computational methods most frequently used associated to time-scales of chemico-physical phenomena.

controlled by particle transport phenomena like diffusion, convection, migration, adsorption and desorption. Chemical reaction happening at the surface of metals, alloys or minerals are many and diverse. The knowledge of the principles ruling transformations in heterogeneous catalysis will allow one to “tailor” a catalyst to carry out the target transformation sought [124]. Figure 1.6 relates schematically the dimensionality of chemico-physical phenomena to the length/time scale of computational methods [125]

1.3.3.1 Statistical thermodynamics

The calculation of the Hessian matrix (second derivative of energy with respect to atomic coordinates) leads to the recovery of force constants, that can be used to the simulate infrared (IR) spectra [89] and intensities; Raman spectra [89] can be obtained this way when Raman intensities are calculated for differentiation of dipole terms with respect to an external electric field (E). Clearly, obtaining first derivatives is less time-consuming than calculating superior orders in the expansion. MP2 and some DFT methods (e. g., GGA like BP86, B97D) in combination with medium-to-large basis sets usually simulate vibrational spectra with a maximum average error of 10–50 cm^{-1} versus spectroscopic experiment, while techniques like CCSD(T) can deliver much better results [49]. Simulations can also include the solvation effect through the methods seen before. These simulations might help synthetic chemists to explain the structure of unknown compounds through selection rules and fingerprint matching and assign calculated atomic motions to experimental fundamental or overtone bands of known compounds. In addition, vibrational analysis of harmonic modes and statistical thermodynamics afford a straightforward and convenient link from single-molecule electronic structure and potential energy to molar thermodynamic properties (zero-point energy (ZPE), thermal energy (E), enthalpy (H), entropy (S) and Gibbs free energy (G)) through partition functions [15]. Since most methods to calculate electronic structures feature analytic second derivatives, this is also a computationally advantageous approach [45]. Partition functions are generated in the framework of the ideal gas approximation as functions of the temperature (T) and volume (V) for the translational partition function. $q(V,T) = q_{ele}(T){\cdot}q_{tra}(V,T){\cdot}q_{rot}(T){\cdot}q_{vib}(T)$. In solvent phase rotations are substituted with liberation contributions and translations with liberational contributions [111]. Quantum harmonic oscillator [89] is an approximation that holds in most of the cases. When an accurate thermochemistry is expected, however, anharmonic corrections to zero-point energy (ZPE) and roto-vibrational coupling [89] are strongly recommended, since the Morse potential curve deviates sensibly from the harmonic curve at higher internuclear distances, near bond-breaking region [89]. Also, floppy molecules that have low-barrier torsions (hindered rotors) and vanishing-barrier vibrations (free rotors) show a dynamic behavior that falls far from HO approximation. Thus, without correction, large errors would be brought into the entropic partition function (consequently, into Gibbs free energies). Few different correction schemes for low frequency vibrations are available in the literature such as the Pitzer-Gwinn [126], Truhlar [127], McClurg [128] and quasi-harmonic approximations [129–131].

1.3.3.2 Mean field theory

Mean field theory is an important branch of condensed matter physics. "Mean field" definition applies to all those theories where the study of a multi-body network of interactions is reduced to an average interaction acting upon the individual under study [44]. In all but the simplest situations (like the Ising model [132]), the mean field simplification is very convenient. Without it, for instance, the calculation of the configuration integral Z_N, thus the estimation of the potential energy term U(r) between particles, would become cumbersome in the study of dense fluids [44]. Mean field theory has been originally applied to fluid model (Van der Waals equation) [44], phase transitions, phase transition and ferromagnetism [133], alloys and superconductivity [134, 135]. Similar theoretical treatments were developed for game [136] and queueing theories [137].

1.3.3.3 Molecular dynamics

Molecular dynamics is a very popular and largely used technique among researchers in many fields. A many-particle system is left evolved in time through phase-space trajectories that respond to classical laws of motion (Newton): $F[x(t)] = -\nabla U[x(t)] = m \bullet \ddot{x}(t)$, with U being the potential energy of the particle interaction. The equations of motion are usually integrated numerically over the selected interval of time ($t+\Delta t$) chosen to describe the system. The initial parameters needed to evolve a system usually are the internal forces among particle, F[x(t)], the composition of the sample (its composing masses, m), the considered ensemble (usually microcanonical) and the time of simulation ($t_{final} - t_{initial}$). The forces between particles are usually derived from parameterized force fields. The standard choice for the integrator falls upon the Verlet algorithm, $x(t+\Delta t) = 2\,x(t) + x(t-\Delta t) + (\Delta t^2/m)\bullet F[x(t)]$. This algorithm is particularly useful because it does not allow error accumulation, thus the simulations can be stable for long time [138]. The systems are normally simulated for few femtosecond under periodic boundary conditions to simplify the constraints of a boundary. MD simulations can be efficiently used to obtain static descriptions of the system as well as dynamic evolution of its physical properties. MD simulations can be also run at constant temperature or pressure [139]. One of the major drawback of MD is that the calculation of interatomic forces does not embed any type of quantum mechanical treatment. Without a proper electronic description, MD is not suitable to describe reactions where bonds are formed or broken. The addition of quanto-mechanical electronic calculations extends the use of MD to bond-breaking and forming events (thus scale), but also reduce the size of the sample treatable with this approach by four orders of magnitude (~10^2 atoms). Quantum dynamics approaches like Born–Oppenheimer molecular dynamics (BOMD) [140] and Car-Parrinello molecular dynamics (CPMD) [140] methods are elegant and rigorous treatment of quantum electronic systems where classical atomic motion is separated, in virtue of Born-Oppenheimer principle, from quanto-mechanic electronic motion. The electronic degrees of freedom are calculated at every step of the optimization in BOMD, but introduced through fictitious variables in CPMD. This allows the simultaneous calculation of both electronic and nuclear degrees of freedom. If from one side the predictive power of these techniques in terms of chemical bonding and its evolution with time is incredible, on the other side, these techniques require great amounts of computational time and resources [140]. This makes them unfeasible for large systems and/or for system where long trajectories and high quality of energy conservation laws are needed. An interesting use of Born-Oppenheimer molecular dynamics (BOMD) to unravel the contribution of entropic disorder on the stabilization of surface polar terminations was published by M. Capdevila-Cortada and N. Lopez [141]. The understanding of surface terminations behavior is very important in the design of new materials. The group performed the study using DFT (PBE+U term) under an NVT ensemble [141]. J.K. Nørskov et al. used the RPBE functional in conjunction with plane-waves, contained in the code DACAPO [142], to model the molecular dynamics of a cheap inorganic surface able to evolve hydrogen in place of the expensive and rare platinum [143]. The evolution of hydrogen is one of the most important electrochemical processes taking place on heterogeneous catalysts for its connection with sustainable energy [144]. Metals like Ni and Mo bind H* too strongly to guarantee a sustainable release of H_2, while Au, for instance, suffers from the opposite problem, being a very poor binder instead (Figure 1.7). The observation of naturally occurring enzymes able to evolve hydrogen, like hydrogenases and nitrogenases, has sparked the desire to reproduce the catalytic properties of their active centers using mimicking synthetic alternatives [145–147]. The investigation pointed out that the sites at the edges of MoS_2 are able to bind H* and release H_2, resembling the catalytic structural motif and related activity of the nitrogenase FeMo cofactor (an enzyme that catalyze the evolution of molecular hydrogen, Figure 1.7). Following the computational lead, the group synthesized nanosized MoS_2 and proved that this material produces molecular hydrogen with an overpotential of 0.1–0.2 V [143].

Molecular dynamics, coupled with other techniques like umbrella sampling, conformational flooding, metadynamics, and adaptive force bias, plays a pivotal role in drug discovery as well [148]. Drug discovery deals with the energetics, affinity and selectivity of binding/unbinding events between naturally occurring biological receptors and synthetic inhibitors [148]. These events, however, span microseconds to milliseconds, while ab initio molecular dynamics can effectively sample a few femtoseconds at best [149]. G. de Fabritiis et al. published a paper [150] on the quantitative reconstruction of the binding process between β-trypsin and

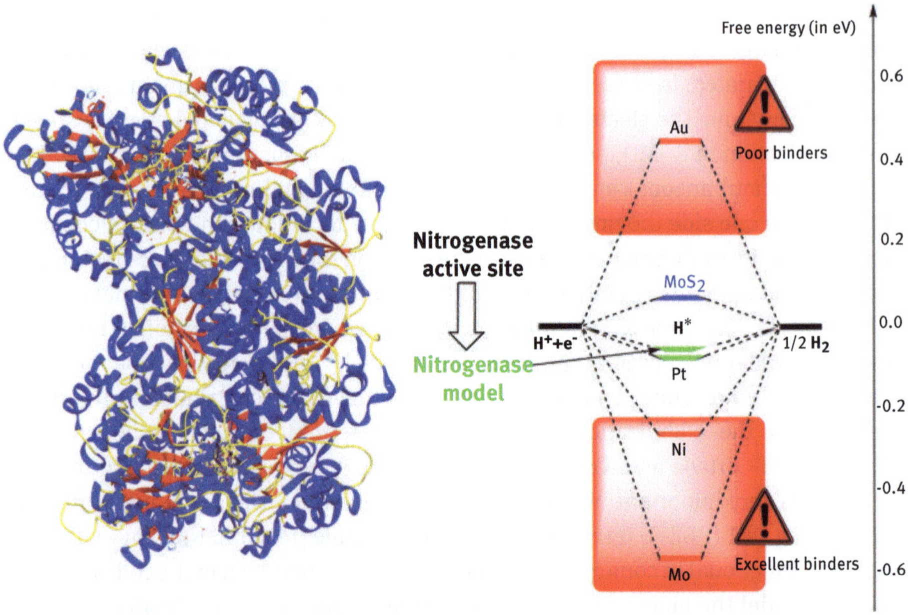

Figure 1.7: Learning from Nature. Mimicking the active site of enzymes like nitrogenase FeMo cofactor (left) leads to rational design of surfaces evolving H_2, like nanosized MoS_2.

benzamidine using a Markov state model (MSM) [151]. High-throughput classical molecular dynamics parallelized on clusters of graphical processor units (GPUs) [152] allowed 495 simulations of this binding event of 100 ns each. 187 simulations gave productive bindings with root-mean-square deviation of atomic positions within 2Å compared to the crystal structure [153], Figure 1.8 left. The estimation of the standard free energy of binding is within 1 kcal/mol from the experimental value [150]. Similar routines are employed to screen and/or improve allosteric drugs (drugs that does not bind the orthosteric pocket of the receptor) and numerous methods to predict allosteric sites and their druggability are described and reviewed in literature [154].

1.3.3.4 Brownian and Langevin dynamics

Brownian dynamics (BD) is another method used to sample the binding/unbinding pathway of associative events [155–157]. The motion of molecules of interest is assumed to be Brownian [158–160], while the solvent is treated stochastically. BD is less time consuming than MD, though its accuracy in reproducing observables strongly depends upon the use of supportive models [161]. An interesting study using BD on the formation of complex barnase-barstar (a ribonuclease and its inhibitor) has been published by A. Spaar et al. in 2006 [162]. The complex barnase-barstar is a well-known case for quantitative studies of protein–protein interactions (wild-type and mutant alike). The computational study focused on the characterization of the electrostatic binding region from diffusional regime to enzyme-inhibitor encounter region. The encounter region was modeled after the X-ray crystal structure [163], Figure 1.8 right. Two regions of preferential binding were authenticated and it

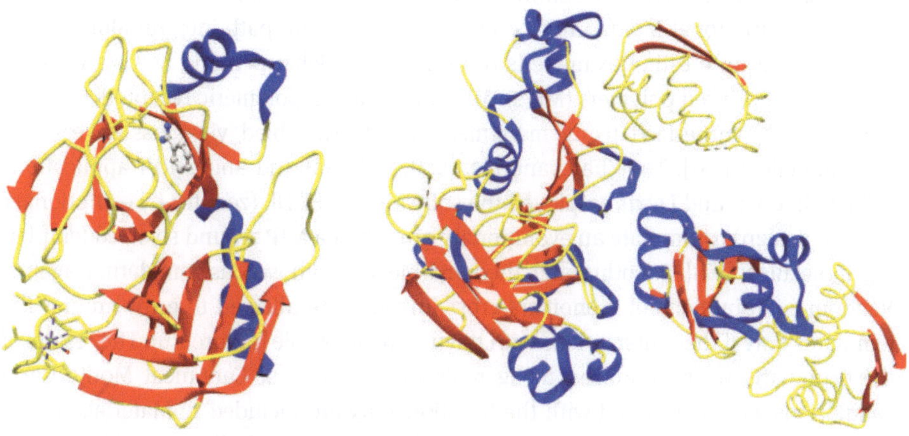

Figure 1.8: Left, X-ray crystal structure of the Ca^{2+}-enzyme-inhibitor complex between benzamidine and serine protease β-trypsin (PDB ID: 3PTB). Right, X-ray crystal structure of the complex barnase-barstar (PDB ID: 1BRS).

was demonstrated that enzymatic mutations alter significantly the populations around these two minima [162]. A particular case of BD is represented by Langevin dynamics. LD complements molecular dynamics calculations with a stochastic treatment of solvent lighter particles perturbing solute heavier particles by the means of averaged frictional forces and collisions [139]. The equation of motion can be written in this case as: $-\nabla U[\ddot{x}(t)] - \gamma\dot{x}(t) + (2\gamma k_b T)^{1/2} \cdot R(t) = m \cdot \ddot{x}(t)$[164]. The term $-\gamma\dot{x}$ (t) represents the frictional forces, while the term $(2\gamma k_b T)^{1/2}$·R(t) is a Brownian "random force", usually distributed as a Gaussian function [139].

1.3.3.5 Monte Carlo methods

These methods own their exotic name to the city of Monte Carlo in Monaco, renowned place where stochastic bets play a huge role in the casinos. MC simulations rely on the simulation of a system based on random or quasi-random numbers and can be divided into three main types: direct Monte Carlo, Monte Carlo integrationand Metropolis Monte Carlo [139]. Direct MC uses random numbers to simulate events. MC integration calculates integrals on random numbers. Metropolis MC is based on building statistically- dependent configurations through Markov chains. The probability of each new configuration is built upon the probability of the previous configuration. Markov chains are normally ergodic: the conditions of irreducibility (every configuration can be accessed from any other configuration within a finite number of steps) and non-periodicity the same configuration does not repeat except after a fixed number of steps) are satisfied [139]. The functions p(X,t) that correlate the probability of occurrence of a certain configuration (X) at time t are called master equations [139]. MC simulations are usually carried out in NVT ensembles, but Metropolis MC can be used in NPT [165] or μVT [166] ensembles as well. MC methods are mostly used to calculate static properties of a system under study. Monte Carlo methods including quantomechanical treatments are called variational Monte Carlo, diffusion Monte Carlo, path-integral Monte Carlo [139]. Another interesting example is represented by Schaupen and Lewis's work on molecular imprinted polymers (MIPs). MIPs are artificial polymeric receptors ("plastic antibodies") synthesized using monomeric units assembled via cross-linkers and template effect [167]. They have antimicrobial, antiviral and anticancer applications [167]. Schaupen and Lewis employed the software ZEBEDDE (zeolites by evolutionary *de novo* design) to simulate an evolutionary growth of a MIP around selected templating molecules [168]. Nicotine and theophylline were chosen as templating agents. Random combinations of monomers/cross-linkers were allowed to grow around the templates only if their interactions with them were favorable (Figure 1.6). The results of these growths were optimized using molecular dynamics. Canonical Monte Carlo simulations were performed with the Sorption software included in Materials Studio and used to characterize various grown imprints and their bindings specificity [168]. The calculations showed that the imprints built this way show a remarkable preference to bind theophylline over similar substrates.

1.3.4 Transition state theory

H. Eyring, M.G. Evans and M. Polanyi proposed this theory in 1933 [169]. Transition structures are mathematical constructs of dynamical instability, dividing surfaces between short-time intrastate (within basin) from long-time interstate dynamics (basin to basin). They possess 3N-7 vibrational degrees of freedom, where the missing degree represents the reaction coordinate. Thus, transition states (TS) are not real structures and many scientists refers to TS-derived thermodynamics as talking of transition state-derived thermodynamics as quasi-thermodynamics [170]. Fundamental requirement of TST is that reactants (or products) and connected transition structures are thermally equilibrated [170]. Several minimization algorithms are available to locate transition structures: steepest descent, Newton-Raphson, rational function optimization, direct inversion of iterative space (GDIIS and GEDIIS) and synchronous transit-guided quasi-Newton (STQN) [171]. The immediate purpose of Eyring-Polanyi equation is to calculate (or estimate at the very least) rate constants for chemical reactive events. Conventional Transition State Theory (CTST) usually provides an upper-bound estimate of experimental reaction rate constants [170, 172]. Within this approach, the transition surface is placed at the saddle point and 100 % crossing-over efficiency is assumed between reactants and products. The variational optimization of the dividing surface in order to minimize re-crossing leads to the Variational Transition State Theory (VTST) [170, 173–176]. VTST and its variants usually produce much better agreement with experimental kinetics than CTST [172]. The program POLYRATE [177] implements several variational approaches (e. g., Canonical VTST, Improved Canonical VTST, Microcanonical VTST), as well as several quantum tunneling corrections for light atoms. As a consequence of the dual particle-wave nature of matter, light particles like electrons or protons can "tunnel" through potential energy barriers [178, 179]. De-Broglie equation predicts a wavelength of ~27 Å associated to an electron and a ~0.6 Å wavelength associated to a proton moving with kinetic energy of 20 kJ/mol. Experimental signs of tunneling effect usually entail strong H/D isotopic effect, non-linear behaviors of the Arrhenius equation and temperature independence of the reaction rate. Examples of electron tunneling can be found in semiconductivity, superconductivity, scanning microscopy and biological phenomena related to charge transfer [180]. The first experimental proof of tunneling in a chemical reaction came from F. Williams's work [181]. There are many experimental evidences claiming tunneling effects of light atoms like protons in elimination reactions [182]. Heavier atoms like carbon or nitrogen or systems with small reduced masses can experience tunneling effect as well, in reactions like conformational changes [183], isomerizations [184] and ring expansions [185]. The computation of tunneling effects is generally accomplished with accurate multi-dimensional derivations of WKB method [186], like the small-curvature (SCT), the large-curvature (LCT), the optimized multidimensional and the least-action tunneling approximations [170, 187]. The inclusion of a multiplicative factor, α_T, called transmission coefficient, takes care of the corrections to the Eyring equation that relate to effects like tunneling, for instance. The inclusion of the transmission factor into the calculation of rate constants (therefore reaction equations) improves the goodness of calculated reaction rates versus experimental [170, 172, 188]. The following equations show the analogy between transition state theory and experimental kinetic laws in calculating rate constants (for standard state of 1 mol/L): **(a)** Eyring-Polanyi equation, **(b)** E-P equation with separated enthalpic and entropic terms and **(c)** Arrhenius law.

$$(\mathbf{a})\ k = \alpha_T \cdot \frac{k_b T}{h} \cdot e^{\frac{-\Delta G^\ddagger}{RT}} \quad (\mathbf{b})\ k = \left(\alpha_T \cdot \frac{k_b T}{h} \cdot e^{\frac{\Delta S^\ddagger}{R}}\right) \cdot e^{\frac{-\Delta H^\ddagger}{RT}} \quad (\mathbf{c})\ k = A \cdot e^{\frac{-Ea}{RT}}$$

Transition state theory provides valuable kinetic discrimination of reaction-controlled events, alias events that need to cross an energy barrier. The first predictive use of transition state theory in organic chemistry involved the electro-cyclic ring opening of 3-formylcyclobutene; computational insights predicted for the latter system an unusual reactivity compared to its congener 3-methylcyclobutene [189]. The Hajos-Parrish reaction was the first organocatalysis to be studied from first principles [190–192]; yet again transition state theory proved to be an invaluable tool in uncovering the energetic of the stereoselection between the more stable s-*trans* geometry (3.5 kcal/mol) and the less stable s-*cis* [193]. The first example of computational design was applied to selective Mannich-type reactions. The reaction between propionaldehyde and N-PMP-protected α-iminoethyl glyoxylate is catalyzed by (S)-proline and affords Mannich products of *syn* configuration (syn:anti 3:1, 99 % enantiomeric excess, entry #1 in Table 1.5). K.N. Houk, C. Barbas et al. produced a computational model of the reaction at HF/6–31G* level in 2006. Transition state theory provided an excellent estimation of the observed diastereoselectivity, 3.5:1, and enantiomeric excess, ~95%[193]. They produced a second computational model where (S)-pipecolic acid replaced (S)-proline as active catalyst of the reaction (Figure 1.9 and entry #2 in Table 1.5) and noticed that a loss of diastereoselectivity in Syn(1) product was taking place in favor of Anti(2) product. Targeted catalytic design, based on the insight built during this strategic approach, lead to the first computational design of an organocatalyst in 2006. The catalyst, an artificial aminoacid (entry #3 in Table 1.5), was designed at HF/6-31G* level of theory to afford selective anti-Mannich reactions, a completely opposite reactive pathway in this type of systems. The predicted outcome of the reaction, 5:95 syn:anti and 98 % ee, were perfectly matched by the successive experimental findings, 6:94 syn:anti and 99 % ee.

Table 1.5: The strategy to catalytic design. Entry #1 shows the data for the Mannich-type reaction catalyzed by (S)-proline, entry #2 shows those for the reaction catalyzed by (S)-pipecolic acid. Computation designs experiment in entry #3.

Catalyst	Type	Diastereomeric Ratio Syn(1): Anti(2)	Enantiomeric excess % Syn (Anti)	Phase
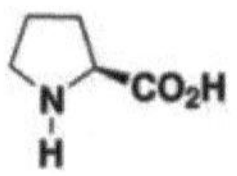	Exp.	3.0:1.0	>99	**Observation I**
	Comp.	3.5:1.0	97	**Model fitting I**
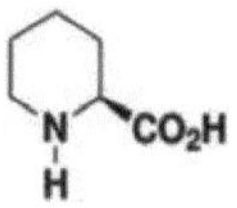	Exp.	2.0:1.0	>99 (>99)	**Observation II**
	Comp.	1.4:1.0	93 (96)	**Model fitting II**
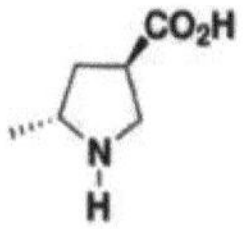	Comp.	**1.0:19.0**	**98**	**Catalytic design**
	Exp.	**1.0:15.7**	**99**	**Exp. proof**

While many other excellent examples of synergistic theory-synthesis approaches in chemistry can be found in literature [194–208], we report here a very interesting paper published by A. Kleij, C. Bo et al. on the use of triazabicyclodecene (TBD) as a organocatalyst of chemoselective ring-opening of cyclic carbonates to afford N-aryl carbamates in cheap and mild conditions [209]. Carbamates are products of CO_2-incorporation. Greenhouse gases (GHG) like CO_2 derived from anthropogenic activities are held responsible for the global warming scenario [210–214]. Sequestrating these gases from the atmosphere and incorporating them in building blocks for fine chemistry is highly sought after [215–217]. DFT-based (B97-D3/6-311G**/SMD) methods as implemented in G09 [98] helped to uncover a favorable proton-relay mechanism as the main source of the fast catalytic effect of TBD at room temperature (Figure 1.10); two hydrogen bonds imprint a 8-membered ring transition state and lower considerably the barrier of activation at the rate-determining step. Two competitive pathways, namely the direct nucleophilic attack of the amine onto the cyclic carbonate and water-assisted pathway, are either kinetically unfeasible or not as effective as TBD (Figure 1.10). The crucial contribution of this proton-relay switch has been counter-checked by experiment: replacing TBD by methylated TBD (MTBD) leads to much poorer catalytic effect [209].

A priori design of enzymatic pockets relies on transition state theory as well. Theozymes are theoretical catalysts tailored by optimizing the transition state of the reactants in a non-catalyzed reaction [218]. Though natural enzymes are molecular

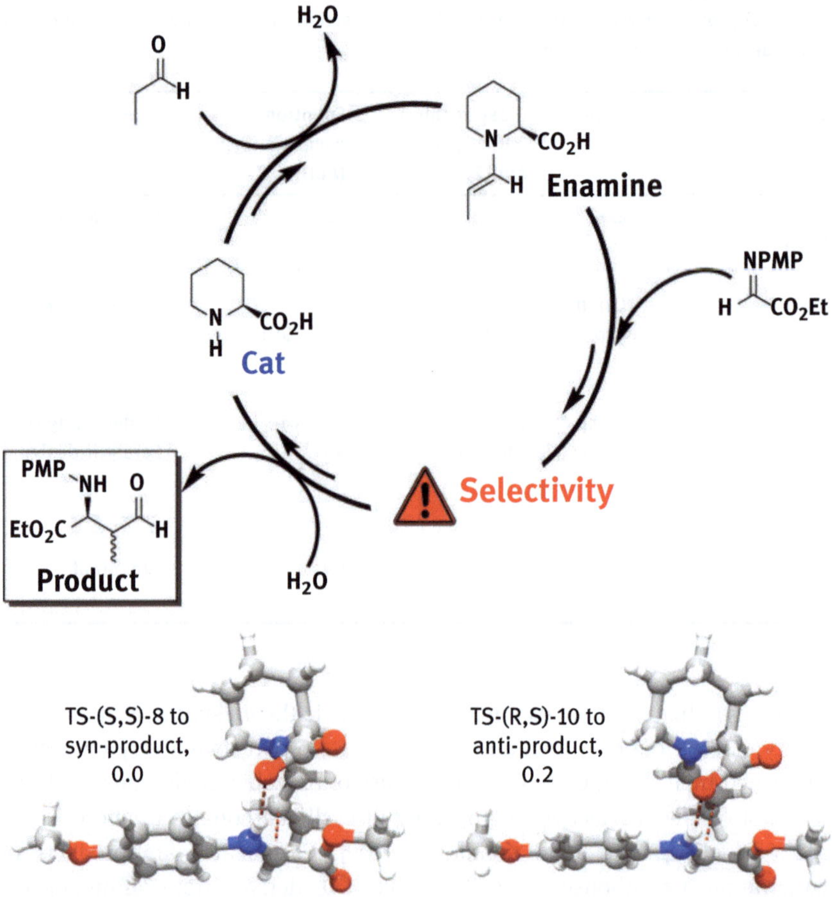

Figure 1.9: Schematic representation of Barbas-Houk's Mannich-type reaction discussed in this section (top). Transition state theory lead to crucial observations (bottom) concerning structures and energetics that sparked computational design of a new catalyst with desired properties.

machines of unmatched efficiency in binding and transforming several substrates with high selectivity, synthetic enzymes can be designed to catalyze reactions unknown to natural enzymes. This is the case of the retro-aldol dissociation reaction of 4-hydroxy-4-(6-methoxy-2-naphthyl)-2-butanone [219], the Kemp rearrangement reaction of a benzisoxazole [220] and the Diels-Alder condensation between 4-carboxylbenzyl *trans*-1,3-butadiene-1-carbamate and *N,N*-dimethylacrylamide [221]. The group employed a sequential methodology involving data-mining, predictive design (Rosetta hashing algorithm [222]), transition state optimizations, selection of best leads and finally expression and directed evolution to generate synthetic enzymes. The strategic approach and the exceptional results of such computationally-assisted design are reported for the first two reactions in Figure 1.11 and Table 1.6,

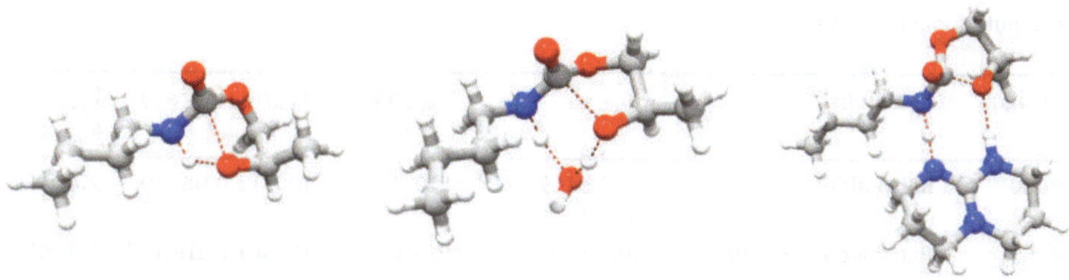

Figure 1.10: Transition structures for ring-opening of a cyclic carbonate enhanced by n-butylamine: left, non-catalyzed ($\Delta G^{\ddagger} = 33.3$ kcal/mol), center, H_2O-catalyzed ($\Delta G^{\ddagger} = 24.2$ kcal/mol) and right, TBD-catalyzed ($\Delta G^{\ddagger} = 18.2$ kcal/mol).

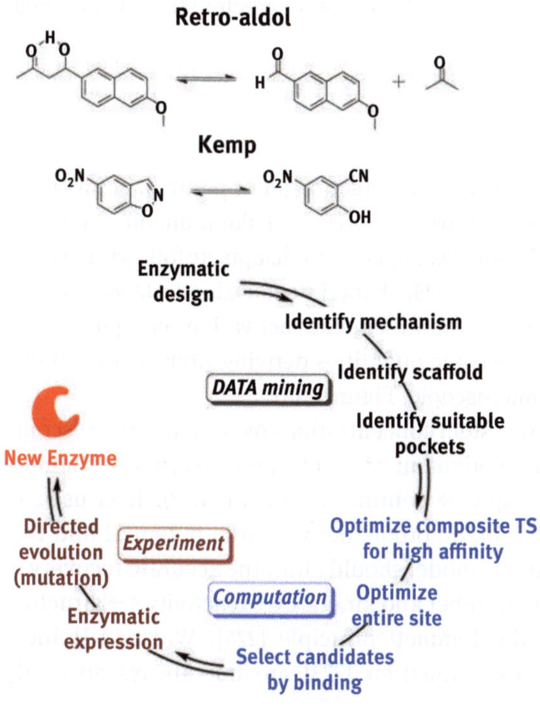

Figure 1.11: Nature does it better, but scientists are closing the gap. The route to "artificial" enzymes is an interplay ground for statistical techniques, theoretical computation and targeted synthesis.

respectively. An experimental technique that could be conceptually linked to transition state theory is femtosecond spectroscopy. This technique allows pulsing wave-packets on a molecule by the means of an ultrafast laser. A vibrational normal mode of interest is excited and the structure rearranges trough an activated complex to the formation of the desired product. This spectroscopy allows the accurate mapping of the entire surface and its dynamics from the reactants to the products,

Table 1.6: Experimental activity and kinetic parameters of mutated artificial enzymes designed by computational methods.

Design	Reaction	k_{uncat}	k_{cat}	K_M (mM)	k_{cat}/K_M ($M^{-1}s^{-1}$)	k_{cat}/k_{uncat}
RA60	Retro-aldol	$3.9 \cdot 10^{-7}$ min^{-1}	$(9.3 \pm 0.9) \cdot 10^{-3}$ min^{-1}	510 ± 33	$(0.30 \pm 0.06) \cdot 10^{-3}$	**$2.4 \cdot 10^{4}$**
RA61	Retro-aldol	$3.9 \cdot 10^{-7}$ min^{-1}	$(9.0 \pm 1.0) \cdot 10^{-3}$ min^{-1}	210 ± 50	$(0.74 \pm 0.11) \cdot 10^{-3}$	**$2.3 \cdot 10^{4}$**
R7 2/5B	Kemp	$1.16 \cdot 10^{-6}$ s^{-1}	(1.20 ± 0.08) s^{-1}	0.86 ± 0.08	1.388 ± 44	**$1.03 \cdot 10^{6}$**
R7 10/11 G	Kemp	$1.16 \cdot 10^{-6}$ s^{-1}	(1.37 ± 0.14) s^{-1}	0.54 ± 0.12	2.590 ± 302	**$1.18 \cdot 10^{6}$**

thus providing important experimental details on the characteristics of the "near transition" region [11].

1.3.5 Kinetic models

Complex multi-step chemical reactions are the direct result of parallel, competitive and consecutive events establishing an intricate network of simultaneous equilibria in the medium: within this network, species appear and disappear following (sometimes) complex phenomenological laws [223]. Experimental kinetic laws use rate constants derived from macroscopic quantities like activation energies (e. g., Arrhenius [223] law) to average microscopic quantities deriving from particle transport and collisions (continuum → microscopic, Figure 1.12).

A kinetic model uses integrated system concentrations over time of microscopic quantities (i. e. transition state thermodynamics) on elemental events to simulate the macroscopic reaction (microscopic → continuum, Figure 1.12): it is used to complement and corroborate mechanistic proofs derived from potential and free energy surfaces. In principles, a kinetic model should allow the accurate recovery of information on selectivity, turnover number and kinetics ↔ reactivity ↔ structure relation in conditions where Curtin-Hammett principle [224], Winstein-Holness equation [224] and steady-state approximation [223] are not always so easily applicable.

Kinetic models abide the principle of microscopic reversibility and represent a strong direct link between pure theory and experiment since they can be compared straightforwardly with plots obtained by experimental kinetics [225]. The higher the number of kinetic equations included in the simulation, the better the agreement of the model versus experiment. Kinetic models are particularly useful to model catalytic reactions, where the rate limiting step(s) may change throughout the reaction following the change in concentrations of the reactants/intermediates [223]. Analytical integration is prohibitive on large systems; therefore one should resort to robust numerical

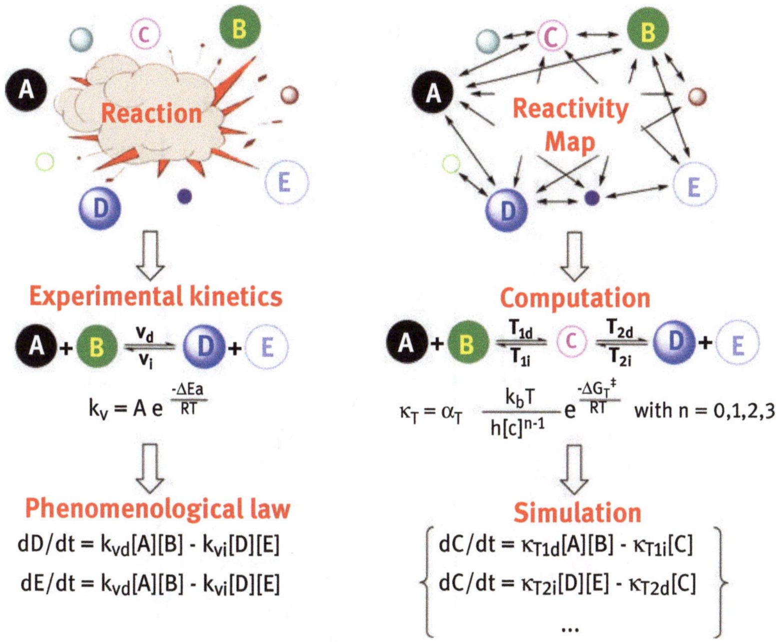

Figure 1.12: The different approach of experimental kinetics versus kinetic model.

algorithms like Euler, 4th-order Runge-Kutta, Runge-Kutta-Fehlberg, Runge-Kutta-Cash-Karp, Runge-Kutta-Dormand-Prince, Bader-Stoer and Adams' methods [223]. "Stiff" equations can be efficiently treated with Bader-Deuflhard algorithm [226]. An elegant example of kinetic models has been reported by F. Maseras et al. It concerns the computational study at B97-D3/6-311G(d)/SMD level of theory of a known 1,3-dipolar cycloaddition between propargylamine and azidoethylamine [227], Figure 1.13. The computed thermodynamic parameters reproduced successfully the experimental findings ($\Delta G^\ddagger = 27.1$ versus 27.3 kcal/mol, for the rate limiting step of the reaction) and the kinetic model helped to re-interpret the experimental mechanism of the reaction in light of the catalytic effect of host-guest encapsulation provided by cucurbit[6]uril (CB6) [227]. Another excellent example has been reported by Maseras et al. The work encompasses the use of non-adiabatic DFT calculations (ωB97X-D) supported by kinetic model in explaining the outcome of photo-activated aromatic perfluoroalkylations [228].

1.3.6 Multi-scale models

The idea of a consistent force-field was born in 1968 [229]. The total potential energy of a molecule can be calculated as a function of simple classical laws involving Hooke's spring law for bonds, periodic laws for dihedral torsions, Lennard-Johns

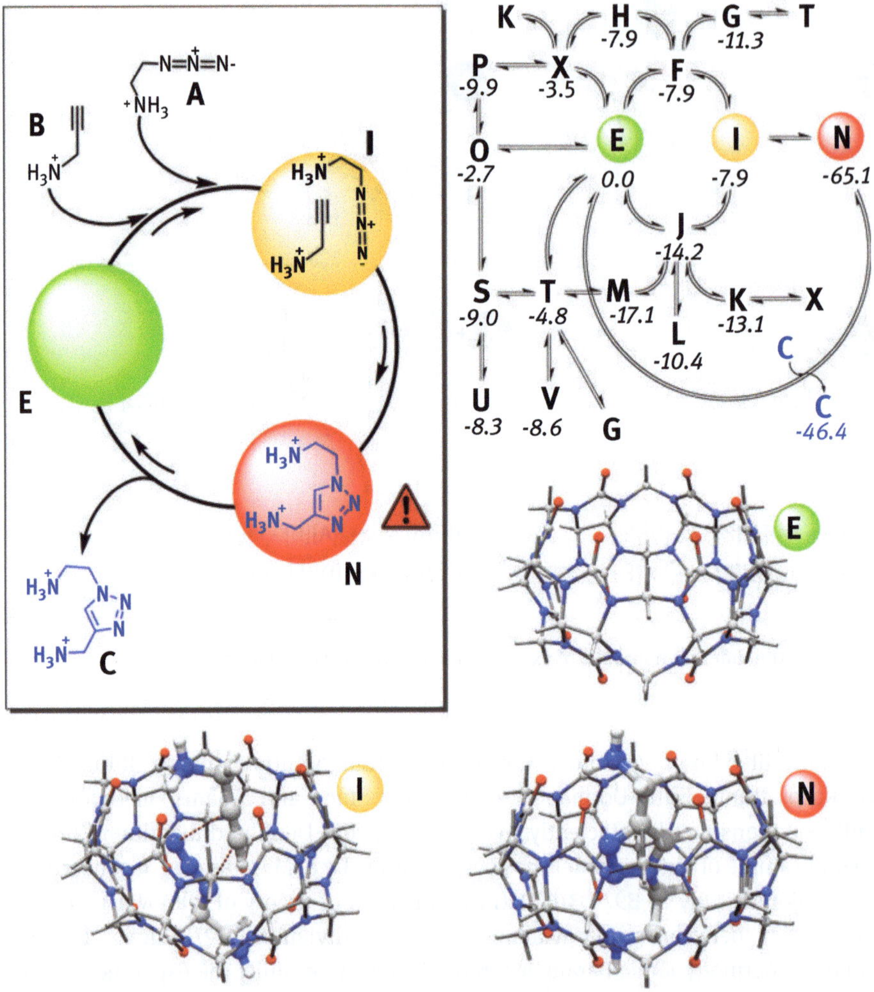

Figure 1.13: Schematic representation of the dipolar cycloaddition discussed in this section, top-left. Maseras's pruned network of calculated competing events proves the "combinatorial" approach behind kinetic model, top right. Molecular structures of the most important intermediates in the reaction (bottom).

equation for Van der Walls interactions and Coulomb law for charge interactions [230]. The potential energy can be further related to energy minimization, intrastate and interstate dynamics and random moves (Monte Carlo) [230]. Multi-scale models were originally proposed in 1976. They allow the optimization and calculation of properties of very large systems (encompassing hundreds of thousands of atoms); thus they have been very useful in the calculations of large peptides, proteins and enzymes like lysozyme [231]. The general idea behind this method is that the chemical system under study can be divided into different layers, calculated separately using different methods. Larger areas that do not take direct part in the reaction (e. g., protein side chains) are calculated using inexpensive classical treatments like molecular mechanics (MM), while active site of reaction can be treated with higher-level quanto-mechanical theory (QM). For such reason these methods are called QM/MM methods. The energy of a total system, E_{Sys}, can be defined as:

$$E_{Sys} = E_{QMQM} + E_{MMMM} + E_{QMinterQM/MM} + E_{MMinterQM/MM}$$

The steric and polarization effects of the classically-treated low-level layer are contained in the term E_{MMMM}, while E_{QMQM} contains the quantum mechanics energy of the high-level layer. $E_{QMinterQM/MM}$ is called electronic embedding, while $E_{MMinterQM/MM}$ is called mechanical embedding. Different methods have different ways to calculate the latter two parameters. Two of these QM/MM methods are available and implemented in computational software: IMOMM [232] and ONIOM [233] methods.

There are excellent reviews [234, 235] and highlights [236] and commentaries [194] on QM/MM calculations in asymmetric catalysis via organometallic species; these works include reactions like hydrogenation, hydroboration, epoxidation and hydroformilation. F. Schoenebeck et al. used ONIOM(B3LYP/6-31+G(d)/LANL2DZ:HF/LANL2MB) level of theory to design a new small-bite phosphine ligand able to enhance the reductive elimination of $PhCF_3$ from Pd(II) complexes. Contrarily to accepted knowledge, the group demonstrated computationally that the reactivity was not correlated with the bite-angle of the phosphine, rather with the repulsive electrostatic interactions between the phosphine and the leaving group (namely CF_3) [237]. Calculations provided a good estimate of the activation enthalpy for elimination of $PhCF_3$ in [(Xantphos)Pd(CF_3)(Ph)], 25.1 kcal/mol versus the experimental value of 25.9 ± 2.6 kcal/mol [238]. Schoenebeck's new complex [(dfmpe)Pd(CF_3)(Ph)] (Figure 1.14), afforded an activation enthalpy range of $\Delta H^{\ddagger} = 20.7–23.5$ kcal/mol, depending on various theories [237]. Driven by the successful computational design, the group synthesized [(dfmpe)Pd(CF_3)(Ph)] and demonstrated that the elimination of $PhCF_3$ is complete after 3 hours at 80 °C after 100 min. Kinetic experiments by NMR measured the activation enthalpy as $\Delta H^{\ddagger} = 27.9 \pm 1.6$ kcal/mol [237]. Another challenge for QM/MM methods is represented by designing chemotherapeutic agents with *in vivo* high affinity and selectivity towards target DNA helix. A synergistic approach between molecular dynamics simulations, QM/MM calculation, force fields and experimental resources (i. e., high-resolution structures of nucleosome-drug adduct) has been applied to further the knowledge on binding location, binding modes and selectivity of metallo-drugs versus DNA and proteins during these last years [239].

1.3.7 Wavefunction analysis

Optimized wavefunctions at the stationary points, $\Psi_i(r) = \sum_i C_{l,i}\Phi_l(r)$, reveal a great deal of information concerning chemical bonds structures and properties like nature,

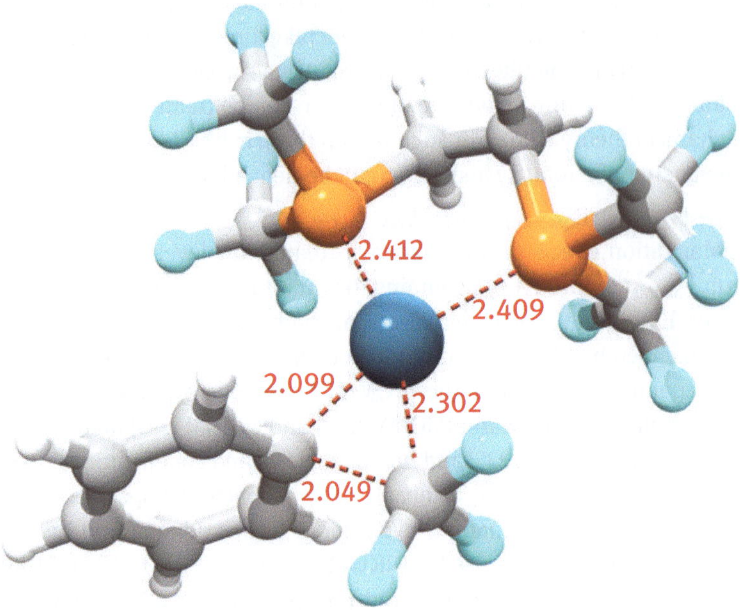

Figure 1.14: Bite-angle is not the issue, says theory. Schoenebeck's designed palladium complex reductive-eliminates $PhCF_3$, despite its small bit-angle phosphine. The figure shows the transition structure for such elimination (bond distances in Å).

directionality, composition, reactivity and selectivity. Among others, some properties derived from wave-function are very useful and routinely used by chemists: molecular orbitals, population analysis, bond order analysis, molecular electrostatic potential, Fukui function [240], charge decomposition analysis [241], natural bond orbital (NBO) analysis [242, 243] and adaptive natural density partitioning analysis [244]. Electron densities obtained from optimized wave-functions, $\rho(\mathrm{r}) = \sum_i \eta_i |\psi_i(\mathrm{r})|^2$ are also excellent tool of investigation to uncover points of interest within distributions of electronic clouds around atoms and molecules. Unlike wave-functions, electron densities are less sensitive to the choice of theory and basis sets and provide a direct comparison to observable electron density phases obtained from x-ray or neutron single-crystal scattering [79], $\rho(\mathrm{xyz}) = 1\mathrm{V} \sum_{hkl} \mathrm{F}_{(\mathrm{hkl})} \mathrm{e}^{-2\pi\, \mathrm{i}(\mathrm{hx+ky+lz})}$. The Quantum Theory of Atoms-in-molecule (QTAIM) analysis is a technique based on the topological analysis (critical points) of $\rho(\mathrm{r})$ and $\nabla^2\rho(\mathrm{r})$ [245]. Figure 1.15 shows the QTAIM of two isoelectronic closed-shell d^{10} complexes, $Ni(\eta^2\text{-}C_2H_4)_3$ and $Cu(\eta^2\text{-}C_2H_4)_3^+$ generated by MULTIWFN program [246, 247] (other software for QTAIM includes AIM2000 [248] and AIMAll [249]). M-C bond pathway is unique in the case of copper and the bonding pattern in $Cu\text{–}\eta^2\text{-}C_2H_4$ is T-shaped, implying a central λ-donation-type interaction between the electron densities of copper and ethylene moieties; in the case of nickel, however, the pathway is

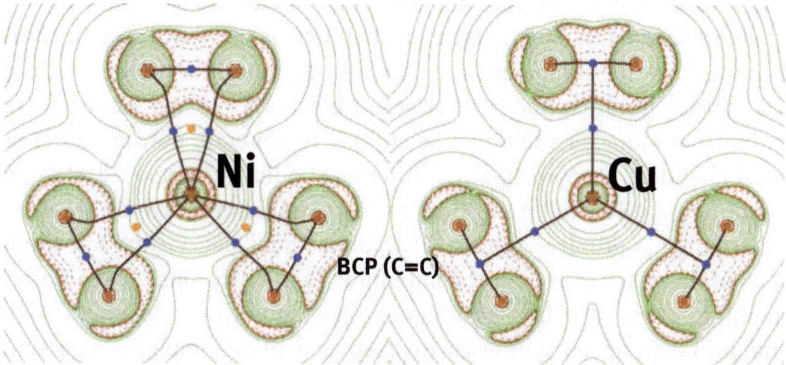

Figure 1.15: Molecular graphs of $Ni(\eta^2\text{-}C_2H_4)_3$, left, and $Cu(\eta^2\text{-}C_2H_4)_3^+$, right, overlaid with their respective Laplacian maps, $\nabla^2\rho(r)$. The structures have been calculated using B3PW91 in conjunction with the non-relativistic full-electron DGDZVP basis set. Blue dots represent bond critical points, or BCP (3,-1), orange dots represent ring critical points, or RCP (3,+1), and brown dots represent nuclear attractors (3,-3). Within green zones kinetic energy density dominates and the electron density is depleted $\{\nabla^2\rho(r) > 0\}$. Within red zones potential energy dominates and the electron density is concentrated $\{\nabla^2\rho(r) < 0\}$.

doubly degenerated and the bonding pattern assumes a more triangular shape, in line with the increased π-back-bonding character in the Ni-C interaction (increased sp^3 character on the carbons). C=C bond critical point (BCP) suggests that ethane loses 10.0 % of C-C internuclear electron density per unit upon complexation with Ni^0 and only 4.5 % upon complexation with Cu^+.

Among others, topological analysis functions are very useful and routinely used by chemists: electron localization function (ELF) [250, 251], localized orbital locator (LOL) [252, 253], reduced density gradient (RDG) [254] and source function [255]. ELF was used by M. Kaupp et al. to explain the bond in another uncommon compound, HgF_4 [256]. The bond pattern, consistent with the representation of a low-spin d^8 complex, defines mercury as real transition metal, rather than a post-transition element. Mercury in high oxidation state was theoretically predicted by Kaupp et al. to be stable as HgF_4 since 1993 [257–259], with weakly-coordinating ligands like SbF_6^- [260] and as HgH_4 by P. Pyykkö et al. [261] and finally isolated and characterized spectroscopically in matrix in 2007 [256].

1.3.8 Chemoinformatics and machine-learning

Under the name of chemoinformatics falls a large and diverse number of different disciplines spanning from computational chemistry, mathematics, and statistics, data-mining to informatics. The accepted goal of chemoinformatics is to fabricate Hit-to-Lead-to-Candidate [262] predictions on unknown molecular systems using validated and statistically significant data obtained from libraries of known properties and chemical behaviors. Computational chemistry is primarily used in pharmaceutical industries to complement, integrate or prioritize targeted drug synthesis via virtual libraries and virtual screening [263]. Virtual libraries are collections of virtual molecules that can be studied to generate new molecule or to hypothesize chemical properties. Such libraries should be composed of large and fitting populations. These libraries can be increased using "trained" Markov chains like in fragment optimized growth (FOG) algorithm [264]. Virtual screening is the analysis of database of chemical compounds to identify viable candidates. Virtual screening uses *in silico* techniques like molecular docking [263]. Combinatorial and high-speed syntheses are used as lead or complementary synthetic methods to chemoinformatics, since they are designed to generate or screen large libraries of compounds [263]. High-throughput computational modeling uses languages like the Chemical Markup Language (CML) and the Simplified Molecular-Input Line-Entry System (SMILES). Physico-chemical properties are normally represented by descriptors: a descriptor is a compact transcription of a certain property within a sample or library of species into mathematical language. The goal of a descriptor is to convey the essence of the quantitative structure-activity relationship (QSAR) [265]. The complexity of the property described by the descriptor increases with the dimensionality of the descriptor itself: 1D-descriptors deal with simple macroscopic bulk properties like molecular weight, while 3D-descriptors capture the essence of the donor-acceptor interaction and binding affinity (polar surface area for instance). 1D-descriptors are used as first-approach filters, while 2D-descriptors (also called fingerprints) seem to work well in reproducing many chemical properties [266]. GRID is an interesting program for drug discovery applications and ADME predictions related to pharmacokinetics [43]. QSAR approach is used extensively in pharmaceutical chemistry [267, 268], but its use in catalytic applications involving small molecules results to be still sparse. An excellent example of QSAR, or in this case QSSR (quantitative structure-selectivity relationship), is represented by M.C. Kozlowski et al.'s work [269, 270]. The group aimed to correlate the structure of in situ-generated zinc catalysts ($ZnBr_2$ + β-aminoacids) with the enantioselectivity of the asymmetric reduction of benzaldehyde to related alcohol. The group generated a grid-based QSAR approach whose descriptors provided a correct replica of the enantioselective footprint of catalysts under study. Since every transition structure have been generated using computationally low-demanding PM3 theory, this model provides important insight at very affordable costs [269].

Machine learning is a branch of computer science that deals with "supervised" or "unsupervised" learning of a machine. The goal of this discipline is the development of artificial intelligences, networks that mimic the human mind's capability to learn, store and relay data using electrochemical signals [271]. Network analysis, for instance, has been applied in organic synthesis since the Sixties, when E. Corey created software called Logic and Heuristics Applied to Synthetic Analysis (LHASA). The software was meant to suggest the sequences of postulated steps in a *de novo* synthesis, based on a database of retrosynthetic rules [272]. In 2005, B. A. Grzybowski et al. developed in 2005 a revolutionary concept of a network connecting molecules ("nodes") by chemical reactions ("arrows") [273]. The network contains about 6 million organic compounds interrelated by 30,000 retrosynthetic rules. The group focused on the evolution of this network, the relations between connections in the network and reactivity and optimization/pruning of the network itself. The software developed after this network was called Chematica. It can successfully generate and optimize synthetic pathways [274, 275] and detect synthetic pathways leading to dangerous chemicals [276]. Other interesting examples on chemoinformatics and machine learning (i. e., artificial neural networks) can be found in literature [277].

1.4 The future: Conclusions and author's perspectives

A brief evolution of the rational way of doing chemistry has been presented throughout this chapter. General principles and descriptions of rational improvements to chemistry have been introduced and described to provide the reader with a general background on this topic. Paradigmatic examples on the synergistic use of theory, spectroscopy and synthesis have been provided to show how the rational approach is more than advantageous in understanding, developing and planning new chemistry. It is certain that the evolution of transistors, semi-conductors and miniaturized circuits will provide a boost in the hardware performances (thus an incentive to study large and more complex systems raising the bar of the overall accuracy) in the next few decades, provided that Moore's law will still be followed faithfully [278]. Furthermore, once the binary code will no longer hold sway, quantum-bits-based architectures might boost computational power and speed to new, far higher plateaus. At the present days, there are reports of commercialization of quantum architectures (e. g., D-Wave 2X) by Canadian company D-Wave System Inc. [279], but still, room is left for scientific debate and criticism [280–284]. Software improvements (i. e., implementation of improved routines, compilers, libraries, parallelization and so on) in commercialized computational programs (an extensive list of the major computational software used in chemistry, physics and material science can be found online [285, 286], as well as a list of the programs and suites to prepare input files and visualize output files [287]) will also play a protagonist role in speeding up the chemistry on screen. Technological improvements may also help experimentalists to keep a better and more functional record of their lab experiments; for instance, electronic laboratory notebooks (ELNs) [288] will become excellent platforms to track and share successful and failed synthetic attempts. These data, particularly the failures, when available on electronic supports for unrestricted mining and sharing, will be of tremendous importance in the validation, refinement and improvement of theoretical models on real systems.

Thus, all these betterments will surely contribute to keep the rational momentum of chemistry evolving in the near future. Truth is, the impact of a higher degree of rationality might profoundly influence the way chemistry is taught, discussed and disseminated as well. One of the greatest limitations in scientific fields is that specialized professionals speak different languages [289]. This loss in translation can sometimes be the cause of scientific misunderstandings, disagreements and loss or disregard of crucial snippets of information in shared projects and assignments. Thus, scientific team-work and collaborations [289] will benefit in efficiency from a new breed of chemist, groomed from early academic stages to blend theory, spectroscopy and synthesis (understanding powers and limits embedded in different approaches!) to pursue research duties and interests. Such a figure may deliver more powerful performances than his/her predecessors in academic and industrial environments alike.

Secondly, rational chemistry might lead to a positive relief of the environmental footprint that is a great concern in the modern era. Chemistry can be a double-edged sword: while in the past it has shown that it can improve the quality of life and fuel technology, it has also contaminated nature and its living creatures. The current drive in research funding is to push for the former without creating the latter as a deleterious effect. Though regulated by national agencies, chemical pollution is a transnational problem that leads inexorably to problems like endocrine disruption due to chronic exposure [290]. Paradoxically, since a clean man-made chemical reaction is a reaction that never took place, the concepts of atom-economy and solvent preservation need to be implemented to their fullest. Validated computational techniques could represent a greener and more sustainable way to design catalysts (or electrodes, membranes, polymers and so on) and tune catalytic effects than indiscriminate trial-and-error synthesis. Finally, there is the issue of the economic sustainability. Pharmaceutical companies are already incorporating some of the computational tools [267] presented in this chapter to design new bioactive molecules and pre-screen their biological activity. High-quality computation techniques are indeed less impacting on R&D budgets compared to trial-and-error synthesis. Furthermore, rational design of reagents could shield enterprises from legal internal liabilities deriving from the use and storage of harmful reagents and the stock and disposal of dangerous waste.

References

[1] Fowler AC. Mathematical models in the applied sciences. Cambridge, UK: Cambridge University Press; 2004.

[2] Melnick R, editor. Mathematical and computational modeling: with applications in natural and social sciences, engineering, and the arts. New York, USA: John Wiley & Sons Inc; 2015.

[3] Ball P. Chemistry: why synthesize? Nature. 2015;528:327–329.

[4] Mortimer CE. Chemistry: a conceptual approach. Belmont, CA, USA: Wadsworth Publishing Co. Inc; 1983.

[5] Francl M. The enlightenment of chemistry. Nat Chem. 2015;7(10):761–762.
[6] Norris JA. The mineral exhalation theory of metallogenesis in pre-modern mineral science. Ambix. 2006;53(1):43–65.
[7] Greenwood NN, Earnshaw A. Chemical periodicity and the periodic table, in chemistry of the elements. Oxford: Butterworth-Heinemann; 1997. p. 20–31.
[8] Kauffman GB. The quiet revolution. Hermann kolbe and the science of organic chemistry. Von A. J. Rocke. Angew Chem. 1994;106(21):2317–2318.
[9] Von Meyer E. Zur erinnerung an hermann kolbe. J Prakt Chem. 1884;30(1):417–467.
[10] Schwarz WHE. 100th anniversary of Bohr's model of the atom. Angew Chem Int Ed. 2013; 52(47):12228–12238.
[11] Atkins P, De Paula J. Atkins' physical chemistry, 8th ed. Oxford, UK: Oxford University Press, 2006.
[12] Einstein A, Podolsky B, Rosen N. Can quantum-mechanical description of physical reality be considered complete? Phys Rev. 1935;47(10):777–780.
[13] Williams DB, Carter CB. Transmission electron microscopy: a textbook for materials science. 2nd ed. New York, NY, USA: Springer-Verlag New York Inc.; 2009.
[14] Goldstein, J., et al. Scanning electron microscopy and x-ray microanalysis. New York, NY, USA: Springer-Verlag New York Inc; 2013.
[15] Jansen F. Introduction to computational chemistry. 2nd ed. Hoboken, USA: John Wiley and Sons Ltd; 2007.
[16] Hensen, B., et al. Loophole-free bell inequality violation using electron spins separated by 1.3 kilometres. Nature. 2015;526(7575):682–686.
[17] De Hoffmann E, Stroobant V. Mass spectrometry: principles and applications. 3rd ed. Hoboken, USA: John Wiley and Sons Ltd; 2007.
[18] Moldoveanu SC, David V. Mass spectrometry: principles and applications. USA: Elsevier Science Publishing Co Inc; 2016.
[19] Poole C. Gas chromatography. Amsterdam, Netherlands: Elsevier Science Publishing Co Inc; 2012.
[20] Corey EJ. Robert robinson lecture. retrosynthetic thinking-essentials and examples. Chem Soc Rev. 1988;17(0):111–133.
[21] Oldenburg H. Epistle dedicatory. Philosophical Trans. 1665;1:1–22.
[22] Huang, B., et al. Chlorinated volatile organic compounds (Cl-VOCs) in environment – sources, potential human health impacts, and current remediation technologies. Environ Int. 2014;71:118–138.
[23] Ni, J.Q., et al. Volatile organic compounds at swine facilities: A critical review. Chemosphere. 2012;89(7):769–788.
[24] Anastas PT, Warner JC. Green chemistry theory and practice. New York, USA: Oxford University Press; 1998.
[25] Mascia, S., et al. End-to-end continuous manufacturing of pharmaceuticals: integrated synthesis, purification, and final dosage formation. Angew Chem Int Ed. 2013;52(47):12359–12363.
[26] Xiang, X.D., et al. A combinatorial approach to materials discovery. Science. 1995; 268(5218):1738.
[27] Lewis GN. The atom and the molecule. J Am Chem Soc. 1916;38(4):762–785.
[28] Gillespie RJ. Fifty years of the VSEPR model. Coord Chem Rev. 2008;252(12–14):1315–1327.
[29] Hückel E. Zur quantentheorie der doppelbindung. Z Phys. 1930;60(7):423–456.
[30] Hückel E. Quantentheoretische beiträge zum benzolproblem. Z Phys. 1931;70(3):204–286.
[31] Woodward RB, Hoffmann R. Stereochemistry of electrocyclic reactions. J Am Chem Soc. 1965; 87(2):395–397.

[32] Figgis BN, Hitchman MA. Ligand field theory and its applications. New York, USA: John Wiley and Sons Ltd; 2000.
[33] Racah G. Theory of complex spectra. II. Phys Rev. 1942;62(9-10):438–462.
[34] Tanabe Y, Sugano S. On the absorption spectra of complex ions. I. J Phys Soc Jpn. 1954; 9(5):753–766.
[35] Tanabe Y, Sugano S. On the absorption spectra of complex ions II. J Phys Soc Jpn. 1954; 9(5):766–779.
[36] Tanabe Y, Sugano S. On the absorption spectra of complex ions, iii the calculation of the crystalline field strength. J Phys Soc Jpn. 1956;11(8):864–877.
[37] Lennard-Jones JE. The electronic structure of some diatomic molecules. T Faraday Soc. 1929; 25(0):668–686.
[38] Mulliken RS. The assignment of quantum numbers for electrons in molecules. I. Phys Rev. 1928;32(2):186–222.
[39] Schweitzer C, Schmidt R. Physical mechanisms of generation and deactivation of singlet oxygen. Chem Rev. 2003;103(5):1685–1758.
[40] Pauling L. Nature of the chemical bond: an introduction to modern structural chemistry. 3rd ed. Ithaca, USA: Cornell University Press; 1960.
[41] Pauling L. The nature of the chemical bond. II. The one-electron bond and the three-electron bond. J Am Chem Soc. 1931;53(9):3225–3237.
[42] Griffith JS, Orgel LE. Ligand-field theory. Quat Rev Chem Soc. 1957;11(4):381–393.
[43] Cruciani G, editor. Molecular interaction fields: applications in drug discovery and adme prediction. Weinheim, Germany: Wiley-VCH Verlag GmbH; 2006.
[44] Andres J, Bertran J. Theoretical and computational chemistry: foundations, methods and techniques. Castellón de la Plana, Spain: Universidad Jaume I. Servicio de Comunicación y Publicaciones; 2007.
[45] Foresman JB, Frisch A. Exploring chemistry with electronic structure methods: a guide to using gaussian. USA: Gaussian, Incorporated; 1993.
[46] Stewart JJP. Optimization of parameters for semiempirical methods VI: more modifications to the NDDO approximations and re-optimization of parameters. J Mol Model. 2013;19(1):1–32.
[47] Hohenberg P, Kohn W. Inhomogeneous electron gas. Phys Rev. 1964;136(3B):B864-B871.
[48] Kohn W, Sham LJ. Self-consistent equations including exchange and correlation effects. Phys Rev. 1965;140x(3B):A1133-A1138.
[49] Koch W, Holthausen MC. A chemist's guide to density functional theory. 2nd ed. Weinheim, Germany: Wiley-VCH Verlag GmbH; 2001.
[50] Perdew, J.P., et al. Prescription for the design and selection of density functional approximations: more constraint satisfaction with fewer fits. J Chem Phys. 2005;123(6): 062201.
[51] Dirac PAM. Note on exchange phenomena in the Thomas atom. Math Proc Cambridge. 1930; 26(3):376–385.
[52] Lieb EH, Simon B. The Thomas-Fermi theory of atoms, molecules and solids. Adv Math. 1977; 23(1):22–116.
[53] Zhao Y, Truhlar DG. The M06 suite of density functionals for main group thermochemistry, thermochemical kinetics, noncovalent interactions, excited states, and transition elements: two new functionals and systematic testing of four M06-class functionals and 12 other functionals. Theor Chem Acc. 2008;120(1):215–241.
[54] Zhao Y, Truhlar DG. Density functionals with broad applicability in chemistry. Acc Chem Res. 2008;41(2):157–167.
[55] Cohen AJ, Mori-Sánchez P, Yang W. Challenges for density functional theory. Chem Rev. 2012;112(1):289–320.

[56] Zhao Y, Lynch BJ, Truhlar DG. Doubly hybrid meta dft: new multi-coefficient correlation and density functional methods for thermochemistry and thermochemical kinetics. J Phys Chem. 2004;108(21):4786–4791.
[57] Karton, A., et al. Highly accurate first-principles benchmark data sets for the parametrization and validation of density functional and other approximate methods. derivation of a robust, generally applicable, double-hybrid functional for thermochemistry and thermochemical kinetics. J Phys Chem. 2008;112(50):12868–12886.
[58] Schwabe T, Grimme S. Theoretical thermodynamics for large molecules: walking the thin line between accuracy and computational cost. Acc Chem Res. 2008;41(4):569–579.
[59] Schwabe T, Grimme S. Towards chemical accuracy for the thermodynamics of large molecules: new hybrid density functionals including non-local correlation effects. Phys Chem Chem Phys. 2006;8(38):4398–4401.
[60] Schwabe T, Grimme S. Double-hybrid density functionals with long-range dispersion corrections: higher accuracy and extended applicability. Phys Chem Chem Phys. 2007;9(26): 3397–3406.
[61] Grimme S. Semiempirical hybrid density functional with perturbative second-order correlation. J Chem Phys. 2006;124(3):034108.
[62] Zhang Y, Xu X, Goddard WA. Doubly hybrid density functional for accurate descriptions of nonbond interactions, thermochemistry, and thermochemical kinetics. Proc Nati Acad Sci USA. 2009;106(13):4963–4968.
[63] Austin, A., et al. A density functional with spherical atom dispersion terms. J Chem Theory Comput. 2012;8(12):4989–5007.
[64] Grimme S. Semiempirical GGA-type density functional constructed with a long-range dispersion correction. J Comput Chem. 2006;27(15):1787–1799.
[65] Grimme, S., et al. A consistent and accurate ab initio parametrization of density functional dispersion correction (DFT-D) for the 94 elements H-Pu. J Chem Phys. 2010;132(15):154104.
[66] Grimme S, Ehrlich S, Goerigk L. Effect of the damping function in dispersion corrected density functional theory. J Comput Chem. 2011;32(7):1456–1465.
[67] Chai J-D, Head-Gordon M. Long-range corrected hybrid density functionals with damped atom-atom dispersion corrections. Phys Chem Chem Phys. 2008;10(44):6615–6620.
[68] Peverati R., Truhlar D. G. Quest for a universal density functional: the accuracy of density functionals across a broad spectrum of databases in chemistry and physics. Philos T Roy Soc A. 2014;372:20120476–20120527.
[69] Sousa SF, Fernandes PA, Ramos MJ. General Performance of Density Functionals. J Phys Chem. 2007;111(42):10439–10452.
[70] Gorin DJ, Toste FD. Relativistic effects in homogeneous gold catalysis. Nature. 2007;446(7134): 395–403.
[71] Pyykko P, Desclaux JP. Relativity and the periodic system of elements. Acc Chem Res. 1979; 12(8):276–281.
[72] Hay PJ, Wadt WR. Ab initio effective core potentials for molecular calculations. Potentials for the transition metal atoms Sc to Hg. J Chem Phys. 1985;82(1):270–283.
[73] Fuentealba, P., et al. A proper account of core-polarization with pseudopotentials: single valence-electron alkali compounds. Chem Phys Lett. 1982;89(5):418–422.
[74] Fuentealba, P., et al. On the reliability of semi-empirical pseudopotentials: simulation of Hartree-Fock and Dirac-Fock results. J Phys Pt B Atom M P. 1983;16(11):L323.
[75] Stevens WJ, Basch H, Krauss M. Compact effective potentials and efficient shared-exponent basis sets for the first- and second-row atoms. J Chem Phys. 1984;81(12):6026–6033.
[76] Stoll H. Large-core vs. small-core pseudopotentials: A case study for Au_2. Chem Phys Lett. 2006;429(1–3):289–293.

[77] Peterson KA, Puzzarini C. Systematically convergent basis sets for transition metals. II. Pseudopotential-based correlation consistent basis sets for the group 11 (Cu, Ag, Au) and 12 (Zn, Cd, Hg) elements. Theor Chem Acc. 2005;114(4):283–296.
[78] Nakajima T, Hirao K. The Douglas–Kroll–Hess approach. Chem Rev. 2012;112(1):385–402.
[79] Guinier A. X-ray diffraction: in crystals, imperfect crystals, and amorphous bodies. New York, USA: Dover Publications Inc; 1994.
[80] Cramer CJ. Essential of computational chemistry: theories and models, 2nd ed. The Atrium, Southern Gate, Chichester, West Sussex PO19 8SQ. England: John Wiley & Sons Ltd, 2004.
[81] Pyykko P. Predicted chemical bonds between rare gases and Au^+. J Am Chem Soc. 1995; 117(7):2067–2070.
[82] Schröder, D., et al. Cationic gold(I) complexes of xenon and of ligands containing the donor atoms oxygen, nitrogen, phosphorus, and sulfur. Inorg Chem. 1998;37(4):624–632.
[83] Seidel S, Seppelt K. Xenon as a complex ligand: the tetra xenono gold(II) cation in $AuXe_4^{2+}$ $(Sb_2F_{11}^-)_2$. Science. 2000;290(5489):117.
[84] Oganov AR, Glass CW. Crystal structure prediction using ab initio evolutionary techniques: principles and applications. J Chem Phys. 2006;124(24):244704.
[85] Dong, X., et al. A stable compound of helium and sodium at high pressure. Nat Chem. 2017;9:440–445.
[86] Zeeman P. VII. Doublets and triplets in the spectrum produced by external magnetic forces. Philos Mag Ser 5. 1897;44(266):55–60.
[87] Ernst RR, Bodenhausen G, Wokaun A. Principles of nuclear magnetic resonance in one and two dimensions. Oxford, UK: Oxford University Press; 1990.
[88] Eaton, G.R., et al. Quantitative EPR. Vienna, Austria: Springer Verlag GMBH; 2010.
[89] Levine IN. Molecular spectroscopy. New York, USA: John Wiley and Sons Ltd; 1975.
[90] Ditchfield R. Self-consistent perturbation theory of diamagnetism. Mol Phys. 1974;27(4): 789–807.
[91] Wolinski K, Hinton JF, Pulay P. Efficient implementation of the gauge-independent atomic orbital method for NMR chemical shift calculations. J Am Chem Soc. 1990;112(23):8251–8260.
[92] Cheeseman, J.R., et al. A comparison of models for calculating nuclear magnetic resonance shielding tensors. J Chem Phys. 1996;104(14):5497–5509.
[93] Keith TA, Bader RFW. Calculation of magnetic response properties using atoms in molecules. Chem Phys Lett. 1992;194(1):1–8.
[94] Keith TA, Bader RFW. Calculation of magnetic response properties using a continuous set of gauge transformations. Chem Phys Lett. 1993;210(1):223–231.
[95] Gauss J. Calculation of NMR chemical shifts at second-order many-body perturbation theory using gauge-including atomic orbitals. Chem Phys Lett. 1992;191(6):614–620.
[96] Gauss J. Effects of electron correlation in the calculation of nuclear magnetic resonance chemical shifts. J Chem Phys. 1993;99(5):3629–3643.
[97] Barone V, in Recent advances in density functional methods, Chong DP, editor 1996, World Scientific Publ. Co.: Singapore.
[98] Gaussian integer G03. Available from: http://gaussian.com/. Accessed 5 June 2017.
[99] Stevens, W.J., et al. Relativistic compact effective potentials and efficient, shared-exponent basis sets for the third-, fourth-, and fifth-row atoms. Can J Chem. 1992;70(2):612–630.
[100] Dias, H.V.R., et al. Synthesis and characterization of the gold(I) tris(ethylene) complex $[Au(C_2H_4)_3][SbF_6]$. Angew Chem Int Ed. 2008;47(3):556–559.
[101] Fianchini, M., et al. Use of $[SbF_6]^-$ to isolate cationic copper and silver adducts with more than one ethylene on the metal center. Organometallics. 2013;32(10):3034–3041.
[102] Fianchini M *Scorpionate Supported And Unsupported Carbonyl And Ethylene Complexes Of Group 11 Metals - Ph.D. thesis*. 2010; Available at: https://uta-ir.tdl.org/uta-ir/handle/10106/2059.

[103] Fianchini, M., et al. A non-classical copper carbonyl on a tri-alkene hydrocarbon support. Dalton Trans. 2009;12:2085–2087.
[104] Moliere A, Vigneron E, eds. New developments in chromophore research. New York, USA: Nova Science Publishers Inc; 2013.
[105] Mayerhöfer TG, Mutschke H, Popp J. Employing theories far beyond their limits—the case of the (boguer-) beer–lambert law. Chem Phys Chem. 2016;17(13):1948–1955.
[106] Elliott P, Furche F, Burke K. Excited states from time-dependent density functional theory. In: Lipkowitz K.B., Cundari T.R., editor(s). Reviews in computational chemistry Vol. 26. Hoboken, NJ, USA: John Wiley & Sons, Inc., 2008:91–165.
[107] Yarkony DR. Modern electronic structure theory: part 1. Singapore, Singapore: World Scientific Publishing Co Pte Ltd; 1995.
[108] Franz K.A., et al. Luminescent materials, in Ullmann's encyclopedia of industrial chemistry. Weinheim, Germany: Wiley-VCH Verlag GmbH & Co. KGaA, 2000.
[109] Marcus RA. On the theory of oxidation-reduction reactions involving electron transfer. I. J Chem Phys. 1956;24(5):966–978.
[110] Marcus RA. Electron transfer reactions in chemistry: theory and experiment (nobel lecture). Angew Chem Int Ed. 1993;32(8):1111–1121.
[111] Marenich, A.V., et al. Computational electrochemistry: prediction of liquid-phase reduction potentials. Phys Chem Chem Phys. 2014;16(29):15068–15106.
[112] Berardi, S., et al. Molecular artificial photosynthesis. Chem Soc Rev. 2014;43(22):7501–7519.
[113] Zeng, Q., et al. Role of ligands in catalytic water oxidation by mononuclear ruthenium complexes. Coord Chem Rev. 2015;304–305:88–101.
[114] Miertuš S, Scrocco E, Tomasi J. Electrostatic interaction of a solute with a continuum. A direct utilizaion of AB initio molecular potentials for the prevision of solvent effects. Chem Phys. 1981;55(1):117–129.
[115] Cossi, M., et al. Energies, structures, and electronic properties of molecules in solution with the C-PCM solvation model. J Comp Chem. 2003;24(6):669–681.
[116] Cances E, Mennucci B, Tomasi J. A new integral equation formalism for the polarizable continuum model: theoretical background and applications to isotropic and anisotropic dielectrics. J Chem Phys. 1997;107(8):3032–3041.
[117] Marenich AV, Cramer CJ, Truhlar DG. Universal solvation model based on solute electron density and on a continuum model of the solvent defined by the bulk dielectric constant and atomic surface tensions. J Phys Chem B. 2009;113(18):6378–6396.
[118] Marenich, A.V., et al. Self-consistent reaction field model for aqueous and nonaqueous solutions based on accurate polarized partial charges. J Chem Theory Comput. 2007;3(6):2011–2033.
[119] Marenich AV, Cramer CJ, Truhlar DG. Universal solvation model based on the generalized born approximation with asymmetric descreening. J Chem Theory Comput. 2009;5(9):2447–2464.
[120] Marenich AV, Cramer CJ, Truhlar DG. Generalized born solvation model SM12. J Chem Theory Comput. 2013;9(1):609–620.
[121] Klamt A, Schuurmann G. COSMO: a new approach to dielectric screening in solvents with explicit expressions for the screening energy and its gradient. J Chem Soc Perk T. 1993;2(5): 799–805.
[122] Klamt, A., et al. Refinement and Parametrization of COSMO-RS. J Phys Chem. 1998;102(26): 5074–5085.
[123] Tomasi J, Mennucci B, Cammi R. Quantum mechanical continuum solvation models. Chem Rev. 2005;105(8):2999–3094.
[124] Norskov, J.K., et al. Towards the computational design of solid catalysts. Nat Chem. 2009;1(1): 37–46.
[125] Stan M. Discovery and design of nuclear fuels. Mater Today. 2009;12(11):20–28.

[126] Pitzer KS, Gwinn WD. Energy levels and thermodynamic functions for molecules with internal rotation I. rigid frame with attached tops. J Chem Phys. 1942;10(7):428–440.
[127] Truhlar DG. A simple approximation for the vibrational partition function of a hindered internal rotation. J Comput Chem. 1991;12(2):266–270.
[128] McClurg RB. "The hindered rotor density-of-states interpolation function". J Chem Phys. 1997;106(16):6675–6680.
[129] Ribeiro, R.F., et al. Use of solution-phase vibrational frequencies in continuum models for the free energy of solvation. J Phys Chem B. 2011;115(49):14556–14562.
[130] Mammen, M., et al. Estimating the entropic cost of self-assembly of multiparticle hydrogen-bonded aggregates based on the cyanuric acid·melamine lattice. J Org Chem. 1998;63(12): 3821–3830.
[131] Grimme S. Supramolecular binding thermodynamics by dispersion-corrected density functional theory. Chem-Eur J. 2012;18(32):9955–9964.
[132] Brush SG. History of the Lenz-Ising model. Rev Mod Phys. 1967;39(4):883–893.
[133] Kadanoff LP. More is the same; phase transitions and mean field theories. J Stat Phys. 2009; 137(5):777.
[134] Bardeen J, Cooper LN, Schrieffer JR. Microscopic theory of superconductivity. Phys Rev. 1957;106(1):162–164.
[135] Bardeen J, Cooper LN, Schrieffer JR. Theory of superconductivity. Phys Rev. 1957;108(5): 1175–1204.
[136] Lasry J-M, Lions P-L. Mean field games. Jpn J Math. 2007;2(1):229–260.
[137] Baccelli, F., et al. A mean-field limit for a class of queueing networks. J Stat Phys. 1992;66(3): 803–825.
[138] Martin RM. Basic theory and practical methods. Cambridge, UK: Cambridge University Press; 2008.
[139] Thijssen JM. Computational physics. Cambridge, UK: Cambridge University Press; 1999.
[140] Marx D, Hutter J. Ab initio molecular dynamics: basic theory and advanced methods. Cambridge, UK: Cambridge University Press; 2010.
[141] Capdevila-Cortada M, Lopez N. Entropic contributions enhance polarity compensation for CeO2(100) surfaces. Nat Mater. 2017;16(3):328–334.
[142] DACAPO. Available at: https://wiki.fysik.dtu.dk/dacapo.
[143] Hinnemann, B., et al. Biomimetic hydrogen evolution: MoS_2 nanoparticles as catalyst for hydrogen evolution. J Am Chem Soc. 2005;127(15):5308–5309.
[144] Dresselhaus MS, Thomas IL. Alternative energy technologies. Nature. 2001;414(6861): 332–337.
[145] Rauchfuss TB. Research on soluble metal sulfides: from polysulfido complexes to functional models for the hydrogenases. Inorg Chem. 2004;43(1):14–26.
[146] Mejia-Rodriguez, R., et al. The hydrophilic phosphatriazaadamantane ligand in the development of H_2 production electrocatalysts: iron hydrogenase model complexes. J Am Chem Soc. 2004;126(38):12004–12014.
[147] Razavet, M., et al. All-iron hydrogenase: synthesis, structure and properties of {2Fe3S}-assemblies related to the di-iron sub-site of the H-cluster. Dalton Trans. 2003;4:586–595.
[148] Ferruz N, De Fabritiis G. Binding kinetics in drug discovery. Mol Inform. 2016;35(6-7):216–226.
[149] Harvey MJ, De Fabritiis G. High-throughput molecular dynamics: the powerful new tool for drug discovery. Drug Discov Today. 2012;17(19–20):1059–1062.
[150] Buch I, Giorgino T, De Fabritiis G. Complete reconstruction of an enzyme-inhibitor binding process by molecular dynamics simulations. Proc Nati Acad Sci USA. 2011;108(25):10184–10189.
[151] Pande VS, Beauchamp K, Bowman GR. Everything you wanted to know about Markov state models but were afraid to ask. Methods. 2010;52(1):99–105.

[152] Buch, I., et al. High-throughput all-atom molecular dynamics simulations using distributed computing. J Chem Inf Model. 2010;50(3):397–403.
[153] Marquart, M., et al. The geometry of the reactive site and of the peptide groups in trypsin, trypsinogen and its complexes with inhibitors. Acta Crystallogr B. 1983;39(4):480–490.
[154] Wagner, J.R., et al. Emerging computational methods for the rational discovery of allosteric drugs. Chem Rev. 2016;116(11):6370–6390.
[155] Votapka LW, Amaro RE. Multiscale estimation of binding kinetics using brownian dynamics, molecular dynamics and milestoning. PLoS Comput Biol. 2015;11(10):e1004381.
[156] Ermak DL, McCammon JA. Brownian dynamics with hydrodynamic interactions. J Chem Phys. 1978;69(4):1352–1360.
[157] Gabdoulline RR, Wade RC. Protein-protein association: investigation of factors influencing association rates by brownian dynamics simulations1. J Mol Biol. 2001;306(5):1139–1155.
[158] Cohen RD. Self similarity in brownian motion and other ergodic phenomena. J Chem Ed. 1986;63(11):933.
[159] Pearle P, Collett B. What Brown saw and you can too. Am J Phys. 2010;78(12):1278–1289.
[160] Von Smoluchowski M. Zur kinetischen Theorie der Brownschen Molekularbewegung und der Suspensionen. Ann Phys-Leipzig. 1906;326(14):756–780.
[161] Greives N, Zhou H-X. BDflex: A method for efficient treatment of molecular flexibility in calculating protein-ligand binding rate constants from Brownian dynamics simulations. J Chem Phys. 2012;137(13):135105.
[162] Spaar, A., et al. Diffusional encounter of barnase and barstar. Biophys J. 2006;90(6): 1913–1924.
[163] Buckle AM, Schreiber G, Fersht AR. Biochemistry-US. 1994;33:8878–8889.
[164] Schlick T. Molecular modeling and simulation: an interdisciplinary guide. 2nd ed. New York, NY, USA: Springer-Verlag New York Inc.; 2010.
[165] Wood WW. Monte carlo calculations for hard disks in the isothermal-isobaric ensemble. J Chem Phys. 1968;48(1):415–434.
[166] Norman GE, Filinov VS. Investigations of phase transitions by a Monte Carlo method. High Temp USSR. 1969;7:216–222.
[167] Cowen T, Karim K, Piletsky S. Computational approaches in the design of synthetic receptors – A review. Anal Chim Acta. 2016;936:62–74.
[168] Schauperl M, Lewis DW. Probing the structural and binding mechanism heterogeneity of molecularly imprinted polymers. J Phys Chem B. 2015;119(2):563–571.
[169] Eyring H. The activated complex in chemical reactions. J Chem Phys. 1935;3(2):107–115.
[170] Truhlar DG. Transition state theory for enzyme kinetics. Arch Biochem Biophys. 2015;582:10–17.
[171] Hratchian HP, Schlegel HB. Chapter 10 - Finding minima, transition states, and following reaction pathways on ab initio potential energy surfaces A2 - Dykstra, Clifford E. In: Frenking G, Kim KS, Scuseria GE, editors. AIP conference proceedings. Amsterdam: Elsevier; 2005. p. 195–249.
[172] Truhlar DG, Garrett BC, Klippenstein SJ. Current status of transition-state theory. J Phys Chem. 1996;100(31):12771–12800.
[173] Wigner E. Calculation of the rate of elementary association reactions. J Chem Phys. 1937; 5(9):720–725.
[174] Horiuti J. On the statistical mechanical treatment of the absolute rate of chemical reaction. B Chem Soc Jpn. 1938;13(1):210–216.
[175] Keck JC. Variational theory of reaction rates. In: Prigogine I., editor(s). Advances in chemical physics Vol. 13. Hoboken, NJ, USA: John Wiley & Sons, Inc, 1967:85–121.
[176] Truhlar DG, Garrett BC. Variational transition state theory. Annu Rev Phys Chem. 1984; 35(1):159–189.

[177] Pu J, Gao J, Truhlar DG. Multidimensional tunneling, recrossing, and the transmission coefficient for enzymatic reactions. Chem Rev. 2006;106(8):3140–3169.
[178] Bell RP. The tunnel effect in chemistry. Boston, MA, USA: Springer; 1980.
[179] Compton RG. Comprehensive chemical kinetics: electron tunnelling in chemistry - chemical reactions over large distances v. 30. Oxford, UK: Elsevier Science & Technology; 1990.
[180] Gray HB, Halpern J. Distant charge transport. Proc Nati Acad Sci USA. 2005;102(10): 3533–3533.
[181] Le Roy RJ, Murai H, Williams F. Tunneling model for hydrogen abstraction reactions in low-temperature solids. Application to reactions in alcohol glasses and acetonitrile crystals. J Am Chem Soc. 1980;102(7):2325–2334.
[182] Miller DJ, Subramanian R, Saunders WH. Mechanisms of elimination reactions. 33. Carbon isotope effects in E2 reactions of (2-phenylethyl-2-14C)trimethylammonium ion. The role of tunneling. J Am Chem Soc. 1981;103(12):3519–3522.
[183] Bettens, F.L., et al. The microwave spectrum, structure, and ring-puckering of the cyclic dipeptide diketopiperazine. J Am Chem Soc. 2000;122(24):5856–5860.
[184] Carpenter BK. Heavy-atom tunneling as the dominant pathway in a solution-phase reaction? Bond shift in antiaromatic annulenes. J Am Chem Soc. 1983;105(6):1700–1701.
[185] Zuev, P.S., et al. Carbon tunneling from a single quantum state. Science. 2003;299(5608):867.
[186] Marcus RA, Coltrin ME. A new tunneling path for reactions such as $H+H_2 \rightarrow H_2+H$. J Chem Phys. 1977;67(6):2609–2613.
[187] Fernandez-Ramos A. Variational transition state theory with multidimensional tunneling. In: Lipkowitz K.B., Cundari T.R., editor(s). Reviews in computational chemistry Vol. 23. Hoboken, NJ, USA: John Wiley & Sons, Inc., 2007:125–232.
[188] Truhlar DG. Tunneling in enzymatic and nonenzymatic hydrogen transfer reactions. J Phys Org Chem. 2010;23(7):660–676.
[189] Rudolf K, Spellmeyer DC, Houk KN. Prediction and experimental verification of the stereoselective electrocyclization of 3-formylcyclobutene. J Org Chem. 1987;52(16):3708–3710.
[190] Bahmanyar S, Houk KN. Transition states of amine-catalyzed Aldol reactions involving enamine intermediates: theoretical studies of mechanism, reactivity, and stereoselectivity. J Am Chem Soc. 2001;123(45):11273–11283.
[191] Rankin KN, Gauld JW, Boyd RJ. Density functional study of the proline-catalyzed direct Aldol reaction. J Phys Chem. 2002;106(20):5155–5159.
[192] Clemente FR, Houk KN. Computational evidence for the enamine mechanism of intramolecular Aldol reactions catalyzed by Proline. Angew Chem Int Ed. 2004;43(43):5766–5768.
[193] Bahmanyar S, Houk KN. The origin of stereoselectivity in Proline-catalyzed intramolecular Aldol reactions. J Am Chem Soc. 2001;123(51):12911–12912.
[194] Houk KN, Cheong PH-Y. Computational prediction of small-molecule catalysts. Nature. 2008;455(7211):309–313.
[195] Mollar, C., et al. Competitive and selective Csp^3-Br versus Csp^2-Br bond activation in palladium-catalysed suzuki cross-coupling: an experimental and theoretical study of the role of phosphine ligands. Chem–Eur J. 2010;16(45):13390–13397.
[196] Bedford, R.B., et al. Exploiting Boron–Zinc transmetallation for the arylation of benzyl halides: what are the reactive species? Angew Chem Int Ed. 2012;51(22):5435–5438.
[197] Harmata, M., et al. Combined computational and experimental studies of the mechanism and scope of the retro-Nazarov reaction. J Am Chem Soc. 2004;126(35):10954–10957.
[198] Butts, C.P., et al. Structure-based rationale for selectivity in the asymmetric allylic alkylation of cycloalkenyl esters employing the trost ‘standard ligand’ (TSL): isolation, analysis and alkylation of the monomeric form of the cationic η^3-cyclohexenyl complex $[(\eta^3\text{-c-}C_6H_9)Pd(TSL)]^+$. J Am Chem Soc. 2009;131(29):9945–9957.

[199] Nielsen MC, Lyngvi E, Schoenebeck F. Chemoselectivity in the reductive elimination from high oxidation state palladium complexes – scrambling mechanism uncovered. J Am Chem Soc. 2013;135(5):1978–1985.
[200] Proutiere F, Aufiero M, Schoenebeck F. Reactivity and stability of dinuclear Pd(I) complexes: studies on the active catalytic species, insights into precatalyst activation and deactivation, and application in highly selective cross-coupling reactions. J Am Chem Soc. 2012;134(1): 606–612.
[201] Sköld, C., et al. Transmetallation versus β-hydride elimination: the role of 1,4-benzoquinone in chelation-controlled arylation reactions with arylboronic acids. Chem–Eur J. 2012;18(15): 4714–4722.
[202] Schreiner, P.R., et al. Methylhydroxycarbene: tunneling control of a chemical reaction. Science. 2011;332(6035):1300.
[203] Wondimagegn, T., et al. Computational design of C_2-symmetric metallocene-based catalysts for the synthesis of high molecular weight polymers from ethylene/propylene copolymerization. Organometallics. 2008;27(24):6434–6439.
[204] Tantillo DJ. Using theory and experiment to discover catalysts for electrocyclizations. Angew Chem Int Ed. 2009;48(1):31–32.
[205] Lin Z. Interplay between theory and experiment: computational organometallic and transition metal chemistry. Acc Chem Res. 2010;43(5):602–611.
[206] Donoghue, P.J., et al. Prediction of enantioselectivity in rhodium catalyzed hydrogenations. J Am Chem Soc. 2009;131(2):410–411.
[207] Conley, B.L., et al. Design and study of homogeneous catalysts for the selective, low temperature oxidation of hydrocarbons. J Mol Cat A-Chem. 2006;251(1–2):8–23.
[208] Alapati SV, Karl Johnson J, Sholl DS. Using first principles calculations to identify new destabilized metal hydride reactions for reversible hydrogen storage. Phys Chem Chem Phys. 2007;9(12):1438–1452.
[209] Guo, W., et al. A metal-free synthesis of N-Aryl carbamates under ambient conditions. Angew Chem Int Ed. 2015;54(40):11686–11690.
[210] Fiore AM, Naik V, Leibensperger EM. Air quality and climate connections. J Air Waste Manage. 2015;65(6):645–685.
[211] Antonakakis N, Chatziantoniou I, Filis G. Energy consumption, CO_2 emissions, and economic growth: an ethical dilemma. Renew Sust Energ Rev. 2017;68:808–824.
[212] Rogelj, J., et al. Paris agreement climate proposals need a boost to keep warming well below 2 °c. Nature. 2016;534(7609):631–639.
[213] Smith, P., et al. Biophysical and economic limits to negative CO_2 emissions. Nat Clim Change. 2016;6(1):42–50.
[214] Gattuso J.P., et al. Contrasting futures for ocean and society from different anthropogenic CO2 emissions scenarios. Science. 2015;349(6243):AAC4722–AAC4722-10.
[215] Aresta M, Dibenedetto A, Angelini A. Catalysis for the valorization of exhaust carbon: from CO_2 to chemicals, materials, and fuels. technological use of CO_2. Chem Rev. 2014;114(3): 1709–1742.
[216] Liu, Q., et al. Using carbon dioxide as a building block in organic synthesis. Nat Commun. 2015;6:5933.
[217] Yu B, He L-N. Upgrading carbon dioxide by incorporation into heterocycles. Chem Sus Chem. 2015;8(1):52–62.
[218] Tantillo DJ, Jiangang C, Houk KN. Theozymes and compuzymes: theoretical models for biological catalysis. Curr Opin Chem Biol. 1998;2(6):743–750.
[219] Jiang, L., et al. De novo computational design of retro-aldol enzymes. Science. 2008; 319(5868):1387.

[220] Rothlisberger, D., et al. Kemp elimination catalysts by computational enzyme design. Nature. 2008;453(7192):190–195.
[221] Siegel, J.B., et al. Computational design of an enzyme catalyst for a stereoselective bimolecular Diels-Alder reaction. Science. 2010;329(5989):309.
[222] Zanghellini, A., et al. New algorithms and an in silico benchmark for computational enzyme design. Protein Sci. 2006;15(12):2785–2794.
[223] Metiu H. Physical Chemistry: kinetics. CT, USA: Taylor & Francis Inc; 2006.
[224] Seeman JI. The Curtin-Hammett principle and the Winstein-Holness equation: new definition and recent extensions to classical concepts. J Chem Ed. 1986;63(1):42.
[225] Rush LE, Pringle PG, Harvey JN. Computational kinetics of cobalt-catalyzed alkene hydroformylation. Angew Chem Int Ed. 2014;53(33):8672–8676.
[226] Bader G, Deuflhard P. A semi-implicit mid-point rule for stiff systems of ordinary differential equations. Numer Math. 1983;41(3):373–398.
[227] Goehry C, Besora M, Maseras F. Computational study on the mechanism of the acceleration of 1,3-dipolar cycloaddition inside cucurbit[6]uril. ACS Catal. 2015;5(4):2445–2451.
[228] Fernández-Alvarez, V.M., et al. Computational study with DFT and kinetic models on the mechanism of photoinitiated aromatic perfluoroalkylations. Org Lett. 2015;17(11):2676–2679.
[229] Lifson S, Warshel A. Consistent force field for calculations of conformations, vibrational spectra, and enthalpies of cycloalkane and n-alkane molecules. J Chem Phys. 1968;49(11): 5116–5129.
[230] Levitt M. Birth and future of multiscale modeling for macromolecular systems (nobel lecture). Angew Chem Int Ed. 2014;53(38):10006–10018.
[231] Warshel A, Levitt M. Theoretical studies of enzymic reactions: dielectric, electrostatic and steric stabilization of the carbonium ion in the reaction of lysozyme. J Mol Biol. 1976;103(2):227–249.
[232] Maseras F, Morokuma K. IMOMM: A new integrated ab initio + molecular mechanics geometry optimization scheme of equilibrium structures and transition states. J Comput Chem. 1995; 16(9):1170–1179.
[233] Vreven, T., et al. Combining quantum mechanics methods with molecular mechanics methods in ONIOM. J Chem Theory Comput. 2006;2(3):815–826.
[234] Balcells D, Maseras F. Computational approaches to asymmetric synthesis. New J Chem. 2007;31(3):333–343.
[235] Bo C, Maseras F. QM/MM methods in inorganic chemistry. Dalton Trans. 2008;22:2911–2919.
[236] Brown JM, Deeth RJ. Is enantioselectivity predictable in asymmetric catalysis? Angew Chem Int Ed. 2009;48(25):4476–4479.
[237] Nielsen MC, Bonney KJ, Schoenebeck F. Computational Ligand design for the reductive elimination of $ArCF_3$ from a small bite angle Pd^{II} complex: remarkable effect of a perfluoroalkyl phosphine. Angew Chem Int Ed. 2014;53(23):5903–5906.
[238] Anstaett P, Schoenebeck F. Reductive elimination of $ArCF_3$ from bidentate Pd^{II} complexes: a computational study. Chem-Eur J. 2011;17(44):12340–12346.
[239] Palermo, G., et al. Fighting cancer with transition metal complexes: from naked DNA to protein and chromatin targeting strategies. Chem Med Chem. 2016;11(12):1199–1210.
[240] Parr RG, Yang W. Density functional approach to the frontier-electron theory of chemical reactivity. J Am Chem Soc. 1984;106(14):4049–4050.
[241] Dapprich S, Frenking G. Investigation of donor-acceptor interactions: a charge decomposition analysis using fragment molecular orbitals. J Phys Chem. 1995;99(23):9352–9362.
[242] Weinhold F, Landis CR, Glendening ED. What is NBO analysis and how is it useful? Int Rev Phys Chem. 2016;35(3):399–440.
[243] Glendening ED, Landis CR, Weinhold F. NBO 6.0: natural bond orbital analysis program. J Comput Chem. 2013;34(16):1429–1437.

[244] Zubarev DY, Boldyrev AI. Developing paradigms of chemical bonding: adaptive natural density partitioning. Phys Chem Chem Phys. 2008;10(34):5207–5217.
[245] Bader RFW. A quantum theory of molecular structure and its applications. Chem Rev. 1991; 91(5):893–928.
[246] Lu T, Chen F. Multiwfn: a multifunctional wavefunction analyzer. J Comput Chem. 2012;33(5): 580–592.
[247] Lu T *MULTIWFN: a multifunctional wavefunction analyzer*. Available at: http://sobereva.com/ multiwfn/. Accessed 5 June 2017
[248] Software, B.f.I. *Aim2000*. Available at: http://www.aim2000.de/impressum.htm. Accessed 5 June 2017
[249] Keith TA *AIMAll*. Available at: http://aim.tkgristmill.com/ Accessed 5 June 2017.
[250] Becke AD, Edgecombe KE. A simple measure of electron localization in atomic and molecular systems. J Chem Phys. 1990;92(9):5397–5403.
[251] Savin, A., et al. Electron localization in solid-state structures of the elements: the diamond structure. Angew Chem Int Ed. 1992;31(2):187–188.
[252] Schmider HL, Becke AD. Chemical content of the kinetic energy density. J Mol Struc-Theochem. 2000;527(1–3):51–61.
[253] Jacobsen H. Localized-orbital locator (LOL) profiles of chemical bonding. Can J Chem. 2008; 86(7):695–702.
[254] Johnson, E.R., et al. Revealing noncovalent interactions. J Am Chem Soc. 2010;132(18): 6498–6506.
[255] Bader RFW, Gatti C. A green's function for the density. Chem Phys Lett. 1998;287(3–4): 233–238.
[256] Wang, X., et al. Mercury is a transition metal: the first experimental evidence for HgF_4. Angew Chem Int Ed. 2007;46(44):8371–8375.
[257] Kaupp M, Von Schnering HG. Gaseous mercury(IV) fluoride, HgF_4: an ab initio study. Angew Chem Int Ed. 1993;32(6):861–863.
[258] Kaupp, M., et al. Oxidation state +IV in group 12 chemistry. Ab initio study of Zinc(IV), Cadmium(IV), and mercury(IV) fluorides. Inorg Chem. 1994;33(10):2122–2131.
[259] Liu W, Franke R, Dolg M. Relativistic ab initio and density functional theory calculations on the mercury fluorides: is HgF_4 thermodynamically stable? Chem Phys Lett. 1999;302(3–4): 231–239.
[260] Riedel S, Straka M, Kaupp M. Can weakly coordinating anions stabilize mercury in its oxidation state +IV? Chem–Eur J. 2005;11(9):2743–2755.
[261] Pyykko P, Straka M, Patzschke M. HgH_4 and HgH_6: further candidates for high-valent mercury compounds. Chem Commun. 2002;16:1728–1729.
[262] Manly, C.J., et al. Strategies and tactics for optimizing the hit-to-lead process and beyond—A computational chemistry perspective. Drug Discov Today. 2008;13(3–4):99–109.
[263] Manly CJ. Computational chemistry, data mining, high-throughput synthesis and screening – informatics and integration in drug discovery. J Autom Meth Manag. 2001;23 (6):191–192.
[264] Kutchukian PS, Lou D, Shakhnovich EI. FOG: fragment optimized growth algorithm for the de novo generation of molecules occupying druglike chemical space. J Chem Inf Model. 2009; 49(7):1630–1642.
[265] Cherkasov, A., et al. QSAR modeling: where have you been? Where are you going to? J Med Chem. 2014;57(12):4977–5010.
[266] Bajorath J. Selected concepts and Investigations in compound classification, molecular descriptor analysis, and virtual screening. J Chem Inf Comp Sci. 2001;41(2):233–245.

[267] Goodford PJ. A computational procedure for determining energetically favorable binding sites on biologically important macromolecules. J Med Chem. 1985;28(7):849–857.
[268] Cramer RD, Patterson DE, Bunce JD. Comparative molecular field analysis (CoMFA). 1. Effect of shape on binding of steroids to carrier proteins. J Am Chem Soc. 1988; 110(18):5959–5967.
[269] Kozlowski, M.C., et al. Quantum mechanical models correlating structure with selectivity: predicting the enantioselectivity of β-amino alcohol catalysts in aldehyde alkylation. J Am Chem Soc. 2003;125(22):6614–6615.
[270] Ianni, J.C., et al. A priori theoretical prediction of selectivity in asymmetric catalysis: design of chiral catalysts by using quantum molecular interaction fields. Angew Chem Int Ed. 2006; 45(33):5502–5505.
[271] Gentili PL. Small steps towards the development of chemical artificial intelligent systems. Rsc Adv. 2013;3(48):25523–25549.
[272] Corey EJ, Howe WJ, Pensak DA. Computer-assisted synthetic analysis. Methods for machine generation of synthetic intermediates involving multistep look-ahead. J Am Chem Soc. 1974; 96(25):7724–7737.
[273] Fialkowski, M., et al. Architecture and evolution of organic chemistry. Angew Chem Int Ed. 2005;44(44):7263–7269.
[274] Gothard, C.M., et al. Rewiring chemistry: algorithmic discovery and experimental validation of one-pot reactions in the network of organic chemistry. Angew Chem Int Ed. 2012;51(32): 7922–7927.
[275] Kowalik, M., et al. Parallel optimization of synthetic pathways within the network of organic chemistry. Angew Chem Int Ed. 2012;51(32):7928–7932.
[276] Fuller, P.E., et al. Chemical network algorithms for the risk assessment and management of chemical threats. Angew Chem Int Ed. 2012;51(32):7933–7937.
[277] Mitchell JBO. Machine learning methods in chemoinformatics. Wires Comput Mol Sci. 2014; 4(5):468–481.
[278] Moore GE. Cramming more components onto integrated circuits, reprinted from electronics, volume 38, number 8, April 19, 1965, pp0.114 ff. IEEE Solid-State Circuits Soc Newsletter. 2006;11(5):33–35.
[279] *D-Wave System Inc.* Available at: https://www.dwavesys.com/.
[280] Boixo, S., et al. Experimental signature of programmable quantum annealing. Nat Commun. 2013;4:2067.
[281] Albash, T., et al. Consistency tests of classical and quantum models for a quantum annealer. Phys Rev. 2015;91(4):042314.
[282] Lanting, T., et al. Entanglement in a quantum annealing processor. Phys Rev X. 2014;4(2): 021041.
[283] Cho A. Quantum or not, controversial computer yields no speedup. Science. 2014; 344(6190):1330.
[284] Steiger, D.S., et al. *Performance of quantum annealing hardware.* in *Proc. SPIE 9648, Electro-Optical and Infrared Systems: Technology and Applications XII; and Quantum Information Science and Technology.* 2015.
[285] Wikipedia, t.f.e. *List of quantum chemistry and solid-state physics software.* Available at: https://en.wikipedia.org/wiki/List_of_quantum_chemistry_and_solid-state_physics_software. Accessed 5 June 2017.
[286] Wikipedia, t.f.e. *Category: computationalchemistry software.* Available from: https://en.wikipedia.org/wiki/Category:Computational_chemistry_software. Accessed 5 June 2017
[287] Wikipedia, t.f.e. *List of molecular graphic systems.* Available at: https://en.wikipedia.org/wiki/List_of_molecular_graphics_systems. Accessed 5 June 2017.

[288] Butler D. Electronic notebooks: A new leaf. Nature. 2005;436(7047):20–21.
[289] Choi BCK, Pak AWP. Multidisciplinarity, interdisciplinarity, and transdisciplinarity in health research, services, education and policy: 2. Promotors, barriers, and strategies of enhancement. Clin Invest Med. 2007;30(6):E224-E232.
[290] Zhang, Y., et al. Biological impact of environmental polycyclic aromatic hydrocarbons (ePAHs) as endocrine disruptors. Environ Pollut. 2016;213:809–824.

Juan Torras, David Zanuy, Oscar Bertran, Carlos Alemán, Jordi Puiggalí, Pau Turón and Guillem Revilla-López

2 Close contacts at the interface: Experimental-computational synergies for solving complexity problems

Abstract: The study of material science has been long devoted to the disentanglement of bulk structures which mainly entails finding the inner structure of materials. That structure is accountable for a major portion of materials' properties. Yet, as our knowledge of these "backbones" enlarged so did the interest for the materials' boundaries properties which means the properties at the frontier with the surrounding environment that is called interface. The interface is thus to be understood as the sum of the material's surface plus the surrounding environment be it in solid, liquid or gas phase. The study of phenomena at this interface requires both the use of experimental and theoretical techniques and, above all, a wise combination of them in order to shed light over the most intimate details at atomic, molecular and mesostructure levels. Here, we report several cases to be used as proof of concept of the results achieved when studying interface phenomena by combining a myriad of experimental and theoretical tools to overcome the usual limitation regardind atomic detail, size and time scales and systems of complex composition. Real world examples of the combined experimental-theoretical work and new tools, software, is offered to the readers.

Keywords: computer simulations, corrosion processes, mica-supported deposition of dendrimers, polymer materials, biominerals

> God made the bulk, the surface was invented by the devil
>
> Wolgang Pauli

Materials are primarily studied in basic undergraduate material science texts books as an infinite bulk with no initial considerations about their surfaces and edges beyond the simple truncation of the bulk. The latter makes good sense for the initial steps in the discipline but any realistic study of the material properties needs to mandatorily deal with the boundaries [1]. In the bulk atoms have a steady regular

This article has previously been published in the journal *Physical Sciences Reviews*. Please cite as: Torras, J., Zanuy, D., Bertran, O., Alemán, C., Puigallí, J, Turón, P., Revilla-López, G. Close contacts at the interface: experimental-computational synergies for solving complexity problems. *Physical Sciences Reviews* [Online] **2018**, *3*. DOI: 10.1515/psr-2017-0135

https://doi.org/10.1515/9783110482065-002

environment, that environment is easily reproduced periodically in the three directions of the space which allows the study of their structures and properties. Yet, surfaces by itself show a myriad of different atom arrangements that do not necessarily follow those of the bulk [1–3]. The bare termination of bulk atom organization is only a realistic approach for the lowest energy facets of some materials under precise physicochemical conditions. The latter makes the "rule" of understanding the surface as the truncation of the bulk an exception: surface reconstruction is reported since decades ago [2] which in combinations with the multiple terminations problem [3] is also able to dramatically tune the physicochemical properties of the boundaries. Even if no major changes are detected in the structure of the bulk after truncation the sheer fact of sharply finishing the bulk mesh alters the electronic structure, and consequently the properties, of the exposed facets [4]. The aforementioned phenomena is even enhanced when no solid surfaces are considered like in the polymer-based surfaces [5], and nanoparticles that may show radically different properties and structure than those found in the bulk [6]. Thus the study of many phenomena of scientific and industrial interest necessarily involves the surface at material boundaries [1]. Whenever we regard any phenomena like heterogeneous catalysis, adsorption, wetting, material growth or surface nanodecoration [7–10], interfaces reveal their utmost importance in their understanding.

In the last decades, huge improvements have been made in the understanding of surfaces itself and the rapport between them and their surrounding environment. This development consist of a paired deployment of new and better experimental and computational chemistry [11]. The combined use of both families of techniques is not always as straightforward as one may initially think: projects combining them often suffer from unaligned progression of both lines that so often tend to work in parallel but separately and with one line, either experimental or computational, lagging behind the other, in a standstill to provide its insight into the unsolved questions. Nevertheless, the always growing need for further detail and for studying new phenomena makes it mandatory for any research project embracing a modern approach to surface science to take into account all available tools whatever their nature is. In this regard, recent approaches hold that there should not be such a differentiation between computational and experimental techniques since the phenomena of interest does not show a dual nature and that when we talk about the combined use of theory and bench work there is only knowledge to be obtained without regard whether experiments are done in a computer or on a bench. The present chapter aims at providing the reader with some "proof of this concept" examples that illustrate how *in silico* and experiment are used to derive useful knowledge.

The cases reported in this chapter deal with phenomena ranging from corrosion prevention to biomolecule-mediated crystallization passing through two examples of surface decoration: one of them achieved by polymer deposition while the other relies on direct polymer chain growth on the surface. The selected situations offer in each case a practical approach to combine the two origins of knowledge so they must be understood as example of what can be done to overcome limitations. The examples focus on four experimental limitations whose overcoming act as the *Zeitgeist* of each of the presented examples: the first reported cases explains the macroscopic protection against corrosion induced by silanes and phosphonic acids on alumina at the atomic level by employing precise quantum mechanical (QM)-based tools; the second example focuses on the adsorption of dendronized polymers on mica, the size of these macromolecules challenges the reasonable limits of QM-based methods so a less computationally demanding classical technique like the molecular dynamics (MD) is employed; the third presented proof of concept focuses on the polythiophene film growth on iron which not only represents a problem due to the size of the polymer chains but also due to the time scale to be studied so an stochastic method like Monte Carlo-Metropolis protocol is required to shed light on the process (the in-house developed software being available to the reader upon request); the final example sums up all the mentioned challenges to offer a genuine solution to the three-dimensional complexity of biomineralization and its size, time and atom detail-related challenges.

Advice must be given to the reader in order to understand this chapter as a provider of proof-of-concept cases that illustrate computational solutions to state-of-the-art problems at levels for which experiment can hardly supply detail-based knowledge. Consequently, the physical and mathematical fundamentals of the employed methods have been deliberately obviate provided they can be found elsewhere in outstanding textbooks [1, 11–13] and cited references. It is up to the reader to develop the provided information and knowledge in these pages into its own wisdom.

2.1 Elucidation of the aluminum corrosion protection mechanism of silanes

Corrosion processes are a well-known threat to facilities specially the industrial ones which causes a huge amount of economic losses every year. The World Corrosion Organization, www.corrosion.org, estimates these losses around 2.2 trillion US$ per year. The scientific community has been facing this problem since many year ago thus trying to slash down these costs. Computational chemistry has come, in the last decades, to provide significant aid to experimental work in the fight against the natural tendency of metallic materials to become rust. The presented case focuses in the field of organic–inorganic interactions of modern hybrid materials and how these can help understand anticorrosion treatments.

Among all new materials, aluminum alloys have been extensively used because of their interesting properties and large applicability. Aluminum and its alloys show an excellent resistance to atmospheric corrosion in marine, urban and industrial environments [14]. One of the most employed aluminum alloys is the AA2024, which is classified as aluminum-copper alloy because of its high content of the latter. This alloy is well-known for its excellent mechanical properties though it has the highest rate of corrosion due its inhomogeneous microstructure [15]. Several surface treatments that can be applied to prolong the life of aluminum alloy-based materials have been developed to address the aforementioned tendency [16, 17]. Recently, sol-gel coating has been developed as a new friendly alternative to traditional aerospace finishing materials [17, 18]. Sol-gel coating structure is based on a reactive mixture of an organo-functionalized silane deposited over the metallic material. This adsorbed layer is able to create a protective that provides durable adhesive substrate for paints, adhesives and sealants. The modification of the inorganic/organic interactions networks leads to several multifunctional coatings with potential in corrosion prevention [19]. Indeed, sol-gel technologies have been spotted as one of the most promising methods for aluminum surfaces protection against corrosion [20].

Sol–gel coating is seriously handicapped when only silicate is used as a reagent thus deploying poor adherence, homogeneity and film thicknesses [21], which enables the access of electrolytes to the metal surface through pores and cracks in the silane films. Phosphonic acids can be added to the sol–gel matrix in order to improve its anticorrosion properties; these acids have been shown as effective corrosion inhibitors for aluminum alloys able to form self-assembled monolayers (SAMs) [22, 23] and to synergically boost the silane films protection to aluminum alloy systems against corrosion [24, 25].

Recently, Dalmoro et al. [26] reported a case of this phosphonic acids effect; the employed acids were the 1,2-diaminoethanetetrakis methylenephosphonic acid (EDTPO) and aminotrimethylenephosphonic acid (ATMP), in combination with tetraethylorthosilicate (TEOS) films prepared by sol–gel synthesis. The found synergy improved the performance of TEOS for preventing the corrosion of the aluminum alloy AA1100 thus obtaining two new modified coating layers denoted as TEDTPO and TATMP, depending on which phosphonic acid was used. FTIR-RA spectroscopy, SEM, EDX analysis and X-ray photoelectron spectroscopy (XPS) in combination with QM calculations were used in that work to study the linking capacity of phosphonic acids to metallic surface and are summarized hereby. The experimental results suggested two main interaction mechanisms for the protective action of the modified silane layer: (i) the formation of very stable six-membered hydrogen bonded rings, involving Si-OH···O=P interactions; (ii) formation of a bidentate surface-complex Al-O-P-O-Al, more stable than the monodentate chemisorbed species, see Figure 2.1. Density functional theory (DFT)-based QM calculations on a short model of organophosphate-surface complex (TATMP) allowed reproducing the two experimental interactions of ATMP with the silica and aluminum surface. Moreover, a stabilization of 6.5 kcal/mol on the six-membered hydrogen-bonded ring when compared with a ring holding a disrupted hydrogen bond was obtained thus corroborating the thermodynamic stability of the proposed six-membered ring.

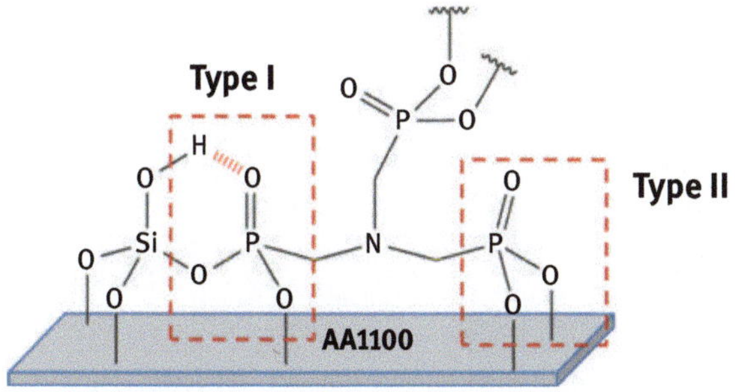

Figure 2.1: Example of monodentate, Type I, and bidentate, Type II, bindings of TATMP coating to the AA1100 surface.

Additionally, the role that phosphonic acid molecules play in the formation mechanism of covalent linkages between the silane layer and the aluminum surface has recently been ascertained through a combination of computation and lab work as well. The work presented a XPS depth profile attack (from 0 to 4 h) obtained on different organosilane-coated aluminum surfaces and the phosphonic acid–treated sample showed that phosphorus is located in a region near the metal surface–silane interphase, establishing stable Al–O–P and P–O–Si linkages. Thus, the concentration of phosphorus increases with the depth profile, evidencing a mechanism where the phosphonic acids actuated as stabilization of the silane coating.

The latter suggestion of the relevant role played by phosphorous still required from further calculations to elucidate the interaction scheme of phosphonic acids within the metallic surface silica. Torras et al. [27] attained the latter by modeling the crystal structure of boehmite (γ-AlOOH) as the substrate of the adsorption of the stable self-assembling monolayers of alcoxisilanes and methylphosphonic acids. This approach is justified since pre-treated aluminum has a natural oxide layer, which presents an amorphous phase but covered with a thin hydroxylate layer of pseudoboehmite (mainly made of hydroxylate γ-Al_2O_3 and γ-AlOOH phase) [28].

To compare the stability between different adsorption complexes and γ-AlOOH surface interactions, the authors of Ref [27] studied the chemisorption of three different molecules on the (010) surface model of γ-AlOOH using different coordination modes, i. e., orthosilicic ($Si(OH)_4$), methylphosphonic acid (MePA), and the condensed molecule of $Si(OH)_3$OMePA. The model was created by repeating the bulk unit cell taken from the optimized bulk of γ-AlOOH and adding a vacuum region of 25 Å along the b-direction (Figure 2.2). All performed calculations were based on the DFT under two-dimensional periodic boundary conditions.

Both species MePA and $Si(OH)_4$ presented a monodentated adsorbed complex as the most stable complex; however, MePa was energetically favored when is compared with the monodentated chemisorption of $Si(OH)_4$. Yet, when the condensed molecule of $Si(OH)_3$OMePA was chemisorbed through a bidentated anchorage an synergistic

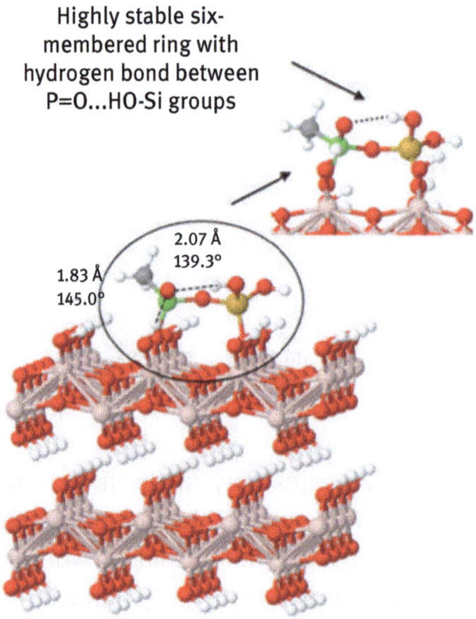

Figure 2.2: Bidentate binding coordination of condensed Si-$(OH)_3$OMePA molecule, Si and P are represented green and orange respectively, chemisorbed on γ-AlOOH (010) surface, showing the stable six-membered ring between phosphonic and silanol groups.

increase of stability was obtained. This overstabilization was attributed to the presence of the previously suggested six-membered hydrogen bonds [26]. Indeed, the computational simulations corroborated the synergic complexation between alkylorthosilicate and alkylphosphonic acid and bolstered the idea that interaction is a source of stabilization of the protective layer on the alumina surface that increases upon the presence of MePA.

The interacting experimental and computational techniques of the study [27] finally allowed to the authors to propose a chemical mechanism for silane deposition and their interaction with phosphorus: initially, hydrolyzed molecules of alkylorthosilicate are attracted by the hydrophilic surface of boehmite. Then, alkylphosphonic acid catalyze the covalent adhesion of alkylorthosilicate to the modified metal surface by hydrogen bond interactions between hydroxyl and P=O linkages. When the surface is stabilized, the cured process takes place and the network is formed. The organophosphonic groups and silane are place in the inner layer and the Si–OH groups in outer layer.

2.2 Computational insights of the mica-supported deposition of dendrimers

Surfaces can accommodate molecules much bigger that the acidic ligands proposed in the previous case. Among these systems there are the dendronized polymers (DPs) that are a special subclass in which hyperbranched sidegroups hang from the backbone thus creating molecules with a huge number of atoms. The interaction of DPs with surfaces represents a challenge with regard to the size of the surface adsorbate since the bigger these are, the more difficult it becomes to extract details from both experiment and computation. At the computational level, DPs cannot be treated with the same set of techniques that were used in the previous example, DFT-based QM, since the computational cost would be immense, molecular mechnics (MM) and its time-dependent derivative the molecular dynamics (MD) are suitable resource-saving oprtions. Computational costs are slashed down thanks to the assumption, by MD, of atoms and bonds to behave as a classical balls and springs system. The second example in this chapter deploys an example of how MM and MD are applied to a system hardly attainable through QM.

In DPs, each monomer is called a dendron which, in turn, is composed of dendritic units that have three functionals fully reacted, through which the monomer grows for different generations (Figure 2.3). This structure produces in the DPs: (i) strong steric repulsions (ii) local cylindrical symmetry and (iii) highly stressed backbone.

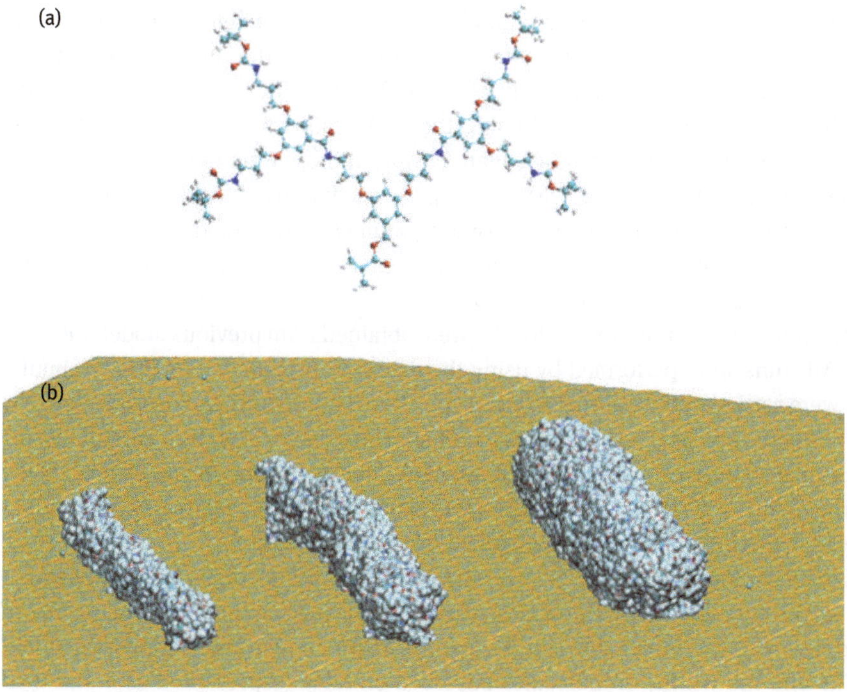

Figure 2.3: (a) Monomer of PG2. (b) Images taken from MD simulations of the polymers PG2, PG3 and PG4 deposited on the attractive mica surface.

As the degree of generation (g) increases so does the size of the polymer, the three mentioned features and the rigidity of the structure of the DPs. Additionally, solubility, aggregation and chemical functionalization can be controlled to a large extent through the large number of active terminals they possess. DPs structural features also significantly differ from those of conventional flexible polymers (which normally have random coil configurations) and from the dendrimers since they have a central unit that confers a fairly cylindrical rigid geometry for the whole molecules.

These characteristics have drawn great attention on DPs as an emerging class of nanomaterials with applications for physicochemical response [29], polymeric fluorophores nanoaggregation [30] and drug delivery ability [31].

In the presented proof of concept we will focus on DPs adsorbed onto attractive mica surfaces and their deviation from their cylindrical shape, a full report of the results can be found in Ref [32]. The presented DPs are composed of a poly(methacrylic acid) backbone whose repeat units are regularly branched dendrons of generation degree, g, to be determined as a number for each example, containing both amide and aromatic groups, see Figure 2.3. The global results in Ref [32] are a product of the interaction between atomic force microscopy imaging (AFM), transmission electron microscopy (TEM) [33–35] and the aforementioned MD.

MD simulations require that the spring constants for bonds and angles altogether with the phase parameters of dihedrals and the parameters of the non-bonding potentials are provided. These are supplied by general force fields or derived manually if not found in them. The presented MD simulations of the DPs absorbed onto an attractive mica surface, hereafter PGg^{mica}, and the surface-free, gas phase, DPs, hereafter PGg^{free}, were performed using the NAMD software package [36] and the AMBER force field parameters when available [37], if these were not found in AMBER, the Generalized AMBER force field (GAFF) [38] was employed. Atomic charges were parametrized from a previous electrostatic parametrization of these systems. The mica $K1.0[Si3Al1O8][Al_2O_2(OH)_2]$ surface was represented using the force-field parameters reported by Heinz and co-workers [39].

The initial geometries for the PGg^{mica} were obtained from previous models [40] and 20 ns MD runs were performed by using them as starting points. The PGg^{free} simulations were carried out in parallel as a control. These simulations can be understood as equilibration steps after which the structure relaxation is checked by a careful inspection of time-evolutions of structural parameters, such as the end to end distance (Lee) and the radius of rotation (Rg). On the other hand, Figure 2.3b evidences adsorption-induced flattening of PG2–PG4 as obtained from MD simulations.

DPs in the free state (PGg^{free}) have circular fairly cross sections, this arrangement is to be partially lost upon interaction with the mica surface thus the total degree of deformation of the molecules can be quantified by comparing the cross-sectional heights and the widths of molecules adsorbed on mica (represented as h and w in Table 2.1). These parameters, h and w, are derived from the distributions of the mass of DP as a function of the radial (i.e., perpendicular to the mica surface) and axial distance (i.e., parallel to the surface) of the center of mass of the polymer chain respectively, see Ref **[32]**. These distributions were obtained by averaging over 300 snapshots taken during the last 15 ns of the 20 ns MD production runs. For each DP, *h* and *w* were estimated as the distance at which the radial and axial mass profiles had dropped to 50 % of their maximum values before reaching the external layer of the DP, values being listed in Table 2.1. The data in Table 2.1 reflects that the values of *w* are ~30 % larger than the values of *h*, independently of generation degree, g, thus corroborating the deformation suffered by the DPs on the mica surface. Conversely, the PGg^{free} simulations led to cylindrical structures. Simulations also allowed disclosing the energetics of the polymer–surface interactions: the total energy decreasing as the generation degree increases, the decomposition renders electro-static component with up to a 75 % contribution to the total energy, whereas the role of van der Waals stands at a 20–30 % which is non-negligible specially for high generation degree, g.

Experimental estimations of the dimensions of second- to fourth-generation DPs adsorbed on mica, PG2–PG4 in Table 2.1, obtained using different microscopy techniques, TEM and SEM for *h* and *w* respectively, are in very good agreement with the computational estimations in the same table. Moreover if we consider that the arbitrary definition of the parameters can produce a slight overestimation of *h* and *w* in the computation. The most relevant proof that computation fits experiment lies on the fact that the *h*/*w* coefficient stands in 0.7–0.8 range for both experiment and simulation thus confirming the deformation of the rigid DPs is faithfully reproduced in all g.

The deformation caused by the mica surface in the cylindrical models of PG2-PG4 can be assessed qualitatively by looking at Figure 2.3b, a quantitative estimate can be obtained by comparing the cross-section circularities of PGg^{mica} and PGg^{free}. To this end, solid images of the sections (equatorial projections) of each model were constructed for subsequent analysis and the cross-section circularity (C) of such two-dimensional shapes was calculated using the known definition

$$C = 4\pi A/P^2$$

Table 2.1: Codes goes for structural parameters as the first letter for height (**h**; in Å), width (**w**; in Å) and diameter (**D**; in Å).PGg^{mica} and PGg^{free} for DPs adsorbed on mica or in the gas phase. Determined for PGg^{mica} using MD simulations and experimental techniques, SEM and TEM microscopy, derived from MD simulations for PGg^{free}. PG*n* denotes the generation degree.

	PG2	PG3	PG4
h^{MD} PGg^{mica}	27.1	35.0	57.5
w^{MD} PGg^{mica}	33.9	48.0	73.6
h^{TEM} PGg^{mica}	23 ±4	34±5	49±3
w^{SEM} PGg^{mica}	31 ±4	45±4	60±4
D^{MD} PGg^{free}	34.4	49	71.8

where A and P refer to the area and perimeter, respectively, obtained by counting the pixels of the different images (i. e. C=1 for an ideal circle). Obviously, in all cases the deformation of the DPs over mica produces a loss of C from PGg^{free} to PGg^{mica}, having the results: $\Delta C = 1.8 \pm 0.2$, 3.0 ± 0.3, and $7.7 \pm 0.2\%$ for PG2, PG3, and PG4, respectively. It should be noted that ΔC involves two contributions: the external mica-induced shape deformation and internal small shape variations seen as surface protuberances of the DPs at the periphery of their cylindrical cross section. This result suggests that the loss of circularity of the consecutives generation degrees PG(g+1) vs PGg doubles through all the investigated range of g.

The behavior of the dendrimers on a mica surface may significantly differ from that of the free DPs, their deformation being much greater (greater flattening) for the same value of g. DPs can adopt this behavior using a small number of monomers (i. e. about 25 monomers). In this case, the end-caps monomers tend to occupy all possible space (spherical shape) and flatten on the central part interacting with the mica surface; the spherical shape dominates over the cylindrical shape of the polymer and its behavior tends to resemble more like a dendrimer. This is particularly evident by comparing the variation of the height of PGg^{free} and PGg^{mica} with g=4, PG4 in Table 2.1, obtained for the same generation degree poly(amidoamine) dendrimer [41]: the height of PG4 decreases ~20 % upon adsorption while that of the dendrimer far over.

The impact of both mica and g on the rigidity of DPs has been estimated using the inverse variance of the end-to-end distance (L_{ee}) distribution which describes the rigidity along the helical backbone axis. Independently of g, the inverse variance is smaller for PGg^{free} than PGg^{mica}, indicating that the behavior as helical rigid rod is enhanced in the latter with respect to the former case. Moreover, the inverse variance increases progressively with g for PGg^{free}, indicating that compression and elongation movements become more difficult. In contrast, the inverse variance decreases with increasing g for PGg^{mica}. On the other hand, the coefficient $(L_{ee}/R_g)^2$ usually stands at 12 for rigid macromolecules, where R_g is the radius of gyration. For PGg^{free} and PGg^{mica} the relative deviation of $(L_{ee}/Rg)^2$ from 12 is around 2% and 0.5 %, respectively. This behavior is fully consistent with the fact that the inverse variance of L_{ee}, and therefore, the rigidity is higher for PGg^{mica} than for PGg^{free}. Therefore, the inherent rigidity of PGs can be enhanced by surface interactions and, consequently, so can be the related properties.

In conclusion, the results reveal that although the widths of DPs upon mica adsorption are not heavily affected, the interaction with the surface produces a slight flattening of DPs and a reduction of the transverse height, a feature that increases with g over the investigated interval of g. This loss predicted by simulation and corroborated by experiment represent an interesting point since and at the maximum generation degree, g_{max}, DPs are expected to be virtually undeformable. Yet, a careful comparison of the computationally obtained minimum and maximum diameters of DPs explains this seeming contradiction: while for low and generations near g_{max} these diameters are very similar the intermediate generations exhibit difference. For the DPs investigated in the present study the maximum difference is in fact expected for g = 4 where the largest flattening on mica was found. It is worth noting that these findings can easily be done and understood thanks to simulation used as an accessory to experiment, the results of the latter also play a role as quality checks as it has been shown.

2.3 An atomistic-scale portrayal of polyethylene film growth

The atomistic details of polymer based materials are seldom available from microscopic observations. Most of the experimental techniques used to characterize the topographic features roughly encompass scales that reach decimals of micrometers. These technical limitations are generally ignored and most of the available literature about the atomistic organization of polymeric materials shows structures that are fitted to low-resolution images based on known systems without much attention to their actual reliability [42]. Moreover, most available structural works are completed without assessing the basic details of those systems, generally by either avoiding the atomistic nature of the polymer ensembles with simplified coarse-grained models, models that do not properly described the system energy [43], or by totally ignoring the atomic organization approaching the problem to finite elements computations [44]. Beyond these simulation limitations, experimental techniques are not generally tuned for the atomistic description of the polymer bulk, lacking the required structural finesse. AFM and other more classical electronic microscopy techniques do not provide yet such details on polymers ultrastructure. These methodological limitations can be overcome by strategically combining the available mathematical algorithms with force fields calculations. In the presented case, a novel modeling strategy is presented for obtaining a fast and reliable atomistic description of a grafting polymer using molecular models based on classical mechanics, MM and MD. The results are validated by the comparison of experimentally found topographic magnitudes of the polymer films deposited on metal surfaces with the simulated ones for the same system. Not only do the presented system challenge the size scale limitation but they also challenge the usually simulated time scale: film formation processes may take up to seconds and these time scale are far away, at the moment, from these achievable through QM- and MM-based techniques. The MM-based Monte Carlo suggested approach for this case can help solve these limitations.

Metal surfaces coated by polymers are commonly found in nowadays literature for micro-engineering applications, since this sort of nanoscale constructs is especially suited for electronic micro-circuitry. Reduced scale circuitry requires the use of low-density materials that can efficiently store and release charge. Electroactive conducting polymers (ECPs) are an excellent choice for becoming important building blocks on nanocircuitry [45]. ECPs are stable macromolecular scaffolds that show induced and reversible electrochemical activity, which is a direct consequence of allowing several oxidation states. They show simplicity and versatility of synthesis as well (e. g. chemical synthesis, vapor-phase oxidation and electrochemical processes) and satisfactory environmental stability. Poly(3,4-ethylenedioxythiophene), abbreviated *PEDOT*, features the majority of those properties: excellent conductivity (up to 500 $S \cdot cm^{-1}$), optimal environmental and chemical stability, fast doping–dedoping processes and exceptional biocompatibility [46, 47]. Moreover, PEDOT shows excellent electroactivity and electrostability, both explaining its high ability to reversibly exchange charge [47–50]. These features makes PEDOT suited for its inclusion as active parts in both rechargeable a batteries and electrochemical ultracapacitors [50–53], symmetric ultracapacitors including very thin PEDOT films (i. e. those with a thickness, ℓ, typically lower than 250 nm) are among the most reliable choices, due to their excellent capacitive properties. The work presented in this section focus on the ultrastructure of coated metal surfaces that were designed for working as capacitor electrodes [51–54].

Although PEDOT on steel constructs have extensively been examined by both scanning electron microscopy and AFM [54–56], structural details about its deposition and growth are yet very scarce. Molecular simulation approaches recently hinted few defining features of these nano-constructs never proven before [57]. Initial attempt to describe the molecular details of the polymer bulk failed to describe the surface properties (e. g. topography and morphology), because they relayed on polymer chains of unique lengths, relinquishing to explore the roughness of the polymer-coated surface. Unfortunately, surface roughness is one of the ions movement determinant during oxidation and reduction cycles, i. e., explains how topography affects the charge exchange in these systems [54–56, 58].

The presented case is a significant improvement with respect the mentioned initial approaches that has enabled the overcoming of their limitations and, conversely, keeping reducing the experimental scales in which surface phenomena can be simulated at reasonable computational costs. The reliability of the computational approach is demonstrated by systematically exploring with AFM the topography of newly synthesized PEDOT on steel electrodes using extremely short polymerization times (symbolized by the Greek letter θ), $\theta \leq 5$ seconds, and comparing the surface topographies with the roughness computed using our atomistic modeling that combined stochastic generation with MD energy relaxation.

All atomistic models were built using an in house developed program available upon request to the corresponding author of the chapter. The code follows several consecutive steps that are listed below.

2.3.1 Deposition of polymer chains on the metal substrate

The deposition of PEDOT chains onto the metallic substrate was simulated using a reported strategy [57], in which the experimental density of PEDOT films doped with ClO_4^- (ρ= 1.665 g/cm) [59] that had grown onto a steel substrate was used as a initial reference for determining the number and length of chains placed on a metallic surface of known area. In that work, the doping level of the EDOT repeat units had been determined as well by assessing the amount of ClO_4^- ions in polymer samples [59, 60]. In the presented results, 60 identical PEDOT chains (made of different number of EDOT units each one) were used to emulate the polymer deposition on a metallic substrate: 10 layers of iron atoms arranged according to the (111) plane, with crystallographic dimensions of a= 7.5 nm and b= 7.5 nm, reproduced the metallic surface, iron atoms being represented as frozen spheres [57]. At this generation phase PEDOT chains were arranged fully extended, represented as rigid conformers, oriented parallel to z-axis and symmetrically distributed on the iron atoms surface. The only restriction imposed to their equatorial distribution is to minimize the steric hindrance between neighboring chains. Lennard-Jones potential adapted to AMBER force field was used to achieve that goal [57, 60].

A realistic equatorial asymmetrical distribution of chains was then obtained using stochastic movements (Metropolis-based algorithm). Accordingly, a polymer chain was randomly selected from its initial location and, subsequently, moved off its current position by applying a translation over the x,y plane and/or by performing a rotation over the z-axis. If the new position does not imply steric overlaps with other chains, an energy-based Metropolis criterion is used to accept or reject the movement. Convergence of this initial stage was reached when no new positions were accepted after 500 attempts. The initial building algorithm [57] finalized at this point.

Counterions (dopant agents) [57] were now placed into the polymer bulk. Previous results using MD relaxation [57] clarified the structural role of the counterions and allowed now reproducing earlier investigations [57]: Along the polymer chains each perchlorate appeared between EDOT repeat units, and generally equidistant to two adjacent chains when their equatorial projection is observed. Consequently, 30 ClO_4^- anions for each PEDOT chain were initially placed amid EDOT units following those topological restrictions.

2.3.2 Growth algorithm

A new set of movements has been included for this current investigation, becoming the second stage of the generation protocol. Different sets of chains are randomly selected and make them grown by adding a random even number of EDOT units. Before explaining this new procedure, several clarifications should be made: at this stage the position of the chains on the substrate is not changed (i. e. their xy position, the position set on their equatorial projection). Despite this apparent bias, it was observed that generating a meaningful amount of starting superficial distributions of chains and then selecting those that energetically were more favored was more efficient [57]. In the current case, the models that are to be simulated with MD are selected after the growing stage is completed.

It is very important not to misplace the meaning of the generation method: the used procedure does not try to emulate the growth mechanism of PEDOT but to build reliable atomistic models of the surface morphology once the modeling steps are over. This clarification relates with the next step for "making grow" the PEDOT chains. In order to assure the random selection of nascent segments, the change on the potential energy is assessed for each set of new added residues. At each growing cycle a fixed amount of residues is added. The differences between structures put to Metropolis trial stem from how those residues are distributed among polymer chains (Figure 2.4).

In order to apply a metropolis algorithm for selecting a growth path, change in the potential energy is evaluated with three different terms, separated in two main components: (i) van der Waals and electrostatic components to assess the effect of new interchain interactions after adding new EDOT units, (ii) an energy penalty that would scale the probabilities of selecting chains once they started to growth, probabilities that initially were independent of the chains height. This later component is computed as an external mathematical scaling factor that ensures that the molecular model reproduces the experimental relationship between roughness and thickness (l/r ratio). If such information is not available, this component is set to 0 and the growth is unbiased.

The strength and the sign of this latter term would depend on the correlation between l/r ratio that has previously been computed for each possible way of

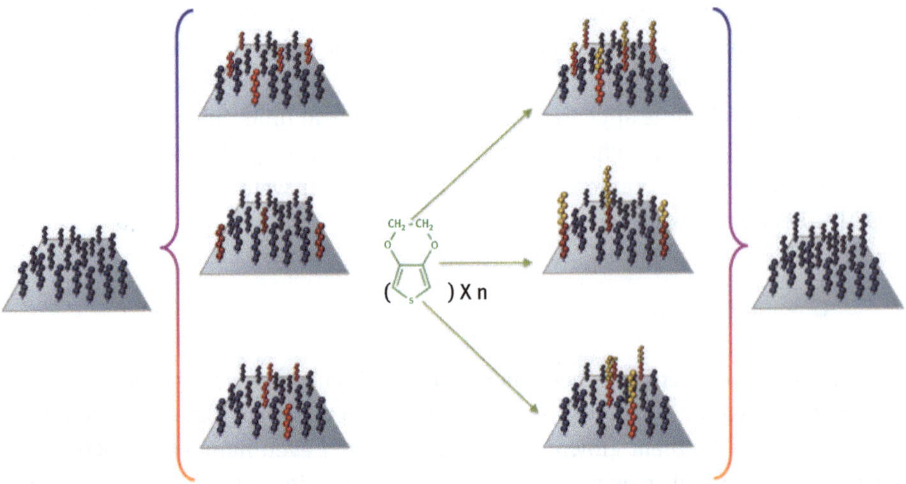

I. stochastic distribution of all considered chains over the parceled area.

II. Selection of a random number of nascent chains to be enlarged. More than one set of number of chains is selected (*paths*)

III. Random selection of a RUs amount to be added onto nascent chains.

IV. Chains enlargement. Each potential path randomly distributes the RU amount among previously selected chains.

V. Selection of one possible growth path on the basis of a modified metropolis algorithm

Figure 2.4: Schematic representation of the metropolis algorithm followed to mimic the growth of polymer chains.

distributing the new residues. For instance, if the aforementioned relationship has decreased, it means that there has been an increase in the polymer roughness, the incorporation of the new segments will be promoted on shorter chains, in order to decrease roughness, whereas its incorporation on longer chains will be disfavored.

2.3.3 Energy relaxation: *Molecular dynamics*

Models derived from the deposition and growth algorithms are rigid and do not take into account the effect of temperature and pressure. Thus, before characterizing the topology and topography, production MD trajectories were run for all generated models. Each model is contained in an orthorhombic cell of 4.95 × 10.035 × 70.0 nm, in which the space not occupied by the substrate and polymer is filled with acetonitrile molecules (Figure 2.5). Before starting the relaxation model, the number of acetonitrile molecules, the solvent, was equalized to a total of 21,627 solvent molecules. Accordingly at this point, the chemical composition of all models was strictly identical.

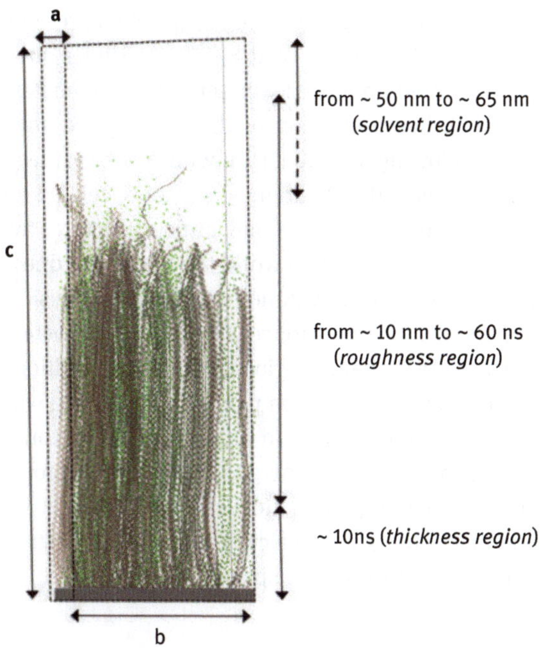

Figure 2.5: Schematic representation of each molecular model generated to mimic PEDOT growth in very thin films. The simulation box dimensions and each box region have been included in the scheme. Regions that appear empty are actually filled with the necessary number of Acetonitrile molecules to reach its density (0.77 g·cm^{-3}). Hydrogen atoms belonging to EDOT units and oxygen atoms belonging to perchlorate molecules have been omitted for clarity.

After 0.80 ns of thermal and baric equilibration (NPT) each 40 models were submitted to short periods of MD (1.5 ns). After this short period, five of those models showed better equilibration than the rest. These five systems were selected for further analysis of longer simulation time (up to 20 ns more). Finally, the lateral summation of these five models, each of them representing almost 50 nm^2 of surface, provided a global model of 250 nm^2 (Figure 2.5).

2.3.4 Force field details

All parameters were extrapolated from AMBER03 libraries [59] with exception of partial charges of the EDOT repeating unit, which had previously been parametrized [60, 61]. Acetonitrile molecules were represented by the standard AMBER03 model [62] and ClO_4^- parameters were obtain from Baaden et al., 2000 [63]. Bond and length distances of PEDOT repeat units were approximately kept at their equilibrium distances using the same strategy employed at Ref [60]. After an extended bibliographic search, no AMBER compatible parameters were found for iron atoms when organized in an FCC lattice. Since the surface was to be kept fixed in our simulations, we extrapolated the latest published parameters for pure iron in a bcc to the geometrical features of a FCC arrangement. Reference 62 shows that solid iron adopting a bcc arrangement presents a cell parameter of a = 2.87 Å and a metallic radius of R = 1.24 Å. Under such geometrical arrangement the authors presented optimized van der Waals parameters of σ = 2.96 Å and ε = 0.945 (kcal·mol^{-1}). Representation of the investigated steel substrate as an iron material organized following an FCC lattice with a = 3.65 Å lead to an apparent metal radius of R = 1.29 Å. If the AMBER parameters of a bcc cell are re-scaled to fit the increment in the atomic radius due to the new lattice structure, the corresponding van der Waals radius is σ = 3.08 Å. For this work, the atomic hardness (ε) will be kept to the reported value [64], because we will not simulate any physical process that implies the active participation of the iron surface. This part of the system will only serve as the anchoring lattice for PEDOT chains. Periodic boundary conditions were applied using the nearest image convention and atom pair cut-off distance was set at 14.0 Å to compute the van der Waals interactions. In order to avoid discontinuities in the potential energy function, non-bonding energy terms were forced to slowly converge to zero, by applying a smoothing factor from a distance of 12.0 Å. Beyond cut off distance, electrostatic interactions were calculated by using Particle Mesh of Ewald, with a points grid density of the reciprocal space of 1Å^3 [65].

2.3.5 Molecular dynamics: simulation details

All simulations were performed using the NAMD 2.9 program [36]. Each system was submitted to 5000 steps of energy minimization (Newton Raphson method) and this was the starting point of several minicycles of optimization and equilibration. The numerical integration step for all performed runs was set at 1 fs, periodic boundary conditions were applied using the nearest image convention and the non-bonded pair list was updated every 1000 steps (1 ps). The first equilibration target involved solvent and ClO_4^- molecules. This stage was integrated by 100,000 steps of heating and equilibration in NVT conditions to thermally stabilize each model at 298 K. The Berendsen thermostat was used [66] with a relaxation time of 1 ps, the fast convergence of the external bath being used to improve the efficiency of the thermal equilibration. During such period of time all the solid phase atoms were kept frozen. After this cycle, the solvent density was equalized to its optimum value using 250,000 steps of NPT simulation at 298 K. The Nose–Hoover piston combined with the piston fluctuation control of temperature implemented for Langevin dynamics [67] was used in this equilibration cycle. Pressure was kept at 1.01325 bars, the oscillation period was set at 1 ps while the piston decay time was set at 0.001 ps. The piston temperature was set at the same value as the thermostat control, 298 K, which used a damping coefficient of 2 ps. Once the simulation box dimensions were stabilized, the final equilibration cycles started. After unfreezing all the atoms present in the solid but the iron atoms, 200,000 more steps of NVT simulation were performed to thermally equilibrate the whole biphasic system. For the latter run the Langevin method [68] was used to maintain the system temperature constant with a damping coefficient of 2 ps. Finally, 250,000 steps of anisotropic pressure control (NP_zT conditions) were run to reach the required simulation conditions. In the NP_zT ensemble, only the box length in the *z*-direction is allowed to change such that the *zz*-component of the pressure tensor is equal to the imposed pressure [69]. Thus, in this last mini run, the metallic surface dimensions previously optimized were kept and only *z*-axis dimension was allowed to change as a response to pressure control. Therefore, the accumulated equilibration time is 0.8 ns, 0.25 ns of which corresponding to the final anisotropic equilibration. The last snapshot of this latter run was the starting point of 15 ns of production time. The production runs were performed under the same conditions previously mentioned for the NPzT equilibration.

2.4 A complementary computational-experimental investigations of DNA-hydroapatite biominerals

Calcium phosphate in the form of hydroxyapatite (HAp), with formula $Ca_{10}(PO_4)_6(OH)_2$, is a mineral of particular importance as it is the main component of hard human tissues [70]. Biomineralization of HAp, which is understood as the production of HAp crystals on organic templates within an organism, is extremely important for many processes like the formation and growth of bone and teeth within the human body. The HAp biomineralization is highly regulated by anionic phosphoproteins with long stretching of acidic amino acids and post-translational phosphorylation [71, 72]. Mineralization represents the ultimate barrier to be overcome in this chapter: the multi-facet interaction. Until now, all presented cases dealt with a single facet the only variation being the size and time scales of the system nevertheless biomineralization challenges the same size and time-scale limitations while at the same time includes the three-dimensional complexity of multi-faceting interaction during the solid solution formation.

The HAp biomineralization concept has led to new bioinspired or biomimetic materials with biomedical applications, for example in tissue repair and regeneration. More specifically, peptides, proteins, self-assembled monolayers, bioglass, and responsive polymers have been used to template HAp mineralization [73–76]. However, despite its therapeutic interest, templated HAp synthesis using DNA has been scarcely studied. Biominerals formed by HAp nanoparticles with embedded DNA chains were successfully obtained by *in situ* precipitation of the inorganic salt in presence of DNA under a narrow range of physic-chemical conditions [77–79] and on the use of a HAp core that was coated using colloidal solutions to give a multi-shell particle [80]. In contrast, Okazaki et al. [81], who investigated the synthesis of HAp in presence of DNA, reported that DNA inhibits the HAp crystal growth, indicating that DNA molecules are not in the HAp structure but adsorbed at the surface of the mineral. In this section, we illustrate how an appropriate design of complementary experimental and *in silico* research has recently provided valuable information and advances in the field of DNA-templated HAp biomineralization.

From a general perspective, one can imagine that DNA should promote the HAp crystal growth because number of reasons: (1) DNA has an extended anionic phosphate backbone that is known to interact with HAp [82, 83]; (2) the period of translation along the *c*-axis of the HAp lattice, 3.4 Å, is relatively similar to the period of the DNA double helix [84] and (3) Mg^{2+} is required for DNA activity *in vivo* [85], but can be replaced by alkali earth metals, including Ca^{2+}, *in vitro* [86]. These considerations motivated a number of combined computational and experimental studies devoted to obtain microscopic details about bioinspired materials formed through the encapsulation of DNA into HAp (hereafter denoted DNA/HAp), in which the biomolecule acted as the nucleating template for the formation of mineral.

This complementary approach started with a preliminary *in silico* study of DNA/HAp biominerals formed by single-stranded (ss) or double-stranded (ds) DNA embedded into HAp nanopores, which was focused on [87]: (i) the nanopore size required for the accommodation of DNA; and (ii) the strain induced by the HAp crystalline field into the DNA structure. Classical force field simulations and quantum mechanics calculations indicated that ds DNA adopting a B double helix can be encapsulated inside nanopores of HAp without undergoing significant conformational distortions (Figure 2.6a). This observation was found to be independent of the DNA sequence, which was attributed to the strong stabilizing interactions between the Ca^{2+} atoms of HAp and the phosphate groups of DNA. Consequently, the mineral completely eliminates the flexibility of sugar rings of DNA used as templates for HAp

(a)

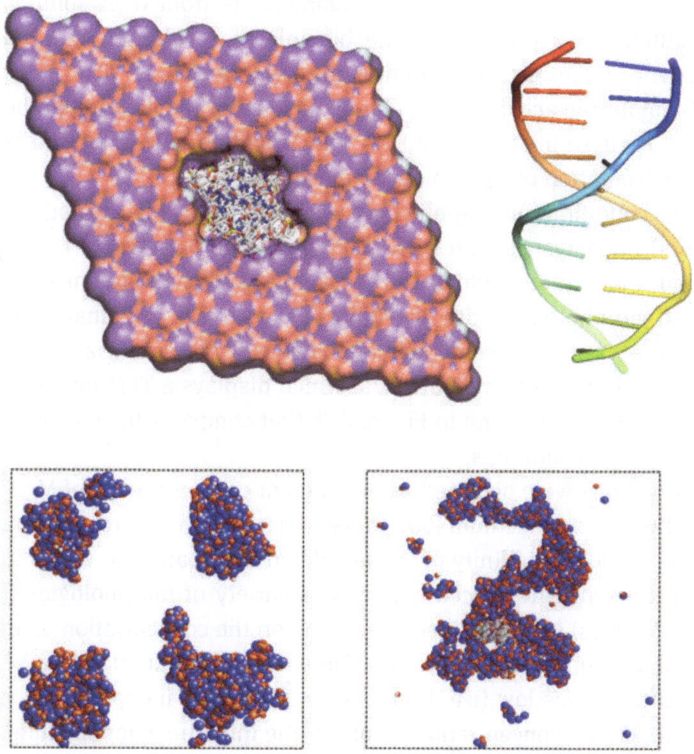

(b)

Figure 2.6: (a) Equatorial perspective of the ds B-DNA embedded into HAp and axial perspective of the biomolecule after simulation. (b) Distribution of Ca^{2+}, PO_4^{3-} and OH^- at last snapshot of a 20 ns MD trajectory in absence (left) and presence (right) of DNA.

crystallization, contributing to preserve the secondary structure of the biomolecule [88]. The minimal dimensions of the pore needed to encapsulate B-DNA were proposed to be ~30 Å × ~21Å. In opposition, ss DNA molecules encapsulated into HAp exhibited drastic conformational changes, which were obviously attributed to the lack of inter-strand hydrogen bonding and the strong DNA–mineral interactions [87].

On the basis of these results, we computationally investigated the mechanism for the growing of HAp crystals using DNA as nucleating agent [89]. For this purpose, a rational design approach based on atomistic MD simulations of DNA immersed in an inorganic solution showed the rapid formation of calcium phosphate clusters at the biomolecule template. Indeed, comparison of the MD trajectories of inorganic Ca^{2+}, PO_4^{3-} and OH^- aqueous solutions in presence and absence of DNA revealed that the templating effect of DNA is accompanied by a change in the aggregation mechanism. Thus, simulations reflected the formation of multiple calcium phosphate clusters in absence of DNA (Figure 2.6b), which incorporate OH^- anions after some nanoseconds transforming into nucleation centers for the formation of HAp. The simultaneous formation of multiple nucleation centers is fully consistent with the mechanism typically found for the crystallization of inorganic salts from their solutions. In opposition, simulations in presence of the biomolecule evidenced that calcium phosphate clusters are formed surrounding the DNA backbone. The phosphate groups of the biomolecule acted as a very large nucleus for the growing of the HAp (Figure 2.6b). As occurred in absence of DNA, the formation of HAp around DNA is limited by the incorporation of OH^- anions to the cluster.

These *in silico* conditions were applied in the laboratory to create experimentally nanoparticles and nanocrystals with DNA inside [89]. The presence of DNA inside nanospheres and nanocrystals, which showed a spherical shape (diameter of ~20 nm) and nanorod morphology (width of ~20 nm), respectively, was unambiguously determined using a variety of techniques after removal of superficially adsorbed DNA. This is clearly illustrated in Figure 2.7a, which displays a TEM micrograph of DNA-containing nanocapsules, and in Figure 2.7b that compares the UV-vis spectra of nanorods in different conditions.

Similar observations were obtained when different concentrations of Mg^{2+} were incorporated to the reaction medium [90]. Indeed, Mg^{2+} was found to play a crucial role in morphology and crystallinity of the templated HAp particles, which in turn also depend on the synthetic procedure. Thus, a variety of morphologies (nanospheres, rods and sheets) were obtained depending on the concentration of Mg^{2+} in the reaction medium and synthetic process. Moreover, the content of Mg^{2+} in the resulting HAp particles was low (i. e. from 0.32 to 3.82 wt%), independently of the synthetic method and the concentration of ions in the initial reaction medium. This was consistent with the composition of hard tissues in nature, which also exhibit very low content of Mg^{2+} in their composition. MD simulations in presence of DNA corroborated that the biomolecule acts as a soft template enabling the formation of inorganic clusters around the polyphosphate backbone, this process being independent of the concentration of Mg^{2+} ions in the inorganic solution [90]. Indeed, the abundance of Mg^{2+}-based interactions increased with the Mg^{2+} concentration, albeit the amount and frequency of Mg^{2+}-phosphate interactions was independent of such concentration. The templating role of DNA in the nucleation of Mg-containing HAp is illustrated by the radial distribution function obtained for $P_{inorg} \cdots P_{DNA}$ pairs (where P_{inorg} and P_{DNA} refer to the phosphate groups of the inorganic solution and DNA backbone, respectively), which is displayed in Figure 2.8a. On the other hand, comparison of radial distribution functions for $Mg^{2+} \cdots P_{DNA}$ pairs derived from trajectories with different Mg^{2+} content in the medium (i. e. 0.5, 1.5, 3 and 6 % wt. Mg^{2+}), which are provided in Figure 2.8b, proves that abundance of $Mg^{2+} \cdots$phosphate interactions is similar in all cases.

Computer simulations also provided relevant information about the microscopic mechanism of dissolution in acidic conditions (i. e. similar to those found in the cellular environment) of HAp nanoparticles templated with DNA [91]. For this purpose, four consecutive (multi-step) MD simulations, involving different temperatures and proton transfer processes, were carried out. Results were consistent with a

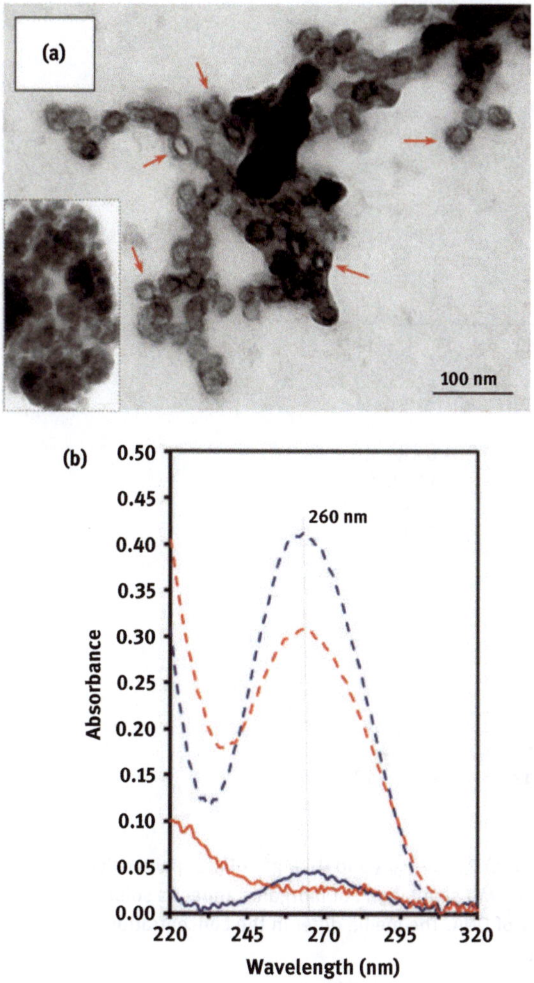

Figure 2.7: (a) TEM micrographs showing fish sperm DNA-containing HAp nanocapsules. Red arrows point out capsules with a clearly distinctive contrast that suggest the incorporation of DNA in their inner part. For comparison purposes the inset shows nanocapsules prepared under the same experimental conditions but in absence of DNA. (b) UV-vis absorption spectra of as prepared DNA-containing nanorods before and after dissolution of the in acid medium (blue solid and dashed lines, respectively): the presence of DNA was detected in both cases by the characteristic absorption peak at 260 nm. UV-vis spectra of samples digested with deoxyribonuclease to eliminate adsorbed DNA before and after dissolution are indicated with red traces (solid and dashed, respectively). As it was expected, the absorbance at 260 nm of solid nanorods disappears after enzymatic treatment, indicating that the absorbance displayed by blue traces correspond to superficially adsorbed DNA. However, the significant absorbance detected after dissolution of the digested sample in acid medium unambiguously demonstrates the incorporation of DNA inside nanoparticles.

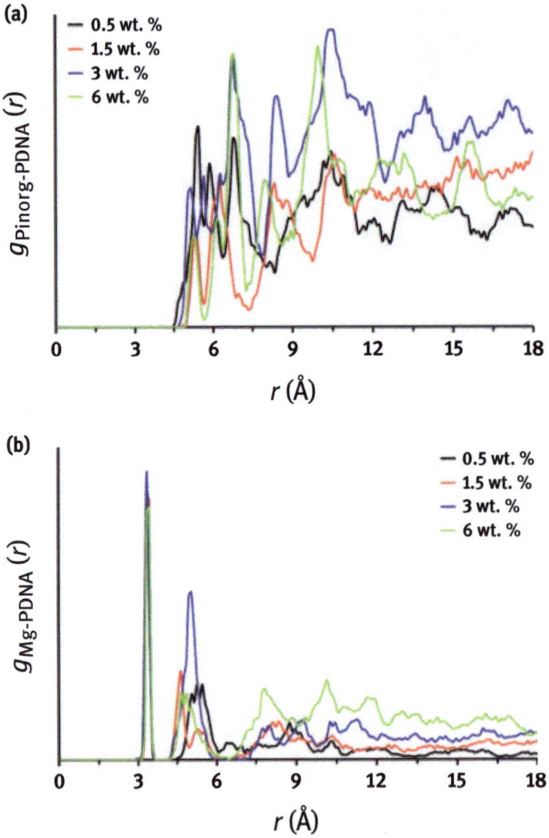

Figure 2.8: Radial distribution functions of (a) $P_{inorg}\cdots P_{DNA}$ and (b) $Mg^{2+}\cdots P_{DNA}$, pairs calculated using all the snapshots recorded during the MD simulations of inorganic aqueous solutions with Ca^{2+}, $Mg^{2+}PO_4^{3-}$ and OH^- in presence of DNA. The media differ in the concentration of Mg^{2+}, which ranges from 0.5 to 6 % wt.

polynuclear decalcification mechanism where Ca^{2+} detachment was triggered by the initial hydration of phosphates, and it was enhanced by their further hydrogenation resulting from hydronium reactivity. Also, results proved that HAp acts as a shield for DNA molecules protecting their functionality from temperature stressing conditions and chemicals. Results from such work had a number of applications in biomedicine and genetics biotechnology and the work itself provides a particular approach to the complex problem of mineral nucleation.

References

[1] Nilsson A, Petterson LGM, Norskov J. Chemical bonding at surfaces and interfaces. Oxford, U.K., Ed: Elsevier Ltd, 2008.

[2] Eiswirth, M., Möller, P., Wetzl, K., Imbihl, R., Ertl, G. Mechanisms of spatial self-organization in isothermal kinetic oscillations during the catalytic CO oxidation on Pt single crystal surfaces. J Chem Phys 1989;90:510.
[3] Higashi GS, Chabal J, Trucks GW, Raghavachari K. Ideal hydrogen termination of the Si (111) surface. Appl Phys Lett 1990;56:656.
[4] Muller DA. The electronic structure at the atomic scale of ultrathin gate oxides. Nature 1999;399:758–761.
[5] Garbassi F, Morra M, Occhiello E. Polymer surfaces: from physics to technology. Weinheim, Germany, Ed: Wiley-VCH, 1997.
[6] Astruc D. Nanoparticles and catalysis. Weinheim, Germany, Ed: Wiley-VCH, 2008.
[7] Carrasco J, Hodgson A, Michaelides A. A molecular perspective of water at metal interfaces. Nat Mater 2012;11:667.
[8] Naoi K, Simon P. New materials and new configurations for advanced electrochemical capacitors. Electrochem J Soc 2008;17:34.
[9] Carchini G, Almora-Barrios N, Revilla-López G, Bellarosa L, García-Muelas R, García-Melchor M. How theoretical simulations can address the structure and activity of nanoparticles.Top Catal 2013;56:1262.
[10] Cölfen H. Biomineralization: a crystal-clear view. Nat Mater 2010;9:960.
[11] Chorkendorff I, Niemanstverdriet JW. Concepts of modern catalysis and kinetics. Weinheim, Germany, Ed: Wiley-VCH, 2003.
[12] Martin RM. Electronic structure: basic theory and practical methods. Cambridge, Ed: Cambridge University Press, 2008.
[13] Rapaport DC. The art of molecular dynamics simulations. Cambridge, Ed: Cambridge University Press, 2004.
[14] Vargel C. Corrosion of aluminium. Oxford, U.K.: Elsevier Ltd; 2004.
[15] Ezuber H, El-Houd A, El-Shawesh F. A study on the corrosion behavior of aluminum alloys in seawater. Mater Des 2008;29:801–805.
[16] Tedim J, Poznyak SK, Kuznetsova A, Raps D, Hack T, Zheludkevich ML, et al. Enhancement of active corrosion protection via combination of inhibitor-loaded nanocontainers. ACS Appl Mater Inter 2010;2:1528–1535.
[17] Jerman I, Vuk AS, Kozelj M, Orel B, Kovac J. A structural and corrosion study of triethoxysilyl functionalized POSS coatings on AA 2024 alloy. A Langmuir 2008;24:5029–5037.
[18] Liu J, Chudhury MK, Berry DH, Seebergh JE, Osborne JH, Blohowiak, KY. Effect of surface morphology on crack growth at a sol-gel reinforced epoxy/aluminum interface. J Adhes 2006;82:487–516.
[19] Warren SC, Perkins MR, Adams AM, Kamperman M, Burns AA, Arora H, et al. Nat Mater 2012;11:460–467.
[20] Zheludkevich ML, Salvado IM, Ferreira MGS. Sol–gel coatings for corrosion protection of metals. J Mater Chem 2005;15:5099–5111.
[21] Dalmoro V, dos Santos JHZ, Azambuja DS. Corrosion behavior of AA2024-T3 alloy treated with phosphonate-containing TEOS. J Solid State Electr 2012;16:403–414.
[22] Holzle LRB, Azambuja DS, Piatnicki CMS, Englert GE. Corrosion behaviour of aluminium in ethyleneglycol–water electrolytes containing phosphonic acid. Mater Chem Phy 2007;103: 59–64.
[23] Lecollinet G, Delorme N, Edely M, Gibaud A, Bardeau J-F, Hindré F, et al. Self-assembled monolayers of bisphosphonates: influence of side chain steric hindrance. Langmuir 2009;25:7828–7835.
[24] Mutin PH, Guerrero G, Vioux A. Hybrid materials from organophosphorus coupling molecules. J Mater Chem 2005;15:3761–3768.

[25] Khramov AN, Balbyshev VN, Kasten LS, Mantz RA. Sol–gel coatings with phosphonate functionalities for surface modification of magnesium alloys. Thin Solid Films 2006;514: 174–181.
[26] Dalmoro V, dos Santos JHZ, Armelin E, Alemán C, Azambuja DS. A synergistic combination of tetraethylorthosilicate and multiphosphonic acid offers excellent corrosion protection to AA1100 aluminum alloy. Appl Surf Sci 2013;273:758–768.
[27] Torras J, Azambuja DS, Wolf JM, Alemán C, Armelin E. How organophosphonic acid promotes silane deposition onto aluminum surface: a detailed investigation on adsorption mechanism. J Phys Chem C 2014;118:17724–17736.
[28] Alexander MR, Thompson GE, Beamson G. Characterization of the oxide/hydroxide surface of aluminium using x-ray photoelectron spectroscopy: a procedure for curve fitting the O 1s core level. Surf Interface Anal 2000;29:468–477.
[29] Zhang A, Okrasa L, Pakula T, Schlüter D. Homologous series of dendronized polymethacrylates with a methyleneoxycarbonyl spacer between the backbone and dendritic side chain: synthesis, characterization, and some bulk properties. J Am Chem Soc 2004;126: 6658–6666.
[30] Kim H-J, Young E-Y, Jin JY, Lee M. Solution behavior of dendrimer-coated rodlike coordination polymers. Macromolecules 2008;41:6066–6072.
[31] Li W, Zhang A, Schlüter AD. Thermoresponsive dendronized polymers with tunable lower critical solution temperatures. Chem Commun 2008;5523–5525.
[32] Bertran O, Zhang B, Schlüter AD, Kröger M, Alemán C. Modeling nanosized single molecule objects: dendronized polymers adsorbed onto mica. J Phys Chem C 2015;119:3746–3753.
[33] Guo Y, van Beek JD, Zhang B, Colussi M, Walsde P, Zhang A, et al. Tuning polymer thickness: synthesis and scaling theory of homologous series of dendronized polymers. J Am Chem Soc 2009;131:11841.
[34] Zhang B, Wepf R, Fischer K, Schmidt M, Besse S, Lindner P, et al. The largest synthetic structure with molecular precision: towards a molecular object. Angew Chem 2011;50:763–766.
[35] Zhang B, Wepf R, Kröger M, Halperin A, Schlüter AD. Height and width of adsorbed dendronized polymers: electron and atomic force microscopy of homologous series. Macromolecules 2011;44:6785–6792.
[36] Phillips JC, Braun R, Wang W, Gumbart J, Tajhordhid E, Villa E, et al. Scalable molecular dynamics with NAMD. J Comput Chem 2005;26:1781–1802.
[37] Cornell WD, Cieplak P, Bayly CI, Gould IR, Merz KM, Ferguson DM, et al. A second generation force field for the simulation of proteins, nucleic acids, and organic molecules. J Am Chem Soc 1995;117:5179–5197.
[38] Wang J, Wolf RM, Caldwell JM, Kollman PA, Case DA. Development and testing of a general amber force field. J Comput Chem 2004;15:1157–1174.
[39] Heinz H, Koermer H, Anderson KL, Vaia RA, Farmer BL. Force field for mica-type silicates and dynamics of octadecylammonium chains grafted to montmorillonite. Chem Mater 2005;17:5658–5669.
[40] Bertran O, Zhang B, Schlüter AD, Halperin A, Kröger M, Alemán C. Computer simulation of dendronized polymers: organization and characterization at the atomistic level. RSC Adv 2013;3:126–140.
[41] Mecke A, Lee I, Baker jR, Banaszak MM, Orr BG. Deformability of poly (amidoamine) dendrimers. Eur Phys J E 2004;14:7–16.
[42] Li D, Zheng Q, Wang Y. Combining surface topography with polymer chemistry: exploring new interfacial biological phenomena. Polym Chem 2014;5:14–24.
[43] Tretyakov N, Müller M. Directed transport of polymer drops on vibrating superhydrophobic substrates: a molecular dynamics study. Soft Matter 2014;10:4373–4386.

[44] Yin J, Yagüe JL, Boyce MC, Gleason KK. Biaxially mechanical tuning of 2-D reversible and irreversible surface topologies through simultaneous and sequential wrinkling. Appl Mater Interfaces 2014;6:2850–2857.
[45] Long YZ, Li M-M-, Gu C, Wan M, Duvail J-L, Liu Z, et al. Recent advances in synthesis, physical properties and applications of conducting polymer nanotubes and nanofibers. Prog Pol Scien 2011;36:1415.
[46] Del Valle LJ, Estrany F, Armelin E, Oliver R, Alemán C. Cellular adhesion, proliferation and viability on conducting polymer substrates. Macromol Biosci 2008;8:1144.
[47] Groenendaal L, Jonas F, Freitag D, Pielartzki H, Reynols JR. Poly (3, 4-ethylenedioxythiophene) and its derivatives: past, present, and future. Adv Mater 2000;12:481.
[48] Kirchmeyer S, Reuter K. Polyfullerenes for organic photovoltaics. J Mater Chem 2005;15:2077.
[49] Tamburri E, Orlanducci S, Toschi F, Terranova ML, Passeri D. Growth mechanisms, morphology, and electroactivity of PEDOT layers produced by electrochemical routes in aqueous médium. Synth Met 2009;159:406.
[50] Aradilla D, Estrany F, Alemán C. Symmetric supercapacitors based on multilayers of conducting polymers. J Phys Chem C 2011;115:8430.
[51] Aradilla D, Azambuja DS, Estrany F, Casas MT, Ferreira CA, Alemán C. Hybrid polythiophene–clay exfoliated nanocomposites for ultracapacitor devices. J Mater Chem 2012;22:13110.
[52] Xuan Y, Sandberg M, Berggren M, Crispin X. An all-polymer-air PEDOT battery. Org Electron 2012;13:632.
[53] Aradilla D, Estrany F, Casellas F, Iribarren JI, Alemán C. All-polythiophene rechargeable batteries. Org Electron 2014;15:40.
[54] Aradilla D, Pérez-Madrigal MM, Estrany F, Azambuja D, Iribarren JI, Alemán, C. Nanometric ultracapacitors fabricated using multilayer of conducting polymers on self-assembled octanethiol monolayers. Org Electron 2013;14:1483.
[55] Aradilla D, Estrany F, Alemán C. Synergy of the redox pair in the capacitive properties of nanometric poly (3, 4-ethylenedioxythiophene). Org Electron 2013;14:131.
[56] Aradilla D, Estrany F, Armelin E, Alemán C. Ultraporous poly (3, 4-ethylenedioxythiophene) for nanometric electrochemical supercapacitor. Thin Solid Films 2012;520:4402.
[57] Zanuy D, Aleman C. Resolving the subnanometric structure of ultrathin films of poly (3, 4-ethylenedioxythiophene) on steel surfaces: a molecular modeling approach. Soft Matter 20139:11634.
[58] Ahumada O, Pérez-Madrigal MM, Ramírez J, Curcó D, Esteves C, Salvador-Mata G, et al. Sensitive thermal transitions of nanoscale polymer samples using the bimetallic effect: application to ultra-thin polythiophene. Rev Sci Instrum 2013;84:053904.
[59] Ocampo C, Oliver R, Armelin E, Alemán C, Estrany F. Electrochemical synthesis of poly (3, 4-ethylenedioxythiophene) on steel electrodes: properties and characterization. J Polym Res 2006;13:193–200.
[60] Teixeira-Dias B, Zanuy D, del Valle LJ, Estrany F, Armelin E, Alemán C. Influence of the doping level on the interactions between poly (3, 4-ethylenedioxythiophene) and plasmid DNA. Macromol Chem Phys 2010;211:1117–1126.
[61] Preat J, Zanuy D, Aleman C. Binding of Cationic Conjugated Polymers to DNA: Atomistic Simulations of Adducts Involving the Dickerson's Dodecamer. J Comput Chem. 2011;12:1304–2012.
[62] Duan Y, Wu C, Chowdhury S, Lee MC, Xiong G, Zhang W, et al. A point-charge force field for molecular mechanics simulations of proteins based on condensed-phase quantum mechanical calculations. J.Comput.Chem. 2003;24:1999–2012.
[63] Baaden M, Burgard M, Boehme C, Wipff G. Lanthanide cation binding to a phosphoryl-calix[4] arene: the importance of solvent and counterions investigated by molecular dynamicsand quantum mechanical simulations. Phys Chem Chem Phys 2001;3:1317–1322.

[64] Savio D, Fillot N, Vergne P, Zaccheddu M. A model for wall slip prediction of confined n-alkanes: effect of wall-fluid interaction versus fluid resistance. Tribol Lett 2012;46:11–22.
[65] Toukmaji C, Sagui C, Board J, Darden T. Efficient particle-mesh Ewald based approach to fixed and induced dipolar interactions. J Chem Phys 2000;113:10913–10927.
[66] Berendsen HJC, Postma PM, van Gunsteren WF, DiNola A, Haak GR. Molecular dynamics with coupling to an external bath. J Chem Phys. 1984;81:3684–3690.
[67] Martyna GJ, Tobias DL. Constant pressure molecular dynamics algorithms. Chem Phys. 1994;101:4177–4189.
[68] Feller SE, Zhang Y, Pastor RW. Constant pressure molecular dynamics simulation: The Langevin method. J Chem Phys. 1995;103:4613–4622.
[69] Toxvaer S. Molecular dynamics calculation of the equation of state of alkanes. J Chem Phys. 1990;93:4290–4295.
[70] Margolis HC, Beniash E, Fowler CE. Role of macromolecular assembly of enamel matrix proteins in enamel formation. J Dent Res 2006;85:775–793.
[71] Uskokovic V, Li W, Habelitz S. Amelogenin as a promoter of nucleation and crystal growth of apatite. J Cryst Growth 2011;316:106–117.
[72] Wang L, Guan X, yin H., Moradian-Oldak J, Nancollas GH. Mimicking the self-organized microstructure of tooth enamel. J Phys Chem C 2008;112:5892–5899.
[73] Gungormus M, Fong H, Kim IW, Evans JS, Tamerles C, Sarikaya M. Regulation of in vitro calcium phosphate mineralization by combinatorially selected hydroxyapatite-binding peptides. Biomacromolecules 2008;9:966–973.
[74] Zhu P, Masuda Y, Yonezawa T, Koumoto K. Investigation of apatite deposition onto charged surfaces in aqueous solutions using a quartz-crystal microbalance. J Am Ceram Soc 2003;86:782–790.
[75] Zhu Z, tong H, Jiang T, Shen X, Wan P, Hu J. Studies on induction of l-aspartic acid modified chitosan to crystal growth of the calcium phosphate in supersaturated calcification solution by quartz crystal microbalance. Biosens Bioelectron 2006;22:291–297.
[76] Ball V, Michel M, Boulmedais F, Hemmerles J, Haikel Y, Schaaf P, et al. Nucleation kinetics of calcium phosphates on polyelectrolyte multilayers displaying internal secondary structure. Cryst Growth Des 2006;6:327–334.
[77] Kakizawa Y, Miyata K, Furukawa S, Kataoka K. Size-controlled formation of a calcium phosphate-based organic–inorganic hybrid vector for gene delivery using poly (ethylene glycol)-block-poly (aspartic acid). Adv Mater 2004;16:699–702.
[78] Urabe M, Kume A, Toibita K, Ozawa K. DNA/calcium phosphate precipitates mixed with medium are stable and maintain high transfection efficiency. Biochem 2000;278:91–92.
[79] Ngourn SC, Butts HA, Petty AR, Anderson JE, Gerdon AE. Quartz crystal microbalance analysis of DNA-templated calcium phosphate mineralization. Langmuir 2012;28:12151–12158.
[80] Sololova V, Radkte I, Heumenn R, Epple M. Effective transfection of cells with multi-shell calcium phosphate-DNA nanoparticles. Biomaterials 2006;27:3147–3153.
[81] Okazaki M, Yoshida Y, Yamaguchi S, Kaneo M, Elliott JC. Affinity binding phenomena of DNA onto apatite crystals. Biomaterials 2001;22:2459–2464.
[82] Chen W-Y, Lin M-S, Lin P-H, Tasi P-S, Chang Y, Yamamoto S. Studies of the interaction mechanism between single strand and double-strand DNA with hydroxyapatite by microcalorimetry and isotherm measurements. Colloids Surf, A 2007;295:274–283.
[83] Del Valle LJ, Bertran O, Chaves G, Revilla-López G, Rivals M, Cassas MT, et al. DNA adsorbed on hydroxyapatite surfaces. J Mater Chem B 2014;2:6953–6966.
[84] Kostetsky EY. The possibility of the formation of protocells and their structural components on the basis of the apatite matrix and cocrystallizing minerals. J Biol Phys 2005;31:607–638.

[85] Bertini I, Gray HB, Lippard SJ, Valentine JS. Bioinorganic Chemistry. Principles of bioinorganic chemistry. Mill Valley, CA: University Science Books, 1994.
[86] Lukeman PS, Stevenson ML, Seeman NC. Morphology change of calcium carbonate in the presence of polynucleotides. Cryst Growth Des 2008;8:1200–1202.
[87] Revilla-López G, Casanovas J, Bertran O, Turón P, Puiggalí J, Alemán C. Modeling biominerals formed by apatites and DNA. Biointerphases 2013;8:10–25.
[88] Casanovas J, Revilla-López G, Bertran O, del Valle LJ, Turón P, Puiggalí J. Restricted puckering of mineralized RNA-like riboses. J Phys Chem B 2014;118:5075–5081.
[89] Bertran O, del Valle LJ, Revilla-López G, Chaves G, Cardús L, Casas MT, et al. Mineralization of DNA into nanoparticles of hydroxyapatite. Dalton Trans 2014;43:317–327.
[90] Bertran O, del Valle LJ, Revilla-López G, Rivas M, Chaves G, Casas MT, et al. Synergistic approach to elucidate the incorporation of magnesium ions into hydroxyapatite. Chem Eur J 2015;21:2537–2546.
[91] Bertran O, Revilla-López G, Casanovas J, del Valle LJ, Turón P, Puiggalí J, et al. Dissolving hydroxyolite: a DNA molecule into its hydroxyapatite mold. Chem Eur J 2016;22:6631–6636.

Paulo J. Costa

3 The halogen bond: Nature and applications

Abstract: The halogen bond, corresponding to an attractive interaction between an electrophilic region in a halogen (X) and a nucleophile (B) yielding a R–X···B contact, found applications in many fields such as supramolecular chemistry, crystal engineering, medicinal chemistry, and chemical biology. Their large range of applications also led to an increased interest in their study using computational methods aiming not only at understanding the phenomena at a fundamental level, but also to help in the interpretation of results and guide the experimental work. Herein, a succinct overview of the recent theoretical and experimental developments is given starting by discussing the nature of the halogen bond and the latest theoretical insights on this topic. Then, the effects of the surrounding environment on halogen bonds are presented followed by a presentation of the available method benchmarks. Finally, recent experimental applications where the contribution of computational chemistry was fundamental are discussed, thus highlighting the synergy between the lab and modeling techniques.

Keywords: *ab initio* calculations, charge-transfer, DFT calculations, electrostatic potential, halogen bond, σ-hole

3.1 Introduction

In 1863 [1], Frederick Guthrie reported the purification and formulation of $NH_3{\cdot}I_2$, which he called "iodide of iodammonium", based on earlier observations of M. Colin in 1814 [2]. Still in the nineteenth century, similar adducts with Br_2 and Cl_2 were also prepared and described [3]. However, the odd interactions existing in those complexes remained essentially unexplored until the late twentieth century, in spite the continuous observation of related phenomena: the formation of adducts between iodoform and quinolone [4], the formation of halogen molecule bridges in chains of 1,4-dioxane and X_2 ($X_2 = Cl_2, Br_2, I_2$) [5], where Bent considered that "there is reason to believe that the O···Br–Br interaction is energetically comparable to a strong hydrogen bond" [6], and complexes formed by halogen molecules with oxygenated solvents [7], among other examples.

This article has previously been published in the journal *Physical Sciences Reviews*. Please cite as: Costa, P. J. The halogen bond: Nature and applications. *Physical Sciences Reviews* [Online] **2017**, *2* (11). DOI: 10.1515/psr-2017-0136

https://doi.org/10.1515/9783110482065-003

In his Nobel Lecture in 1970 [8], Hassel pointed out the importance of halogens in self-assembly but the phenomena at the time was not fully understood, falling in the generic classification of "donor–acceptor interaction" [6] or "charge–transfer bonding" [8]. In this interaction, which is now called halogen bond, the covalently-bound electronegative halogen atoms are able to establish attractive interactions with electron-rich entities (nucleophiles), i. e., a R–X···B interaction (X = Cl, Br, I; B = Lewis base/nucleophile).

The first systematic theoretical explanation for this apparently counterintuitive phenomena was put forward by Brinck, Murray and P. Politzer [9]. In that seminal paper the authors noticed that, in covalently-bonded halogens, their electrostatic potential is anisotropic, possessing positive regions at the tip of X (X = Cl, Br, I). This positive region was later coined as σ-hole [10], corresponding to a maxima of the electrostatic potential mapped on a surface ($V_{S,max}$). Therefore, halogen bonds were described as an electrostatically-driven noncovalent interaction between that positive σ-hole and a nucleophile (see Figure 3.1). The electrostatic potential characteristics elegantly explain the main features and directional preferences of the interactions with halogen atoms: side-on interactions are observed with electrophiles whereas head-on interactions, corresponding to halogen bonds with nucleophiles, possess R–X···B angles of ~180° (Figure 3.1). Indeed, halogen bonds and hydrogen bonds are considered orthogonal molecular interactions since they are geometrically perpendicular and energetically independent when sharing a common acceptor [11]. It must be pointed out that, since σ-holes originate from the polarization of the electronic charge toward the R–X covalent bond, if X is less polarizable and very electron-attracting, the σ-hole might be neutralized or even absent [10]. Therefore, typically, the halogen bond strength increases from chlorine to iodine, whereas fluorine is not usually considered a halogen bond donor.

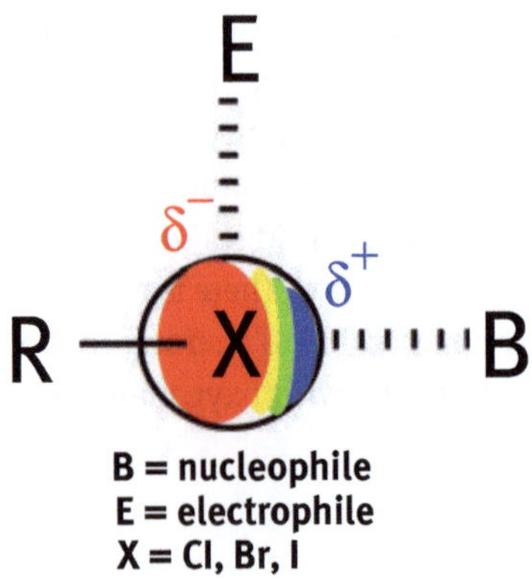

Figure 3.1: Schematic representation of an R–X···B halogen bond showing the anisotropic distribution of charge around the X atom. A possible orthogonal side-on interaction with an electrophile is also shown.

The theoretical explanation was accompanied by an explosive growth of research on halogen bonds, which found applications in diverse fields such as supramolecular chemistry [12], anion recognition [13], crystal engineering [14], and medicinal chemistry/chemical biology [15]. This is promptly seen by the increasing amount of research papers per year with the topic "halogen bonding" or "halogen bond" (Figure 3.2).

All theoretical and experimental results together with the increasing interest eventually led to a rationalization and unification process promoted by the International Union of Pure and Applied Chemistry (IUPAC), which started a project in 2009 aiming at a "comprehensive look at intermolecular interactions involving halogens as electrophilic species" and "give a modern definition of halogen bonding, which takes into account all current experimental and theoretical pieces of information" [16]. The final report issued in 2013 [17] states that "a halogen bond occurs when there is evidence of a net attractive interaction between an electrophilic region associated with a halogen atom in a molecular entity and a nucleophilic region in another, or the same, molecular entity."

This work intends to provide the reader with a succinct overview of the most recent theoretical and experimental developments, typically from 2015-onwards, in

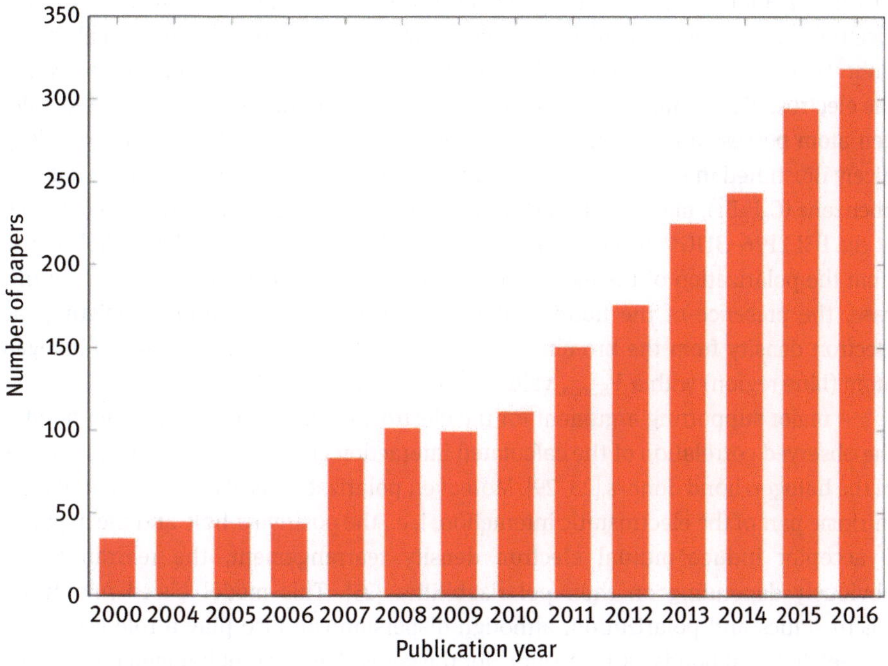

Figure 3.2: Number of publications containing the keywords "halogen bonding" or "halogen bond" (Source: Thomson Reuters Web of Science, 2017/03).

the field of halogen bonding and its applications. The most representative works and areas were selected. For a more systematic assessment, covering a broader range of research papers and also a larger period of time, the reader is directed to the recent reviews and perspectives on the topic [12–15, 18–25].

3.2 On the nature of the halogen bond: A longstanding debate

The nature of halogen bonds is a subject of a heated debate and in this point, theoretical calculations are of paramount importance. The halogen bonds are normally included in the realm of noncovalent interactions and their generic features are largely explained by the electrostatic interaction between a region of depleted electron density at the tip of a covalently bonded halogen, which is, as mentioned earlier, designated as σ-hole [10], and a negative site.

The electrostatic potential that the electrons and nuclei produce at any point r is given by (eq. 3.1).

$$V(r) = \sum_A \frac{Z_A}{|R_A - r|} - \int \frac{\rho(r')dr'}{|r' - r|} \tag{3.1}$$

In this equation, Z_A is the charge of the nucleus A located at R_A and $\rho(r')$ is the electronic density of the molecule. It is common to evaluate $V(r)$ on the 0.001 electrons Bohr^{-3} contour of the density $\rho(r)$ [26] labeling it $V_s(r)$. In halogen bond donors, when the electrostatic potential surface extrema are evaluated, the covalently-bonded halogen atom possess a maxima, $V_{S,max}$, corresponding to the σ-hole [10]. This σ-hole is nicely illustrated in Figure 3.3(a) where the electrostatic potential of bromopentafluorobenzene (C_6F_5Br), mapped on the 0.001 au contour of the electronic density obtained at the B3LYP/6–311G** level of theory is depicted. As mentioned earlier, σ-holes arise from the polarization of the electronic charge toward the R–X covalent bond. In this case, the presence of the fluorine atoms enhance that polarization, depleting the electron density from the bromine atom, thus forming a positive tip at the halogen atom (blue region) with a $V_{S,max}$ value of 26.06 kcal mol^{-1} [27].

A major supporting argument for the electrostatic nature of the halogen bond is the observed correlation of the calculated interaction energies with the $V_{S,max}$ values of the halogen bond donors [28, 29]. Moreover, polarization is also involved as it is an intrinsic part of the electrostatic interaction, i. e., the positive σ-hole and the negative B acceptor induce mutual electron density rearrangement, the rearrangement obviously depending on their polarizabilities [25]. This model based mainly on electrostatics and polarization, although dispersion can also play a role especially in week halogen bonds [30], accounts for the general features of halogen bonds such as the R–X···B bond angles of ~180° and the possibility of a side-on interaction with an electrophile (Figure 3.1).

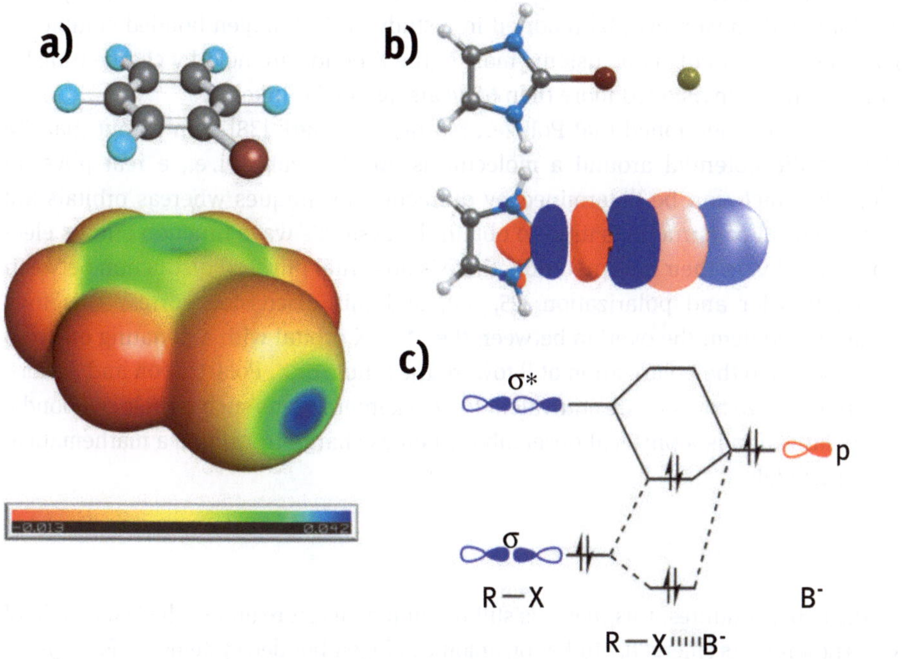

Figure 3.3: On the nature of halogen bonds: a) electrostatic potential of bromopentafluorobenzene (C_6F_5Br) mapped on the 0.001 au contour of the electronic density obtained at the B3LYP/6-311G** level of theory. Notice that the fluorines are negative whereas the bromine σ-hole is shown as a blue positive region, corresponding to a $V_{S,max}$ value of 26.06 kcal mol^{-1}; b) Natural Bond Orbital (NBO) donor-acceptor interaction corresponding to a donation from a chloride anion (light orbital) to a σ* C–Br orbital of 2-bromo-1*H*-imidazol-3-ium in a charge-assisted halogen bond (adapted from reference 36); c) simplified orbital interaction diagram for an R–X⋯B^- halogen bond (R, X, B = F, Cl, Br, and I), adapted from reference 23.

The electrostatic nature of the halogen bond is not consensual and some authors have pointed out a dominant charge-transfer nature, i. e., a donor–acceptor interaction [23] to explain the characteristics of this interaction. This charge-transfer nature, sometimes called covalent character [23], is supported by several theoretical and experimental studies [31–34]. In this interpretation, there is a stabilization arising from HOMO–LUMO interactions between a lone pair of B and the σ* orbital of C–X [35]. This interaction is illustrated in Figure 3.3(b) which shows the donation from a chloride anion to a σ* C–Br orbital in a charge-assisted ion-pair halogen bond between 2-bromo-1H-imidazol-3-ium and Cl^- [36]. This system will be analyzed more thoroughly below. From a molecular orbital theory perspective, schematically represented in Figure 3.3(c), there is mixing between the antibonding σ* R–X orbital of the halogen bond donor and a lone pair of the halogen bond acceptor B. In the case of a low-lying σ* C–X orbital, acting as a Lewis acid, and a strong Lewis base, the interaction is enhanced.

Ab initio valence bond (VB) theory also endorses charge-transfer as a key player as Shaik and co-workers [37] reported in a study on 55 halogen-bonded complexes reaching a "clear cut" conclusion: "most of the X-bonds are held by charge transfer interactions as envisioned more than 60 years ago by Mulliken".

It must be mentioned that Politzer, Murray and Clark [38] pointed out that the electrostatic potential around a molecule is an observable, i. e., a real physical property which can be determined by diffraction techniques whereas orbitals are mathematical constructions used to obtain the system's wave function and its electronic density. In their interpretation, there is no actual physical distinction between charge-transfer and polarization [25, 38], and this discussion corresponds to a semantic problem; the overlap between the σ* C–X orbital with a donating one of B corresponds to the polarization of B toward the σ-hole of X. Polarization and donor–acceptor charge transfer are equivalent for weak interactions such as halogen bonds, but polarization is a physical observable whereas charge-transfer is a mathematical construct [39].

3.2.1 Latest theoretical insights

As the debate endures, this theme is still hot in the literature and in the last couple of years new reports emerged. Studies on anionic halogen-bonded systems $CX(F_nH_{3\text{-}n})\cdots X^-$ (with X = F, Cl, and $n = 0–3$), based on MP2/aug'-cc-pVTZ calculations [40], showed the existence of tetrel, hydrogen, and halogen bonds. For instance, the interaction energy for $CF_3Cl\cdots Cl^-$ was –43.1 kJ mol^{-1}, corresponding to a charge-transfer interaction of –33.1 kJ mol^{-1} from the Cl^- lone pair to the antibonding C–Cl orbital of the substituted methane. Indeed, charge-transfer values correlated with intermolecular distances and with the interaction energies of these complexes.

Coupled cluster theory, in particular, CCSD(T)/aug-cc-pVTZ calculations were performed in order to study the nature and strength of a series of $YX\cdots BR_m$ dimers (where X = F, Cl, Br; Y = donor group; BR_m = acceptor group) comprising 36 neutral and anionic halogen bonded complexes together with eight complexes bearing hydrogen, pnicogen, or chalcogen bonds [41]. In this study, Natural Bond Orbital (NBO) charges, charge-transfer energies, electrostatic potentials, vibrational frequencies, local stretching and bending force constants, and relative bond strength orders, among other properties were analyzed. Bond strength orders (BSO), derived from local stretching force constants, differentiate between weak and predominantly electrostatic halogen bonds, normal halogen bonds and strong, predominantly covalent halogen bonds, indicating that the majority have both electrostatic and covalent nature. The covalent bond is due to a 3c-4e (three-center-four-electron) interaction, giving rise to the characteristic calculated charge-transfer values. The electrostatic term is dependent on the polarizing strength of B together with the polarity of the X–Y bond and the polarizability of X. The same authors expanded this study to 202 halogen-bonded complexes, using ωB97XD/aug-cc-pVTZ and CCSD(T)/aug-cc-pVTZ calculations [42], showing that most of the halogen bonded systems possess sizable covalent contributions and that they are more covalent than its counterparts, i. e., hydrogen bonds and pnicogen bonds. Nonetheless, electrostatic contribution arising from the interaction of the σ-hole with B is also important. However, the authors point out that the analysis of the σ-hole can provide some insight into the electrostatic terms but not the potentially covalent character of a halogen bond since the covalent contributions depend on both potential and kinetic energy.

Halogen bonded complexes involving phosphines, $H_3P\cdots ClF$, $H_3P\cdots BrF$, and $H_3P\cdots IF$, were studied at the CCSD(T)-F12c, M06-2X, and ωB97X-D levels of theory. All systems presented relatively short intermolecular P···X distances and strong interaction energies, as high as -19.61 kcal mol^{-1} for $H_3P\cdots ClF$, leading the authors to the designation "*Mulliken inner complex*". Indeed, NBO analysis portioned the complex $H_3P\cdots ClF$ as H_3PCl^+ and F^-. SAPT analysis indicates that, for instance, in the mentioned $H_3P\cdots ClF$ complex, induction (which includes charge-transfer) contributes almost with the same amount as electrostatics.

In bifurcated halogen bonds formed between dihalogen molecules (ClF, BrF, and BrCl) and the methoxy groups of several 4-substituted 1,2-dimethoxybenzenes [43], M06-2X/def2-TZVPPD calculations show that the asymmetric interaction is preferred whereas the symmetric arrangement is a first-order saddle point. This was ascribed to a polarization effect. A QTAIM analysis indicates closed-shell interactions and the interaction energies correlate linearly with the sum of the delocalization indices between the halogen and the two oxygen atoms obtained by this type of calculations, displaying the contribution from exchange–correlation to the binding energy. Energy decomposition analysis corroborates this data; in weak halogen bonds the main stabilizing term is the exchange–correlation whereas in strong halogen bonds the electrostatic component becomes dominant. Nonetheless, it was evident that "electron sharing" contribution correlates with the halogen bond interaction energy. A second-order stabilization energy calculation from a NBO analysis indicated that the most important stabilization arises from a charge-transfer from the oxygen lone pairs toward the σ* orbital of the dihalogen. Again, it seems that covalency (as represented by the exchange–correlation) is extremely important in the description of halogen bonds.

The role of charge-transfer has been very recently studied by Řezáč and coworkers [44] using the novel constrained DFT method for noncovalent complexes [45]. In this scheme, a charge transfer-free reference state corresponding to the superimposition of the electron densities of the non-interacting fragments is used, thus avoiding the vanishing of the charge-transfer at the complete basis set limit. Using the X40 dataset [46] that covers a wide range halogen bonds and the B-LYP, B3-LYP, BH-LYP, PBE, and PBE0 functionals, the amount of charge-transfer was evaluated. For weak halogen bonds, charge-transfer contributes only with 3 % whereas for all halogen bonded complexes, the average contribution is 10 % of the interaction energy. It was however surprising that the charge-transfer energies correlated with the interaction energies (R^2 = 0.85) which was attributed to a synergy between this character, electrostatics (i. e., the magnitude of the σ-hole), and dispersion. Nonetheless, the nature of B is important and when B involves nitrogen, the halogen bonds are stronger than the ones involving oxygen or sulfur, possessing also the larger values of charge-transfer but this is accompanied by stronger electrostatic interactions. Overall, the authors concluded that charge-transfer is undeniably involved in halogen bonding but cannot be considered its driving force.

Thirman and Head-Gordon [47] developed an energy decomposition analysis (EDA) of intermolecular interactions based on absolutely localized molecular orbitals (ALMOs) which can be applied to MP2 calculations (MP2 ALMO EDA). In this scheme, the binding energy is decomposed into frozen interaction, polarization, charge-transfer, and dispersion. The method was tested on the $FCl{\cdots}OH_2$ system (in analogy with the water dimer) and in this particular halogen bond, the interaction is driven by charge-transfer and is not particularly directional; electrostatics favor a planar geometry but charge-transfer effects are able to overcome this geometry in line with a more favorable alignment with the oxygen lone pair.

As stated before, according to *ab initio* valence bond (VB) theory, halogen bonds are "held by charge-transfer interactions" [37]. Very recently [48], block-localized wave function calculations, which are based on VB theory, tried to offer a unified theory for blue- and red-shifts in the vibrational frequencies of hydrogen and halogen bonds. Typically, hydrogen and halogen bonds exhibit red-shifts upon the formation of the complexes; however, blue-shifts occur in certain systems. The red-shifts are explained by the n(B) → σ*(R–X) charge transfer that, by populating the antibonding orbital, stretches the R–X bond, with a concomitant decrease in the HOMO-LUMO gap due to the interaction with the electrostatic field of B. This also indicates a more effective orbital mixing within the acid, which corresponds to polarization and therefore, both charge-transfer and polarization stabilize halogen bonds. In other words, stretching of the R–X bond leads to a lower energy of the σ* R–X orbital which in turn allows a more efficient charge-transfer from B. The red-shifts are thus explained by the polarization and charge-transfer nature of these interactions. On the other hand, the shrinking of the R–X bond, leading to blue-shifts, occurs when the frozen energy term (composed of the electrostatic and Pauli repulsion energies) is dominant. Overall, the preference toward R–X bond elongation or shrinkage upon halogen bond formation is determined by a competition between this frozen energy term and charge-transfer/polarization. The same holds for the formation of a hydrogen bond.

Another unified picture of hydrogen bonds and halogen bonds was provided in a MP2(full)/aug-cc-pVDZ(-PP) study performed on 108 halogen and hydrogen bonded complexes of the $R{-}X{\cdots}NH_3$ family (X = H, Cl, Br, and I) [49]. The interaction energies correlate linearly with the $V_{S,max}$ values but also with second-order perturbation stabilization energies and the amount of charge transferred (Q_{CT}), indicating that both factors (electrostatics *vs*. charge-transfer) are important. Indeed, $V_{S,max}$ and Q_{CT}

are complementary in the description of the interaction energies and a two-parameter linear regression was proposed allowing a better prediction of the interaction energies of both hydrogen and halogen bonds. This type of two-parameter regression was also recently proposed for charge-assisted ion-pair halogen bonds between imidazolium derivatives and Cl^- [36] as will be discussed in more detail below.

The role of polarizability in halogen bonds was studied recently [50] using the model complexes F–Br⋯X–R (X = F, Cl, Br, I and R = H, F) and the calculation of molecular electrostatic potential (MEP), atoms in molecules (AIM) analysis, energy decomposition analysis (EDA), and molecular polarizability based on MP2/aug-cc-pVTZ optimizations. The topological analysis of the electron density and the analysis of the atomic quadrupole moments reveal that the Br⋯X interactions are electrostatic in nature, however, the analysis of the MEPs indicate that considering only the static electrostatic interactions is insufficient to explain all features of the interactions. The calculated molecular polarizabilities show a deformation of the electron density of X and this deformation capability plays an important role in the stability of these complexes.

A distinction in the character of halogen bonds formed with n-type (e. g. pyridine) and p-type (e. g. benzene) halogen bond acceptors was found in C_6F_5X, C_6H_5X, and CF_3X model systems using CCSD(T) calculations [51]. For halogen-π interactions (p-type), dispersion interactions are quite significant whereas for in n-type acceptors, strong halogen bonds are formed with a larger contribution from electrostatic interactions.

A coordinative nature was attributed to a strong I^+⋯S interaction in the S–I^+–S iodonium complex [52] as depicted in Figure 3.4. The authors assigned this interaction to a halogen bond between an iodonium cation (halogen bond donor) and the sulfur atom in a thione group (halogen bond acceptor) based on a QTAIM analysis complemented with NMR spectroscopy and mass spectrometry studies. They concluded that the interaction is quite strong (−60 kJ mol^{-1}) and even though a considerable electrostatic nature exists, the I^+⋯S interaction is not purely electrostatic as

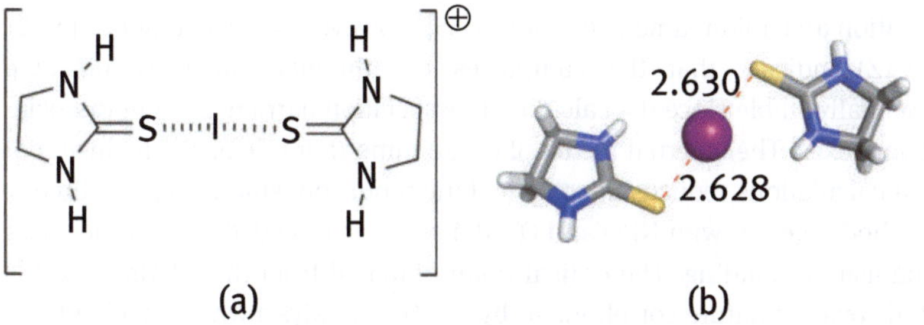

Figure 3.4: Schematic representation of the $[I(2\text{-imidazolidinethione})_2]^+$ complex (left) together with the X-ray structure [52] (right) with the I^+⋯S interaction highlighted in red.

a considerable amount of electron sharing between the iodine and the two sulfur atoms was observed and therefore, according to the authors, this interaction has a dual nature with considerable coordinative character. There is however a question whether if this interaction falls in the halogen bond category given the current IUPAC definition that mentions a halogen in a molecular entity.

Overall, the latest developments afford contradictory information given the multitude of calculations, analysis schemes, and their interpretation. It is not unlikely that the relative importance of the electrostatic, charge-transfer and dispersion contributions are dependent on the system, interacting partners and surrounding environment [13], as will be seen next. However, the controversy on the nature of the halogen bond will surely endure and a very recent paper on NBO and hydrogen bonds [53] might fuel new discussions. In that paper, AJ Stone argues that the charge-transfer component of the interaction energy of a hydrogen bond is contaminated by the basis set superposition error and therefore, NBO analysis of the interaction energies "*are meaningless in the context of intermolecular interactions*". Additionally, the values obtained for the charge-transfer energy were described as "*worthless*" whereas symmetry-adapted perturbation theory (SAPT) methods provide a better description of the interaction. Although the studied system was the hydrogen-bonded hydrogen fluoride dimer, the same conclusions might also hold if this methodology is applied to halogen-bonded dimers.

3.2.2 Surrounding environment effects on the nature and strength of halogen bonds

Halogen bonds are influenced by the surrounding media and the solvent indeed plays a role, opening the door for a multitude of applications [19] including for instance, the design of macromolecular hosts capable of recognizing and binding anions. In this scope, halogen bonds can be superior when compared to hydrogen bonds in the binding of anions in water as demonstrated by the experimental binding studies, complemented with molecular dynamics simulations, on a simple acyclic mono-charged receptor [54]. Therefore, a theoretical understanding of how the external media modulates the halogen bond is of paramount importance.

Halogen bonds are present in dimers of charged molecules with the same sign, i. e., cation–cation and anion–anion complexes [55], however, DFT calculations (M06-2X/aug-cc-pVTZ) indicate that the complexes are thermodynamically unstable although kinetically stable since the calculated dissociation barriers prevent dissociation of the complexes. The external media plays an important role as the inclusion of solvent in the calculations reduces the electrostatic repulsion, stabilizing the dimers. The SAPT method together with NBO and QTAIM analysis provided some indications on the mechanism of bonding. The critical point obtained from the QTAIM analysis show a partial covalent nature corroborated by the NBO analysis which yielded large charge-transfer stabilizing energies in both cation–cation and anion–anion systems. From the SAPT analysis, induction contributions are important.

Calculations on a series of I···O halogen bonds in complexes of iodobenzene derivatives with formaldehyde [56] showed that the solvent, in this case diethylether (ε = 4.2400) or water (ε = 78.3553), modeled using the conductor-like polarizable continuum model, had a destabilizing effect on the interaction. Moreover, in $C_2F_3I\cdots B$ and $C_6F_5I\cdots B$ halogen bonded complexes (B = Cl^-, Br^-, I^-, NH_3, OH_2) calculated in chloroform (ε = 4.7113), acetone (ε = 20.493) and water [57], the solvent had a larger impact on the charged systems whereas for the neutral systems the interaction energies are changed to a lesser extent, although they are also weakened. This is illustrated by the drop of the interaction energy calculated at the B3LYP/aug-cc-pVDZ level of theory in the gas-phase (–19.71 kcal mol^{-1}) and in water (–0.55 kcal mol^{-1}) in the charged system $C_6F_5I\cdots Cl^-$.

The solvent is not the only surrounding factor playing a role in modulating the strength of halogen bonds. For instance, it was soon recognized that halogen bonds are present in biological molecules, stabilizing inter- and intramolecular interactions, ligand bonding, and folding [58]. A putative example of the importance of these type of interactions is the structure of phospho-CDK2/cyclin A in complex with the inhibitor 4,5,6,7-tetrabromobenzotriazole (Figure 3.5).

Recently, it was demonstrated by means of DFT calculations (M06-2X) that the protein environment is not an innocent factor in stabilizing halogen bonds, even the oddest ones involving negatively charged halogen bond donors with acceptors in proteins (B = O, N, S) as depicted in Figure 3.6(a) [60]. Calculations on model systems consisting of $^-OOC\text{–}(C{=}C)_n\text{-}C_6H_5\text{–}X\cdots NH_3$ dimers (n = 0–5; X = Br, I) were performed in different dielectric media in order to predict the variation of the binding energies. Indeed, proteins possess an anisotropic dielectric environment (ε = 4–80) and to mimic that variation, a PCM model was applied to the calculations using chlorobenzene (ε = 5.6), dichlorobenzene (ε = 10.4), acetone (ε = 20.5), dimethyl sulfoxide (ε = 46.8), and water (ε = 78.4) as solvents. The distance between

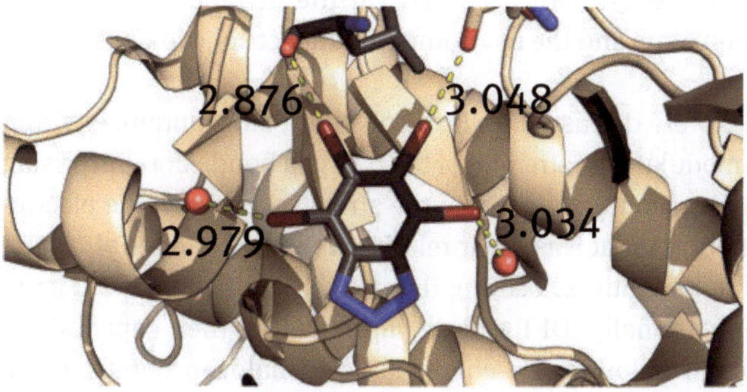

Figure 3.5: Example of a protein-ligand complex highlighting the presence of halogen bonds: the X-ray structure of phospho-CDK2/cyclin A in complex with the inhibitor 4,5,6,7-tetrabromobenzotriazole (PDB ID code 1P5E) showing four Br···O halogen bonds (dashed yellow lines) [59].

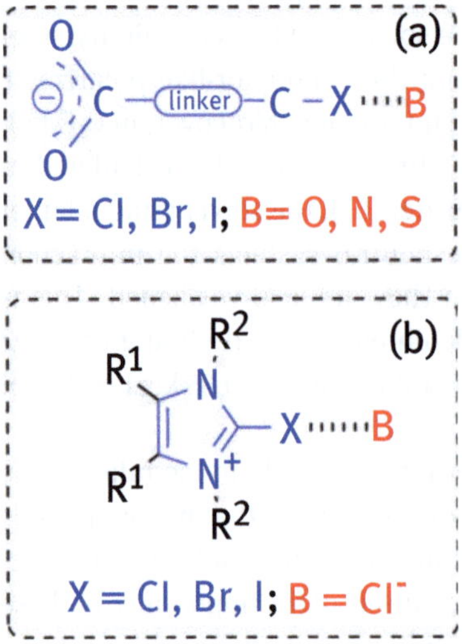

Figure 3.6: (a) Scheme of halogen bonds involving negatively charged donor systems as reported in reference 60. (b) Interaction of a series of 2-halo-functionalized imidazolium derivatives with chloride as reported in reference 36.

the negative COO^- group and the C-X moiety (halogen bond donor) is important and the interaction is strengthened as that distance is increased. More interestingly is the fact that the solvent enhances the halogen bond strength at short distances ($n \leq 2$) more significantly that at large distances ($n > 2$) when compared with vacuum calculations. An ETS-NOCV approach, which is a combination of the extended transition state (ETS) method with the natural orbitals for chemical valence (NOCV) scheme [61], showed that the electrostatic interaction has the major role, however, orbital interactions account for 42–36 % of the halogen bond, which is quite significant. The NBO analysis indicates that the main intermolecular interaction is a charge-transfer between the lone pair of NH_3 and the empty σ* C–X orbital. Overall, the halogen bond strength can be tuned *via* the distance between the negative center and halogen atom and the environment (dielectric), which is indeed relevant for drug design.

Above, some examples were discussed on how solvent or surrounding environment, e. g., a protein, can modulate the strength of the halogen bond. More surprising is when the solvent is able to modulate not only the strength, but also the electrostatic *vs.* charge transfer nature (or at least their relative weights). A series of 2-halo-functionalized imidazolium derivatives, bearing different R^1 and R^2 groups (Figure 3.6(b)) were studied computationally (DFT and *ab initio*) [36] for their capability of binding anions, in particular chloride, *via* charge-assisted halogen bonds. These haloimidazolium motifs are particularly suitable for anion binding [62] and supramolecular assembly, e. g., anion-templated assembly of pseudorotaxanes [63] and catenanes [64]. For this special kind of halogen bonds, MP2/aug-cc-pVDZ calculations gave the best agreement when compared with the state-of-the-art CCSD(T) results. Amongst the tested DFT functionals (B3LYP, B97-1, M06, and the variation M06-2X), the M06-2X gave the best performance. In the gas-phase, unrealistically large interaction energies were calculated owing to the ion-pair nature of the interaction. One might anticipate an enormous contribution from electrostatics given the opposite charge of the imidazolium cation and the chloride anion, however, very large bond orders (Mayer or Wiberg) and high contributions from $n_{Cl^-} \rightarrow \sigma^*$ C–X charge-transfer, were obtained by a second-order perturbation stabilization energy of donor–acceptor interactions E(2). These values are more consistent with a dative/covalent nature of the halogen bond. In these systems, the solvent effect on the interaction energies is dramatic. Indeed, calculations in both chloroform ($\varepsilon = 4.7$) and water ($\varepsilon = 78.4$) showed that the interaction energies and the charge-transfer are highly reduced. However, the variation of the interaction energy for the tested series can be predicted by a two-parameter linear regression which optimizes the weights of charge-transfer and electrostatic interactions *via* the values of the computed E(2) and $V_{S,max}$ values, respectively. Surprisingly, the media modulates not only the strength but also the weight of these contributions; in the gas-phase, the weight of the electrostatic interactions is dominant whereas in water, charge-transfer has the greater coefficient. In chloroform both contributions are more balanced showing the importance of the solvent polarity in the rational design of new chloride receptors based on halogen bonds.

The solvent effects in halogen bond strengths and character are not straightforward. For instance, Brammer, Hunter, and co-workers [65] studied by UV/vis absorption and ^{1}H NMR titration experiments the influence of the solvent polarity on the stability of both hydrogen and halogen bonds using as model systems 4-(phenylazo) phenol (hydrogen bond donor), I_2 (halogen bond donor), and tetramethylurea/tetramethylthiourea (acceptors). Surprisingly, the halogen bonded complex I_2···tetramethylthiourea is stable even in polar alcohol solvents contrasting with the hydrogen bonded complexes which were sensitive to the solvent polarity. This, according to the authors, indicates that simple electrostatic arguments are not sufficient to explain the behavior and stability of the halogen bond and therefore, charge-transfer is an important factor to account for the remarkable stability of these interactions.

3.3 Method benchmarks

In spite the existence of several computer modeling techniques for the description of halogen bonds [66] herein the focus will be on the latest studies concerning quantum mechanical methods. Normally, the CCSD(T) method is the "gold standard" to evaluate the quality of the computational methods and in the particular case of halogen bonds, the X40 data set [46] was developed, being extremely useful as it provides a non-biased molecular set containing both halogen a non-halogen bonded complexes. Other data sets containing solely halogen bonded complexes such as the XB18 and XB51 [67] are also useful. An analysis of the root-mean-square errors (RMSEs) of different post-HF methods results when compared with the CCSD(T)/CBS reference values applied to the X40 data set indicates that MP2/aug-cc-pVDZ yields an error of 0.55 kcal mol^{-1} for the binding energies whereas the best performance was achieved for SCS-MI-CCSD/CBS (0.06 kcal mol^{-1}). BLYP-D3/def2-QZVP calculations outperform MP2 (0.39 kcal mol^{-1}) thus providing a reliable tool for the description of noncovalent interactions [66]. Concerning the XB18 and XB51 sets [67], M06-2X and ωB97XD provided the best performance among the tested hybrid DFT methodologies while for DFT GGAs, only M06-L gave reasonable accuracies. The authors did not recommended MP2 as it overestimates binding. Additionally, they recognized the difficulty of calculating accurate halogen bond energies since the chosen method has to account for electrostatic, dispersion, polarization, and charge transfer, in what the authors defined has a mixture of bonding "unicorns".

Later the X40 data set, together with other data sets for noncovalent interactions, was used to evaluate the accuracy of popular QM methods [68]. Besides the popular DFT functionals B3LYP, B97-D, M06-2X, and ωB97X-D, semiempirical PMx methods were also tested. M06-2X and B97-D3 achieve high accuracy whereas PMx methods were generally inaccurate for halogenated and ionic molecules. The capability of semiempirical methods in describing halogen bonds is important since they could be used in large systems such as the ones involving proteins. In this scope, a review on the applicability of semiempirical quantum mechanical (SQM) methods for noncovalent interactions was recently published [69]. Controversies apart, *i. e.*, whether the interaction is purely electrostatic, the σ-hole is only described by QM methods with reasonable sized basis sets whereas SQM methods use minimal basis sets, and therefore, are not able to correctly account for that positive region at the tip of the halogen. Curiously, the PM6 method affords good interaction energies when compared with QM calculations but their potential energy surface is completely inaccurate; an empirical correction was therefore proposed [70] and the latest version of this method is called PM6-D3H4X [71]

Another assessment of the accuracy of DFT functionals in the description of halogen bonds was performed for the specific case of R–X⋯π interactions involving benzene as halogen bond acceptor [72]. In this case, T-shaped R–X⋯benzene dimers (R = H, F, and HCC; X = Cl, Br) were calculated at the CCSD(T)/CBS level and compared with 34 DFT functionals along with ab initio MP2 and MP2.X methods. Again, MP2 appears to overestimate the binding and double hybrid functionals B2PLYPD and mPW2PLYPD, which account for dispersion contributions, provide the best results concerning energies and equilibrium geometry although M06-2X also performs very well.

The fact that some of the above-mentioned studies refer that MP2 overestimate binding is in contrast with the calculations performed on charge-assisted halogen bonds present in 2-bromo-1*H*-imidazol-3-ium with chloride [36]. Here, MP2/aug-cc-pVDZ gave the best agreement when compared with CCSD(T)/aug-cc-pVDZ calculations. However, it must be mentioned that this was not extended to other systems and therefore, cannot be generalized.

The use of expensive methods with large basis sets to study halogen bonds might not be feasible for large systems such as biochemical or supramolecular entities. Recently, a benchmark of DFT methods with small basis sets was performed [73] using the XB18 data set together with a larger data set comprising 33 complexes from the X40 dataset with iodobenzene derivatives and pyridine (termed iodobenzene data set). Diverse combinations of DFT functionals and basis sets were evaluated, yielding a total of 46 methods (15 functionals and 16 basis sets). For the XB18 data set, the M06-2X/DGDZVP combination had the best performance although PBE0 also gave quite good results. Concerning the larger iodobenzene data set, PBE0 outperforms M06-2X. Nonetheless, the small to moderate sized DGDZVP basis set is a good choice to study halogen bonds in medium to large size systems were the computational cost of more expensive methods could be prohibitive, such as supramolecular and biochemical systems.

Indeed, large supramolecular systems offer particular challenges for the theoretical study of halogen bonds. Although one could in principle apply force field methods [62–64], it is also important to test the applicability of QM methods to these systems. In this scope, and using the first supramolecular capsule based solely on halogen bonds reported by Diederich [74, 75] (Figure 3.7) as test case, Grimme and co-workers [76] reported on the applicability of DFT to predict supramolecular thermochemical parameters. Unquestionably, the existence of experimental thermochemical data allows an alternative approach to test the accuracy of the theoretical methods. In this case, the experimental binding free energies were −4.83 kcal mol^{-1} and −3.60 kcal mol^{-1} for the iodine and bromine analogues, respectively [74]. Calculations were performed with dispersion corrected density functional theory

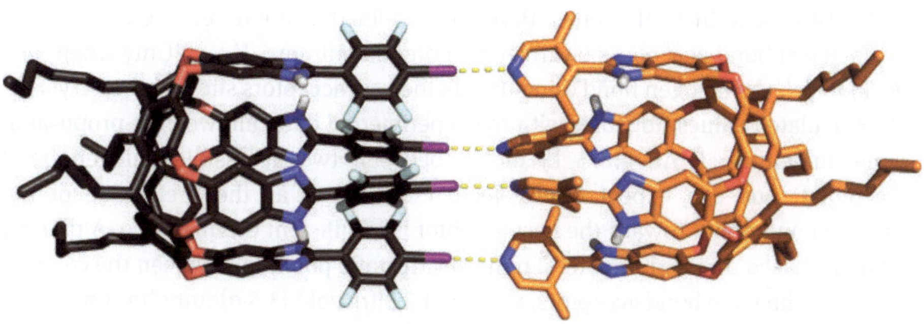

Figure 3.7: X-ray structure of an halogen bond capsule reported in reference 75 assembled *via* two resorcin[4]arene cavitands. Encapsulated benzene molecules, solvent, and non-polar hydrogen atoms were omitted for clarity. The halogen bond donor is represented in dark grey whereas the halogen bond acceptor is depicted in orange.

using the TPSS-D3/TZ level of theory for geometry optimization and the PW6B95-D3/QZ for single-point energies. The statistical corrections to the free energy in the gas phase were calculated with the HF-3c method, which is a fast-minimal basis set Hartree–Fock based method whereas the solvent effects were included *via* the COSMO-RS model. For iodine, PBEh-3c//PBEh-3c (−2.6 kcal mol^{-1}) and PW6B95-D3//TPSS-D3/TZ (−5.7 kcal mol^{-1}) produced the best results. The authors also tested the capability of the methods in predicting the binding affinities for two guest molecules inside the iodine capsule (1,4-dioxane and 1,4-dithiane) and both the less expensive PBEh-3c//PBEh-3c and PW6B95-D3/QZ//TPSS-D3/TZ provided good accuracy.

3.4 Applications

The continuous interest in the theoretical understanding of halogen bonding is accompanied by new developments in the experimental field. Indeed, theoretical calculations are often used to guide experiments or interpret experimental data. Even the simplest calculations such as the evaluation of the $V_{S,max}$ in a series of small halogen bond donors (Figure 3.8 (a)), combined with structural chemistry experiments and vibrational spectroscopy studies, can provide useful insights for crystal engineering applications [77]. This study used as premise the fact that more pronounced σ-holes, i. e., higher $V_{S,max}$ values, yield a more effective halogen bond donor and therefore, will be more efficient at forming co-crystals with a complementary halogen bond acceptor, B. Indeed, the results from the co-crystallization experiments showed the influence of the halogen bond capability as determined by the QM calculations. A similar rationale was later used [78] to guide the selectivity of halogen bond interactions in the co-crystallization of 9 perfluorinated halogen bond donors and 12 ditopic acceptors (Figure 3.8(b)). They were able to achieve co-crystallization (monitored by IR spectroscopy) of 89 compounds and obtained 35 new crystal structures. From the calculated values, the halogen bond acceptors can be categorized according to their ΔE values. Here, ΔE is defined as the difference between the two halogen bond acceptor's electrostatic potential minima, $V_{S,min}$. Only when ΔE > 167 kJ mol^{-1}, the halogen bond donor favors the best acceptors site for all co-crystals. The calculated values, together with the experimental data, allowed the proposal of some guidelines: if ΔE >; 75 kJ mol^{-1} (between two accessible halogen bond acceptors), one can expect intermolecular selectivity as the preference of the halogen bond donor toward the best acceptor is a sufficient driving force. A parallel rationale based on the differences in the electrostatic potential between the competing sites of halogen bond acceptors, namely [1,2,3]triazalo-[3,5-α]quinoline and [1,2,3]triazalo-[3,5-α]pyridine, was also reported [79]. In this case, electrostatic potential differences > 75 kJ mol^{-1} lead to a preference toward the best acceptor whereas no selectivity was observed when ΔE is smaller.

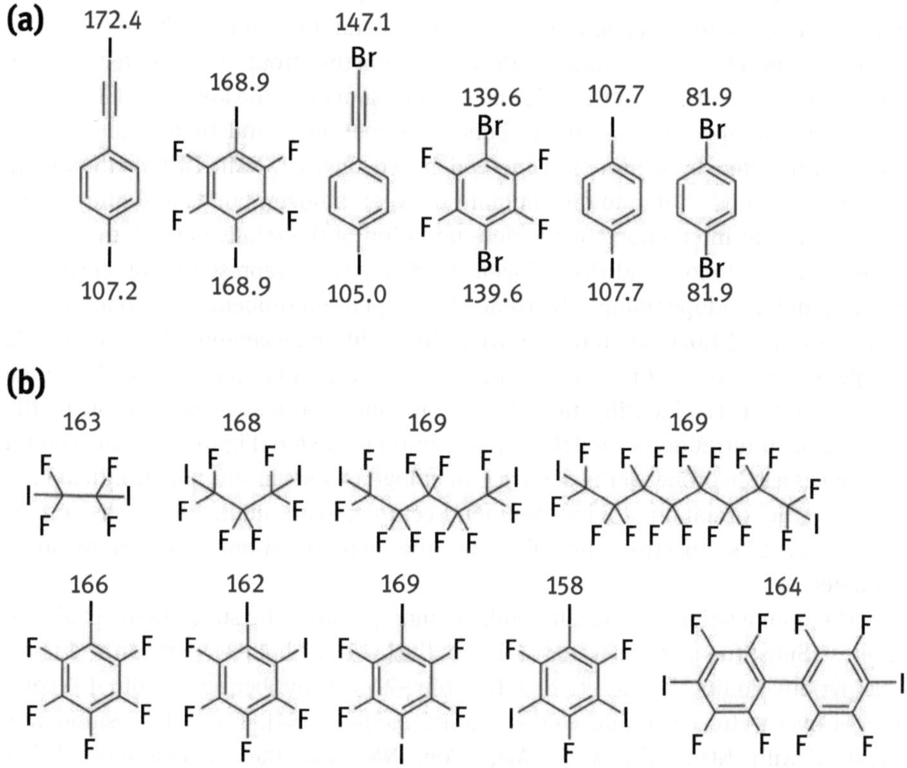

Figure 3.8: Small library of halogen donors together with their respective $V_{S,max}$ values (blue, kJ mol^{-1}) calculated at the B3LYP/6-311+ G** level of theory as reported in reference 77 (a) and reference 78 (b).

Halogen bonds are also responsible for interesting reactivity. For instance, molecular iodine (I_2) acts as a catalyst for many organic reactions but the origin of this phenomena is not completely clear. DFT calculations were performed [80], complemented with experimental work, trying to shed some light into this question. For that purpose, four model reactions involving α,β-unsaturated carbonyls or nitrostyrenes were studied. The catalytic effects of I_2 are ascribed to the possibility of molecular iodine to form halogen bonds, activating the electrophiles, substantially lowering the activation free energies. Halogen σ-holes are also responsible for the reactivity of chlorine in the reduction of a trichloromethyl group by sulfur nucleophiles [81]. In this case, the negative sulfur atom, acting as the halogen bond acceptor, interacts with the chlorine σ-hole, eventually leading to the abstraction of the halogen and the formation of a carbanion. This hypothesis was studied using a model reaction involving the reduction of a trichloromethyl pyrimidine derivative with thiophenol and sodium thiophenolate by means of DFT calculations (BMK/6-31G**). The electrostatic potential was calculated at the MP2/cc-pVQZ level of theory.

This was complemented by the experimental study of the reaction at −78 °C, quenched at different reaction times (1, 5, 10, 30 and 90 minutes). The nucleophilic attack occurs at the chlorine atom in the trichloromethyl group. This is potentiated by the existence of a σ-hole at the chlorine atom as shown by the MP2 calculations.

Other example of the synergy between experiment and theoretical work is nicely illustrated by a proposed method based on NOE NMR/DFT methodology [82]. This method is able to discriminate between halogen bonds and other types of noncovalent interactions by the determination of the relative orientation of the halogen bond donors and the halogen bond acceptors represented by perfluorohexyl iodide, iodopentafluorobenzene, bromopentafluorobenzene (donors), 1,4-diazabicyclo[2.2.2]octane, and 2,4,6-trimethylpyridine (acceptors). Halogen bonds are generally stronger than the alternative noncovalent interactions, such as lone pair/π and π/π. The identification of halogen bonds was also made possible by the use of 2,3,5,6-tetrafluoro-4-iodobenzyl *tert*-butyl nitroxide (Figure 3.9) which could be used as a spin probe for the detection of halogen bonds in solution by EPR [83]. In this case the formation of a halogen bond complex with quinuclidine was accompanied by EPR spectroscopy, allowing the determination of thermodynamic parameters.

The optimization of halogen bonds by manipulating the substituents is also of interest. Substituent effects were studied on the $[N–I–N]^+$ halogen bond on a series of $[bis(pyridine)iodine]^+$ and $[1,2\text{-bis}((pyridine\text{-}2\text{-}ylethynyl)benzene)\text{-}iodine]^+$ complexes by spectroscopic and computational methods [84]. The *para*-position was modified with -NO_2, -CF_3, -H, -F, -Me, -OMe, -NMe_2 and the ^{15}N NMR showed that the chemical shifts are governed by the π population on the ring instead of the electron population at the nitrogen atoms, *i. e.*, its magnitude does not reflect the strength of the halogen bond. Both NMR and DFT studies (B3LYP) indicate the existence of static and symmetric $[N–I–N]^+$ halogen bonds. It is also possible to play with the substituents in 2-halo-functionalized imidazolium derivatives (Figure 3.6(b)) in order to enhance their binding capability toward chloride [36]. When electron-withdrawing groups, *e. g.*, -NO_2, -F, are placed in the 4,5-positions, the halogen bond strength increases both the gas phase and solvent, including water. Bis(triazole)pyridinium receptors were also optimized toward the binding of halide

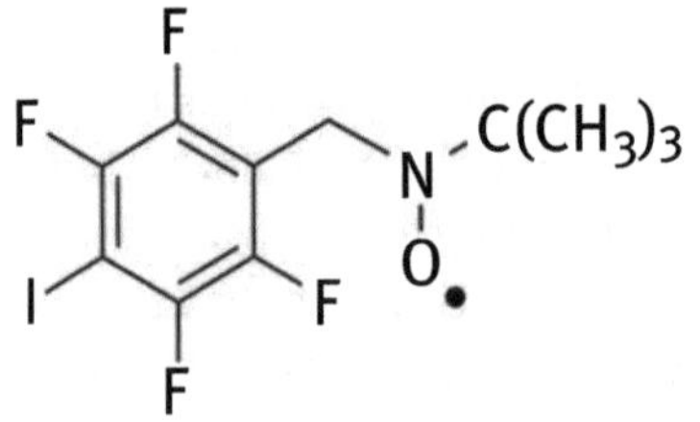

Figure 3.9: Structure of 2,3,5,6-tetrafluoro-4-iodobenzyl *tert*-butyl nitroxide.

anions [85]. Chirality is an important issue and halogen bond receptors can be designed for the enantiodiscrimination of acceptor molecules in solution as the ones based on the halotriazole and halotriazolium core bearing chiral substituents [86]. In particular, the mentioned chiral triazolium receptor was able to discriminate between enantiomeric thioureas *via* halogen bonds. Chiral (*S*)-BINOL-based receptors bearing halotriazolium units capable of halogen bonding enhance the enantioselective recognition of chiral anions (*N*-Boc-alanine, *N*-Boc-leucine, *N*-Boc-tryptophan, and BINOL-PO_4) [87]. The use of halogen bonds was superior to the hydrogen bonding analogues and MD simulations indicate that the intrinsic linearity of halogen bonds together with host–guest steric interactions play a dominant role in the recognition process. Neutral iodotriazoles, in particular 1,4-diaryl-5-iodo-1,2,3-triazole derivatives can also be used as scaffolds for halogen bond receptors [88].

The versatility of halogens bonds is outstanding and this type of interaction is able to potentiate even the oddest structures like supramolecular helixes [89]. Using *N*-amidothioureas bearing iodophenyl groups, β-turns are identified in the crystal structure. Moreover, C–I···π halogen bonds between adjacent molecules are also observed and help to propagate the helicity of the fragments as depicted in Figure 3.10, left. The supramolecular assembly exhibits enhanced CD signals, being the first example of chiral amplification in a supramolecular helix triggered by halogen bonds. The first triple helicate able to encapsulate iodine in organic and aqueous media as well as in the solid state by means of halogen bonds was also reported [90]. This triple helicate is shown in Figure 3.10 right, and is assembled from three tricationic arylethynyl strands around two iodine anions, being stabilized by multiple linear halogen bonds and π-stacking interactions.

In supramolecular arrangements, the halogen bond can be surprisingly short as the one present in a bridged triarylphoshine oxide and an aryl iodide in the solid state [91]. Here the *ortho*-bridged triarylphosphine oxide bearing sterically demanding fluorenyl bridging moieties interacts with 1,4-diiodotetrafluorobenzene establishing a very short P=O···I halogen bond (2.683 Å) in the solid state. Experimental and computational methods were used to study this unusual structure. The PBEh-3c and HSE-3c methods (mentioned earlier) were used for the dimer and the crystal structure, respectively. The halogen bond is not significantly stronger in the crystal when compared to the dimer and therefore, the stacking interactions of the fluorenyl flanks are the main factor governing co-crystal formation, being responsible for the short P=O···I distance.

Molecular machines gained special recognition with the award of the Nobel Prize to Jean-Pierre Sauvage, J. Fraser Stoddart and Bernard L. Feringa [92]. In this scope, halogen bonds are among the interactions that can be exploited for the design of new devices. A two-station rotaxane was designed [93] based on a isophthalamide-containing macrocycle and an axle, represented in Figure 3.11 in pale orange and dark grey, respectively. The axle comprises two stations, a naphthalene diimide group and a iodotriazolium group. These are able to interact with the hydroquinone

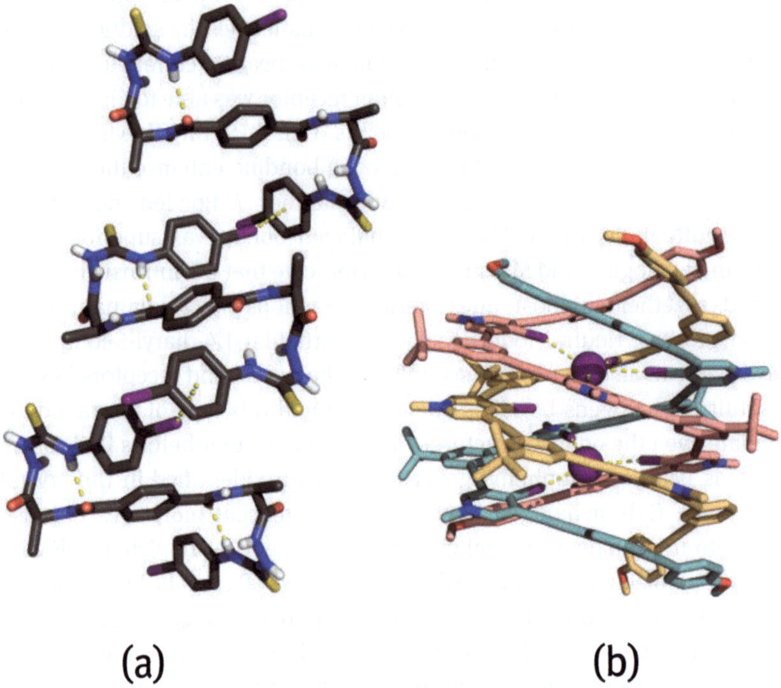

Figure 3.10: Left: X-ray structure of a supramolecular helix of bilateral *N*-amidothioureas bearing β-turns and intermolecular C–I···π interactions as reported in reference 89. Right: X-ray structure of triple helicate assembled from three tricationic arylethynyl as reported in reference 90.

moiety of the macrocycle and anions, respectively. In $CDCl_3$ solution, the macrocycle translocates from the naphthalene diimide station toward the iodotriazolium station by the addition of halide anions. Indeed, the iodotriazolium station recognizes the anion *via* halogen ($C–I···Cl^-$) and hydrogen ($N–H···Cl^-$) bonds, as shown in Figure 3.11. This shuttling behavior is accompanied by a color change in the solution, enabling naked-eye detection. Another example of halogen bonding supramolecular assembly driven by an external stimulus is the pH controlled assembly of halogen-bonded dimers [94]. This system takes advantage of the different orders of magnitude of the halogen binding acceptor capability of phenoxide (good) versus phenol (poor). In this system, an iodotriazole triazole derivative bearing also a phenoxide anion moiety is able to self-assemble via $C–I···I^-$ halogen bonds, forming a stable dimer. Upon protonation, no assembly is observed in solution or solid state, therefore, self-assembly can be controlled by a change in the protonation state of the system. The system was also investigated computationally using DFT calculations (PSSh/def2-TZVP); the protonated dimer yielded an entropy of interaction of $-4.26\ kJ\ mol^{-1}$ (in an acetonitrile PCM model) whereas deprotonated phenoxide dimer afforded an enthalpy of dimerization of $-50.4\ kJ\ mol^{-1}$.

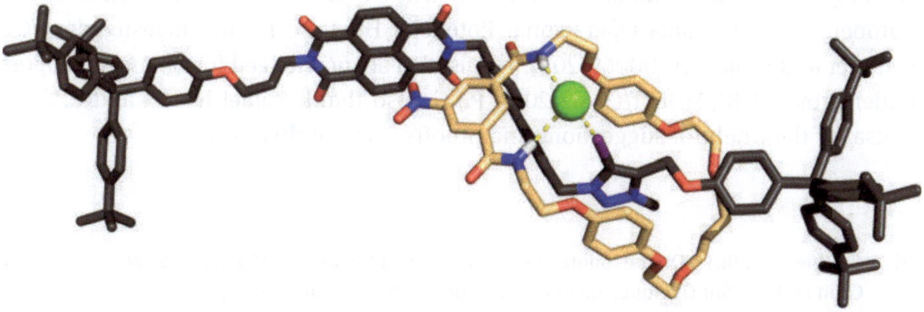

Figure 3.11: X-ray crystal structure of a two-station rotaxane as reported in reference 93. The macrocycle depicted in pale orange is located at the iodotriazolium station on the axle (depicted in dark grey). This structure is assembled by halogen (C–I···Cl$^-$) and hydrogen (N–H···Cl$^-$) bonds.

As mentioned earlier, halogen bonds are important in biological molecules [58] and can certainly be further explored. Indeed, their structural features might be overlooked as it was recently showed [95]. Earlier surveys of the PDB database indicate that halogen bonds were preferentially formed with the protein backbone (~65 %), and the majority of drug design applications focus on the backbone [96]. In this new study [95], the authors showed that the proportion of halogen bonds formed between halogenated molecules and the protein backbone or the protein side-chains changes with the structure resolution and as the resolution becomes worse, the side-chain halogen bonds percentage decreases. It is important to note that since the classic force fields, generally used to refine these structures, do not account for halogen bonding, these lower resolution areas (side-chains) loose the halogen bonds upon refinement. This was also supported by the fact that the effect of the resolution (main chain vs. side-chain interaction) is not observed for hydrogen bonds which the classical force fields can handle. These conclusions should not be ignored when designing new protein inhibitors exploiting halogen bonds. Curiously, the versatility of halogen bonds also enables the modulation of peptide–receptor interactions as shown by the halogenation of two native opioid peptides that bind opioid receptors [97]. If in one hand the replacement of a hydrogen of the peptide by a bulkier halogen atom can produce steric clashes reducing the binding, especially in small cavities, it can also increase the binding if the halogen (chlorine, bromine, or iodine) interacts with negatively charged atoms of the protein *via* halogen bonds. Modifications on the amino acids of a protein to observe the halogenation effect can also be performed. The replacement of the hydroxyl of the tyrosine by an iodine in a model T4 lysozyme [98] caused the displacement of the aromatic side chain toward an oxygen acceptor, showing that halogen bonds can circumvent destabilizing effects that other substitutions cause.

Funding: P.J.C. thank Fundação para a Ciência e a Tecnologia (FCT), Fundo Social Europeu, and Programa Operacional Potencial Humano for the Investigador FCT contract and project IF/00069/2014. FCT is also acknowledged for financial support under project UID/MULTI/00612/2013. P.J.C. also thank Rafael Nunes and C. S. R. Sousa for their helpful suggestions and proofreading of this work.

References

[1] Guthrie F. XXVIII. – On the iodide of iodammonium. J Chem Soc. 1863;16:239–244.
[2] Colin M. Note Sur Quelques Combinaisons de L'iode. Ann Chim. 1814;91:252.
[3] Remsen I, Norris JF. Action of the halogens on the methylamines. Am Chem J. 1896;18:90.
[4] Bjorvatten T, Hassel O. Crystal structure of the 1:3 addition compound iodoform-quinoline. Acta Chem Scand. 1962;16:249.
[5] Hassel O, Rømming C. Direct structural evidence for weak charge-transfer bonds in solids containing chemically saturated molecules. Q Rev Chem Soc. 1962;16:1–18.
[6] Bent HA. Structural chemistry of donor-acceptor interactions. Chem Rev. 1968;68:587–648.
[7] Mulliken RS. Structures of complexes formed by halogen molecules with aromatic and with oxygenated solvents. J Am Chem Soc. 1950;72:600–608.
[8] Hassel O. Nobel Lectures, Chemistry 1963-1970. Amsterdam: Elsevier Publishing Company; 1972.
[9] Brinck T, Murray JS, Politzer P. Surface electrostatic potentials of halogenated methanes as indicators of directional intermolecular interactions. Int J Quantum Chem. 1992;44:57–64.
[10] Clark T, Hennemann M, Murray JS, Politzer P. Halogen bonding: The σ-hole. J Mol Model. 2007;13:291–296.
[11] Voth AR, Khuu P, Oishi K, Ho PS. Halogen bonds as orthogonal molecular interactions to hydrogen bonds. Nature Chemistry. 2009;1:74–79.
[12] Gilday LC, Robinson SW, Barendt TA, Langton MJ, Mullaney BR, Beer PD. Halogen bonding in supramolecular chemistry. Chem Rev. 2015;115:7118–7195.
[13] Brown A, Beer PD. Halogen bonding anion recognition. Chem Commun. 2016;52:8645–8658.
[14] Metrangolo P, Resnati G. Halogen bonding: Where we are and where we are going. Cryst Growth Des. 2012;12:5835–5838.
[15] Wilcken R, Zimmermann MO, Lange A, Joerger AC, Boeckler FM. Principles and applications of halogen bonding in medicinal chemistry and chemical biology. J Med Chem. 2013;56:1363–1388.
[16] Project No. 2009-032-1-100. Categorizing halogen bonding and other noncovalent interactions involving halogen atoms. https://iupac.org/projects/project-details/?project_nr=2009-032-1-100). Accessed 8 Mar 2017
[17] Desiraju GR, Ho PS, Kloo L, Legon AC, Marquardt R, Metrangolo P, et al. Definition of the halogen bond (IUPAC recommendations 2013). Pure Appl Chem. 2013;85:1711–1713.
[18] Cavallo G, Metrangolo P, Milani R, Pilati T, Priimagi A, Resnati G, et al. The halogen bond. Chem Re. 2016;116:2478–2601.
[19] Beale TM, Chudzinski MG, Sarwar MG, Taylor MS. Halogen bonding in solution: Thermodynamics and applications. Chem Soc Rev. 2013;42:1667–1680.
[20] Wang H, Wang W, Jin WJ. σ-hole bond vs π-hole bond: A comparison based on halogen bond. Chem Rev. 2016;116:5072–5104.
[21] Ford MC, Ho PS. Computational tools to model halogen bonds in medicinal chemistry. J Med Chem. 2016;59:1655–1670.
[22] Jentzsch AV, Hennig A, Mareda J, Matile S. Synthetic ion transporters that work with anion–π interactions, halogen bonds, and anion–macrodipole interactions. Acc Chem Res. 2013;46:2791–2800.

[23] Wolters LP, Schyman P, Pavan MJ, Jorgensen WL, Bickelhaupt M, Kozuch S. The many faces of halogen bonding: A review of theoretical models and methods. WIREs Comput Mol Sci. 2014;4:523–540.
[24] Politzer P, Murray JS, Clark T. Halogen bonding and other σ–hole interactions: A perspective. Phys Chem Chem Phys. 2013;15:11178–11189.
[25] Politzer P, Murray JS. Halogen bonding: an interim discussion. Chem Phys Chem. 2013;14: 278–294.
[26] Bader RFW, Carroll MT, Cheeseman JR, Chang C. Properties of atoms in molecules: Atomic volumes. J Am Chem Soc. 1987;109:7968–7979.
[27] Values calculated by the author for this Chapter.
[28] Riley KE, Murray JS, Fanfrlik J, Rezac J, Sola RJ, Concha MC, et al. Halogen bond tunability I: The effects of aromatic fluorine substitution on the strengths of halogen-bonding interactions involving chlorine, bromine, and iodine. J Mol Model. 2011;17:3309–3318.
[29] Riley KE, Murray JS, Politzer P, Concha MC, Hobza P. Br···O complexes as probes of factors affecting halogen bonding: Interactions of bromobenzenes and bromopyrimidines with acetone. J Chem Theory Comput. 2009;5:155–163.
[30] Riley KE, Murray JS, Fanfrlík J, Řezáč J, Solá RJ, Concha MC, et al. Halogen bond tunability II: The varying roles of electrostatic and dispersion contributions to attraction in halogen bonds. J Mol Model. 2013;19:4651–4659.
[31] Palusiak M. On the nature of halogen bond – The Kohn–Sham molecular orbital approach. J Mol Struct Theochem. 2010;945:89–92.
[32] Rosokha SV, Traversa A. From charge transfer to electron transfer in halogen-bonded complexes of electrophilic bromocarbons with halide anions. Phys Chem Chem Phys. 2015;17:4989–4999.
[33] Cabot R, Hunter CA. Non-covalent interactions between iodo-perfluorocarbons and hydrogen bond acceptors. Chem Commun. 2009;0:2005–2007.
[34] Lieffrig J, Jeannin O, Frą Ckowiak A, Olejniczak I, Świetlik R, Dahaoui S, et al. Charge-assisted halogen bonding: Donor–acceptor complexes with variable ionicity. Chem Eur J. 2013;19:14804–14813.
[35] Wolters LP, Bickelhaupt FM. Halogen bonding versus hydrogen bonding: a molecular orbital perspective. ChemistryOpen. 2012;1:96–105.
[36] Nunes R, Costa PJ. Ion-pair halogen bonds in 2-halo-functionalized imidazolium chloride receptors: substituent and solvent effects. Chem Asian J. 2017;12:586–594.
[37] Wang C, Danovich D, Mo Y, Shaik S. On the nature of the halogen bond. J Chem Theory Comput. 2014;10:3726–3737.
[38] Politzer P, Murray JS, Clark T. Mathematical modeling and physical reality in noncovalent interactions. J Mol Model. 2015;21:52.
[39] Clark T, Politzer P, Murray JS. Correct electrostatic treatment of noncovalent interactions: The importance of polarization. WIREs Comput Mol Sci. 2015;5:169–177.
[40] Del Bene JE, Alkorta I, Elguero J. Anionic complexes of F– and Cl– with substituted methanes: hydrogen, halogen, and tetrel bonds. Chem Phys Lett. 2016;655-656:115–119.
[41] Oliveira V, Kraka E, Cremer D. The intrinsic strength of the halogen bond: Electrostatic and covalent contributions described by coupled cluster theory. Phys Chem Chem Phys. 2016;18:33031–33046.
[42] Oliveira V, Kraka E, Cremer D. Quantitative assessment of halogen bonding utilizing vibrational spectroscopy. Inorg Chem. 2017;56:488–502.
[43] Novák M, Foroutan-Nejad C, Marek R. Asymmetric bifurcated halogen bonds. Phys Chem Chem Phys. 2015;17:6440–6450.
[44] Řezáč J, De La Lande A. On the role of charge transfer in halogen bonding. Phys Chem Chem Phys. 2017;9:791–803.

[45] Řezáč J, De La Lande A. Robust, basis-set independent method for the evaluation of charge-transfer energy in noncovalent complexes. J Chem Theory Comput. 2015;11:528–537.

[46] Řezáč J, Riley KE, Hobza P. Benchmark calculations of noncovalent interactions of halogenated molecules. J Chem Theory Comput. 2012;8:4285–4292.

[47] Thirman J, Head-Gordon M. An energy decomposition analysis for second-order Møller Plesset perturbation theory based on absolutely localized molecular orbitals. J Chem Phys. 2015;143:084124.

[48] Wang C, Danovich D, Shaik S, Mo Y. A unified theory for the blue- and red-shifting phenomena in hydrogen and halogen bonds. J Chem Theory Comput 2017. in press. DOI: 10.1021/acs.jctc0.6b01133).

[49] Zou J-W, Huang M, Hu G-X, Jiang Y-J. Toward a uniform description of hydrogen bonds and halogen bonds: Correlations of interaction energies with various geometric, electronic and topological parameters. RSC Adv. 2017;7:10295–10305.

[50] Duarte DJR, Sosa GL, Peruchena NM, Alkorta I. Halogen bonding. The role of the polarizability of the electron-pair donor. Phys Chem Chem Phys. 2016;18:7300–7309.

[51] Tsuzuki S, Uchimaru T, Wakisaka A, Ono T. Magnitude and directionality of halogen bond of benzene with C6F5X, C6H5X, and CF3X (X = I, Br, Cl, and F). J Phys Chem A. 2016;120:7020–7029.

[52] Koskinen L, Hirva P, Kalenius E, Jääskeläinen S, Rissanen K, Haukka M. Halogen bonds with coordinative nature: Halogen bonding in a S–I+–S iodonium complex. Cryst Eng Comm. 2015;17:1231–1236.

[53] Stone AJ. Natural bond orbitals and the nature of the hydrogen bond. J Phys Chem A. 2017;121:1531–1534.

[54] Langton MJ, Robinson SW, Marques I, Félix V, Beer PD. Halogen bonding in water results in enhanced anion recognition in acyclic and rotaxane hosts. Nat Chem. 2014;6:1039–1043.

[55] Quiñonero D, Alkorta I, Elguero J. Cation–cation and anion–anion complexes stabilized by halogen bonds. Phys Chem Chem Phys. 2016;18:27939–27950.

[56] Forni A, Rendine S, Pieraccini S, Sironi M. Solvent effect on halogen bonding: The case of the I···O interaction. J Mol Graph Model. 2012;38:31–39.

[57] Lu Y, Li H, Zhu X, Zhu W, Liu H. How does halogen bonding behave in solution? A theoretical study using implicit solvation model. J Phys Che. 2011;115:4467–4475.

[58] Auffinger P, Hays FA, Westhof E, Ho PS. Halogen bonds in biological molecules. Pnas. 2004;101:16789–16794.

[59] De Moliner E, Brown NR, Johnson LN. The structure of phospho-CDK2/cyclin A in complex with the inhibitor 4,5,6,7-tetrabromobenzotriazole (TBS). Eur J Biochem. 2003;270:1–8.

[60] Chen Z, Wang G, Xu Z, Wang J, Yu Y, Cai T, et al. How do distance and solvent affect halogen bonding involving negatively charged donors? J Phys Chem B. 2016;120:8784–8793.

[61] Mitoraj M, Michalak A, Ziegler T. A combined charge and energy decomposition scheme for bond analysis. J Chem Theor Comput. 2009;5:962–975.

[62] Zapata F, Caballero A, White NG, Claridge TDW, Costa PJ, Félix V, et al. Fluorescent charge-assisted halogen-bonding macrocyclic halo-imidazolium receptors for anion recognition and sensing in aqueous media. J Am Chem Soc. 2012;134:11533–11541.

[63] Serpell CJ, Kilah NL, Costa PJ, Félix V, Beer PD. Halogen bond anion templated assembly of an imidazolium pseudorotaxane. Angew Chem Int Ed. 2010;49:5322–5326.

[64] Caballero A, Zapata F, White NG, Costa PJ, Félix V, Beer PD. A halogen-bonding catenane for anion recognition and sensing. Angew Chem Int Ed. 2012;51:1876–1880.

[65] Robertson CC, Perutz RN, Brammer L, Hunter CA. A solvent-resistant halogen bond. Chem Sci. 2014;5:4179–4183.

[66] Kolář MH, Hobza P. Computer modeling of halogen bonds and other σ-hole interactions. Chem Rev. 2016;116:5155–5187.

[67] Kozuch S, Martin JM. Halogen bonds: Benchmarks and theoretical analysis. J Chem Theory Comput. 2013;9:1918–1931.
[68] Li A, Muddana HS, Gilson MK. Quantum mechanical calculation of noncovalent interactions: A large-scale evaluation of PMx, DFT, and SAPT approaches. J Chem Theory Comput. 2014;10:1563–1575.
[69] Christensen AS, Kubař T, Cui Q, Elstner M. Semiempirical quantum mechanical methods for noncovalent interactions for chemical and biochemical applications. Chem Rev. 2016;116:5301–5337.
[70] Řezáč J, Hobza P. A halogen-bonding correction for the semiempirical PM6 method. Chem Phys Lett. 2011;506:286–289.
[71] Brahmkshatriya PS, Dobeš P, Fanfrlík J, Řezáč J, Paruch K, Bronowska A, et al. Quantum mechanical scoring: structural and energetic insights into cyclin-dependent kinase 2 inhibition by Pyrazolo[1,5-a]pyrimidines. Curr Comput-Aid Drug. 2013;9:118–129.
[72] Forni A, Pieraccini S, Rendine S, Sironi M. Halogen bonds with benzene: an assessment of DFT functionals. J Comput Chem. 2014;35:386–394.
[73] Siiskonen A, Priimagi A. Benchmarking DFT methods with small basis sets for the calculation of halogen-bond strengths. J Mol Model. 2017;23:50.
[74] Dumele O, Trapp N, Diederich F. Halogen bonding molecular capsules. Angew Chem Int Ed. 2015;54:12339–12344.
[75] Dumele O, Schreib B, Warzok U, Trapp N, Schalley CA, Diederich F. Halogen-bonded supramolecular capsules in the solid state, in solution, and in the gas phase. Angew Chem Int Ed. 2017;56:1152–1157.
[76] Sure R, Grimme S. Halogen bonded supramolecular capsules: A challenging test case for quantum chemical methods. Chem Commun. 2016;52:9893–9896.
[77] Aakeröy CB, Baldrighi M, Desper J, Metrangolo P, Resnati G. Supramolecular hierarchy among halogen-bond donors. Chem Eur J. 2013;19:16240–16247.
[78] Aakeröy CB, Wijethunga TK, Desper J, Đakovic M. Electrostatic potential differences and halogen-bond selectivity. Cryst Growth Des. 2016;16:2662–2670.
[79] Perera MD, Desper J, Sinha AS, Aakeröy CB. Impact and importance of electrostatic potential calculations for predicting structural patterns of hydrogen and halogen bonding. CrystEngComm. 2016;18:8631–8636.
[80] Breugst M, Detmar E, Von Der Heiden D. Origin of the catalytic effects of molecular iodine: A computational analysis. ACS Catal. 2016;6:3203–3212.
[81] Caballero-Garćia G, Romero-Ortega M, Barroso-Flores J. Reactivity of electrophilic chlorine atoms due to σ-holes: A mechanistic assessment of the chemical reduction of a trichloromethyl group by sulfur nucleophiles. Phys Chem Chem Phys. 2016;18: 27300–27307.
[82] Ciancaleoni G, Bertani R, Rocchigiani L, Sgarbossa P, Zuccaccia C, Macchioni A. Discriminating halogen-bonding from other noncovalent interactions by a combined NOE NMR/DFT approach. Chem Eur J. 2015;21:440–447.
[83] Gualandi L, Mezzina E, Franchi P, Lucarini M. Nitroxide radical spin probes for exploring halogen-bonding interactions in solution. Chem Eur J. 2016;22:16017–16021.
[84] Carlsson A-C-C, Mehmeti K, Uhrbom M, Karim A, Bedin M, Puttreddy R, et al. Substituent effects on the [N–I–N] + halogen bond. J Am Chem Soc. 2016;138:9853–9863.
[85] Nepal B, Scheiner S. Competitive halide binding by halogen versus hydrogen bonding: Bis-triazole pyridinium. Chem Eur J. 2015;21:13330–13335.
[86] Kaasik M, Kaabel S, Kriis K, Järving I, Aav R, Rissanen K, et al. Synthesis and characterization of chiral triazole-based halogen bond donors: Halogen bonds in solid state and in solution. Chem Eur J 2017. in press. DOI: 10.1002/chem0.201700618.

[87] Lim JYC, Marques I, Ferreira L, Félix V, Beer PD. Enhancing the enantioselective recognition and sensing of chiral anions by halogen bonding. Chem Commun. 2016;52:5527–5530.

[88] Maugeri L, Asencio-Hernández J, Lébl T, Cordes DB, Slawin AMZ, Delsuc M-A, et al. Neutral iodotriazoles as scaffolds for stable halogen-bonded assemblies in solution. Chem Sci. 2016;7:6422–6428.

[89] Cao J, Yan X, He W, Li X, Li Z, Mo Y, et al. C–I···π halogen bonding driven supramolecular helix of bilateral N-amidothioureas bearing β-Turns. J Am Chem Soc 2017. in press. DOI: 10.1021/jacs0.6b13171.

[90] Massena CJ, Wageling NB, Decato DA, Rodriguez EM, Rose AM, Berryman OB. A halogen-bond-induced triple helicate encapsulates iodide. Angew Chem Int Ed. 2016;55:12398–12402.

[91] Schaub TA, Sure R, Hampel F, Grimme S, Kivala M. Quantum chemical dissection of the shortest P=O···I halogen bond: The decisive role of crystal packing effects. Chem Eur J 2017. in press. DOI: 10.1002/chem0.201701234.

[92] The Nobel Prize in Chemistry 2016 http://www.nobelprize.org/nobel_prizes/chemistry/laureates/2016/). Accessed 29 Mar 2017

[93] Barendt TA, Robinson SW, Beer PD. Superior anion induced shuttling behavior exhibited by a halogen bonding two station rotaxane. Chem Sci. 2016;7:5171–5180.

[94] Maugeri L, Jamieson EMG, Cordes DB, Slawin AMZ, Philp D. pH controlled assembly of a self-complementary halogen-bonded dimer. Chem Sci. 2017;8:938–945.

[95] Zhang Q, Xu Z, Zhu W. The underestimated halogen bonds forming with protein side chains in drug discovery and design. J Chem Inf Model. 2017;57:22–26.

[96] Xu Z, Yang Z, Liu Y, Lu Y, Chen K, Zhu W. Halogen bond: Its role beyond drug-target binding affinity for drug discovery and development. J Chem Inf Model. 2014;54:69–78.

[97] Rosa M, Caltabiano G, Barreto-Valer K, Gonzalez-Nunez V, Gómez-Tamayo JC, Ardá A, et al. Modulation of the Interaction between a peptide ligand and a G protein-coupled receptor by Halogen Atoms. ACS Med Chem Lett. 2015;6:872–876.

[98] Scholfield MR, Ford MC, Carlsson A-C-C, Butta H, Mehl RA, Ho PS. Structure-energy relationships of halogen bonds in proteins. Biochemistry 2017. in press. DOI: 10.1021/acs.biochem0.7b00022.

Xavier López

4 Effect of protonation, composition and isomerism on the redox properties and electron (de)localization of classical polyoxometalates

Abstract: This publication reviews some relevant features related with the redox activity of two inorganic compounds: $[XM_{12}O_{40}]^{q-}$ (Keggin structure) and $[X_2M_{18}O_{62}]^{q-}$ (Wells-Dawson structure). These are two well-known specimens of the vast Polyoxometalate (POM) family, which has been the subject of extensive experimental and theoretical research owing to their unmatched properties. In particular, their redox activity focus a great deal of attention from scientists due to their prospective related applications. POMs are habitually seen as 'electron sponges' since many of them accept several electrons without losing their chemical identity. This makes them excellent models to study mechanisms of electrochemical nature. Their redox properties depend on: (i) the type and number of transition metal atoms in the structure, (ii) the basicity of the first reduced species and, occasionally, of the fully oxidized species; (iii) the size of the molecule, (iv) the overall negative charge of the POM, and (v) the size of the central heteroatom. In the last years, important collaboration between the experimental and theoretical areas has been usual on the development of POM science. In the present chapter three of these synergies are highlighted: the influence of the internal heteroatom upon the redox potentials of Keggin anions; the dependence of the redox waves of Fe-substituted Wells-Dawson compounds with pH; and the role of electron delocalization and pairing in mixed-metal Mo/W Wells-Dawson compounds in their ability to accept electrons. In these three cases, a complete understanding of the problem would not have been possible without the mutual benefit of experimental and computational data.

Keywords: density functional theory, calculations, polyoxometalates, transition metals, electronic structure, redox properties

4.1 Introduction

In the area of Inorganic Chemistry, polyoxometalates [1, 2] (POMs) or polyoxoanions – owing to their anionic nature in solution – comprise a growing family of metal oxide

This article has previously been published in the journal *Physical Sciences Reviews*. Please cite as: López, X. Effect of protonation, composition and isomerism on the redox properties and electron (de)localization of classical polyoxometalates. *Physical Sciences Reviews* [Online] **2017**, *2* (11). DOI: 10.1515/psr-2017-0137

https://doi.org/10.1515/9783110482065-004

molecules. They are primarily made of oxygen and early transition metals such as M = W, Mo and V (called *addenda* or peripheral atoms in the present context), although many other elements can be present as well in the main framework. The smallest members of this family are subnanometric, whereas the largest structures can reach sizes of the nanomaterials domain, close to 4–5 nm [3]. Classical POMs are compact structures that can be classified into isopoly- and heteropolyanions, the latter being characterised by the general formula $[X_aM_bO_c]^{q-}$ (typically $a < b < c$). The internal region of these molecules is occupied by the *heteroatom* (X), typically a p-block or transition metal element but, in principle, there are no restrictions to the type of atoms occupying the cavity. The most common structures of the heteropolyanion class are: $[XM_{12}O_{40}]^{q-}$ or Keggin, $[X_2M_{18}O_{62}]^{q-}$ or Wells-Dawson, $[M'X_5M_{30}O_{110}]^{q-}$ or Preyssler. For the sake of compactness, it is commonplace in the literature to use short-hand formulae, without oxygen or charge, which is used throughout the present text when justified. For instance, XM_{12} and X_2M_{18} are used for the Keggin and Wells-Dawson systems, respectively. In the fully oxidised form of tungstates (M = W) and molybdates (M = Mo), all the metal centres feature the formal oxidation state VI. Therefore, the total charge, $q-$, is determined by the internal heteroatom, X.

POM structures usually exhibit many (occasionally some at the same time) properties that make them attractive in wide-ranging fields of Chemistry [4–6]. Although the first POM was reported in the early 19th century, the first great surge of this field took place in the 1960s. In 1998, a reviewing work by Baker and Glick [7] boosted the study of POMs, which entered an era of systematic investigation which continues in constant academic and technological development. Among the plethora of properties and applications, electronics and magnetism [8–10], electrochemistry [11], catalysis [12], electro- and photochromic systems [13, 14], sensors [15], supramolecular organization [16, 17], (nano)materials science [18–20] and medicine [21] are the most remarkable ones. Among the singular phenomena that have been reported, some were not fully explained just by experiments and needed theoretical support. Since the early 1990s and especially in the 21st. century, the growing use of Computational Chemistry applied to the study of POMs has permitted a better understanding of their properties. Especially, their electronic structure and redox behaviour were soon recognised as essential for many applications. Also, the reactivity, magnetism or the solution behaviour have been tackled theoretically [22].

In the redox area, POMs are often called 'electron sponges' for many of them can easily gain several electrons with no appreciable geometrical changes, either at an electrode surface or in the liquid bulk, making them excellent models to study mechanisms of electrochemical nature [20, 23–33]. The following main characteristics govern their redox properties: (i) the type and number of transition metal atoms in the structure, (ii) the basicity of the first reduced species and, occasionally, of the fully oxidised species; (iii) the size of the molecule, (iv) the overall negative charge of the POM, and (v) the size of the central heteroatom. This chapter reviews three examples with strong synergies between the results obtained from experimental and theoretical techniques that came up with plausible explanations on intricate phenomena in the field of POM redox chemistry.

The results herein summarised and discussed are the outcome of a long-lasting collaboration between two research groups: an experimental Electrochemistry group in the Laboratory of Physical Chemistry of the Université de Paris-Sud (Orsay, France) and a theoretical Quantum Chemistry group in the Department of Physical and Inorganic Chemistry of the Universitat Rovira i Virgili (Tarragona, Spain).

4.2 Background

4.2.1 Electrochemistry

In electron exchange processes, the redox potential (Φ) and the reaction free energy (Δ*G*) are formally linked by the number of electrons (*n*) exchanged in the process and the Faraday constant (*F*) through the Nernst equation:

$$\Delta G = -nF\Phi$$

In an oxidation-reduction couple, the formal apparent redox potential is defined as $\Phi^o = (\Phi_{pa} + \Phi_{pc})/2$, where Φ_{pa} and Φ_{pc} are the anodic (oxidation) and cathodic (reduction) peak potentials, respectively.

In this chapter, reduction energies defined as RE = $E(\text{POM}_{n\text{-red}}) - E(\text{POM}_{ox})$ are presented for reactions $\text{POM}_{ox} + ne^- \rightarrow \text{POM}_{n\text{-red}}$. For this purpose, we computed electronic energies for (*n*-fold) reduced and oxidised forms with the energy of the free electron taken as zero. Assuming that the electronic energy change during the reduction process is practically equal to the Gibbs free energy change (entropic *change* term negligible), RE ≈ Δ*G*, the computed REs may be seen as theoretical estimates of the experimental reduction potentials:

$$\text{RE} \approx -nF\Phi \tag{4.1}$$

The last expression shows that a species with a more negative RE than another will consequently have a more positive Φ, and vice versa. The computational results are mostly discussed as *differences* between REs (in eV) or Φ (in V).

4.2.2 Density functional theory

Herein I present a piece of theoretical work performed within the Kohn-Sham formalism in the framework of density functional theory (DFT) [34]. This is, nowadays, a widespread theoretical approach because of its straightforward use and excellent accuracy/cost ratio [35–41] for most POM chemical phenomena. On the basis of many encouraging results, computational analyses have a general acceptance and remarkable significance among the scientific community. DFT calculations provide molecular geometries and energies, orbital shapes and energies, dipole moments, electron distributions and many other properties that help us understanding the behaviour of molecules. The calculations herein discussed reproduce and explain electrochemical measurements, delivering molecular orbitals and total molecular energies that, combined, give us reaction energies and other relevant properties. The anion charges of the reduced and oxidised forms differ, and so the RE must be computed in the presence of a solvent model. So, to make the calculations on redox properties reliable, the stabilizing effects of the molecular environment must be included in the model. Otherwise, the energies would be unreliable for comparison, as in the gas phase approximation [42]. It is worth mentioning that the theoretical results are mostly aimed at explaining the trends rather than the experimental absolute redox potentials with high accuracy.

For each particular study presented, slightly different DFT setups were utilised although all of them based on the same principles. The reader is referred to the original papers for the computational details.

4.3 Influence of the heteroatom size on the redox properties of Keggin anions

Most Keggin $[XW_{12}O_{40}]^{q-}$ anions studied possess electrochemically inactive X heteroatoms. Therefore, their redox behaviour is based on the addenda atoms. During the mid-1960s, Pope and co-workers [24, 26, 27] proved that the one-electron redox potentials of a Keggin anion, in conditions of no protonation, is a linear function of its overall molecular charge, $q-$. Later on, electron addition in Keggin (Figure 4.1) and other POM anionic species [43, 44] was carefully investigated both from the experimental [6] and the theoretical [22] points of view. However, other physical and/or electronic factors governing the energetics of the first electron transfer process in POMs needed extra explanation, such as the role of the heteroatom size. In a 2010 work, an answer to this question was proposed for one-electron transfer by analysing the electrochemical behaviour of a series of $XW_{12}O_{40}{}^{q-}$ compounds (with X = B, Al, Ga, Si, Ge, P, As) [45]. The goal of these calculations was to find the physical origin for the observed redox potentials. Furthermore, theory might help to identify the better of two parameters, equivalent in principle, to describe this physical origin, in terms of electrical charge or of electrostatic potential.

Cyclic voltammetry (CV) and controlled potential coulometry were used to analyse the electrochemical behaviour of each compound in aqueous solution at controlled pH 5 medium to guarantee stability (0.4 M CH_3COONa + CH_3COOH). The first CV wave for each compound features a one-electron reversible process, the electron transfer not being perturbed by protonation. Therefore, the corresponding apparent potential values, $\Phi°$, could be used to assess the influence of the central heteroatom

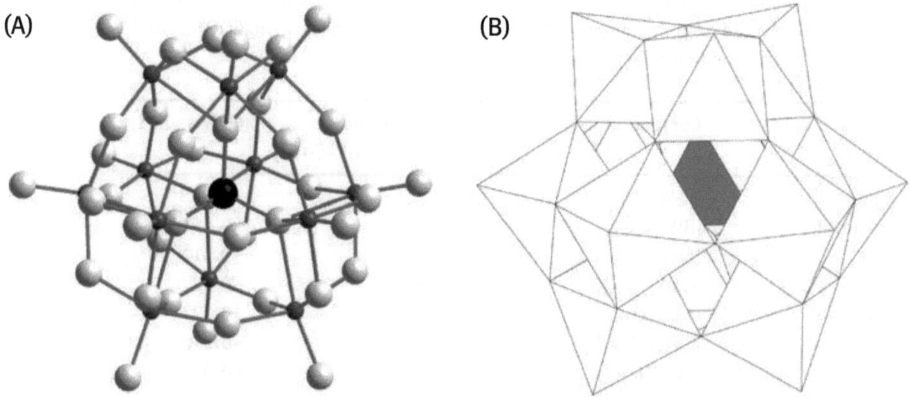

Figure 4.1: (A) Ball-and-stick and (B) polyhedral views of the Keggin structure, $[XW_{12}O_{40}]^{q-}$. Color code: light grey – oxygen; dark grey – tungsten; black – heteroatom, X. In the fully oxidised state, the internal tetrahedron, XO_4^{q-}, is responsible for the negative charge of the structure whereas the external cage, $W_{12}O_{36}$, is formally neutral. Octahedra in (B) are MO_6 units.

size on the reducibility of the POM. This section compares the variations in the redox potentials of Keggin anions in the mentioned conditions upon changes in X of the same group (III: B, Al, Ga; IV: Si, Ge and V: P-As). Table 4.1 gathers the apparent formal potentials Φ^o for the first one-electron redox process of selected Keggin compounds. The molecular charge per volume unit (volumic charge densities) issued from the DFT calculations are added and will be commented later. A general trend emerges from the Φ^o quoted in this table. The reader may notice that apparent Φ^o values get more negative (more difficult to reduce clusters) as the size of the central heteroatom decreases within a given family of Keggin compounds with the same overall negative charge. Within each group, heteroatoms with smaller atomic numbers are smaller in size. Figure 4.2 illustrates the CVs for the series of Keggin anions with X = B, Al, Ga, where the B derivative is the most difficult to reduce of its group, about 100 mV more than the Ga-derivative. What is the subjacent reason explaining these observations?

As has been previously proposed [46–49] and applied [50–53], many close-packed POMs may be seen as an internal anionic fragment encapsulated by a neutral metal oxide cage. For Keggin tungstates it is customary to write $[XO_4]^{q-}@W_{12}O_{36}$ to denote this concept. This assumption can simplify the interpretation of some chemical properties [53]. In the present case, it was taken for granted that the internal XO_4^{q-} unit is the responsible of the observed variations, ruling out the geometrical differences between $W_{12}O_{36}$ cages as a determinant factor. Actually, X-ray characterization revealed that $W_{12}O_{36}$ cages of compounds in a group are nearly equal. In the same line, DFT calculations show that the volume of Keggin molecules fluctuates by 0.4 to 0.9% within a group, showing that the overall size of the Keggin anion can be considered constant.

Table 4.1: Apparent potentials, $\Phi^o = (\Phi_{pa} + \Phi_{pc})/2$, for the first one-electron redox process of selected Keggin compounds at pH 5 (0.4 M CH_3COONa + CH_3COOH),[a] and volumic charge density for each compound.

Family	Compound	Volumic charge density[b][10^{22} C·Å^{-3}]	Φ^o[V vs SCE]
Keggin-I	$[H_2W_{12}O_{40}]^{6-}$		−0.608
Keggin-III	$[BW_{12}O_{40}]^{5-}$	1.882	−0.491
	$[AlW_{12}O_{40}]^{5-}$	1.868	−0.410
	$[GaW_{12}O_{40}]^{5-}$	1.866	−0.387
Keggin-IV	$[SiW_{12}O_{40}]^{4-}$	1.506	−0.227
	$[GeW_{12}O_{40}]^{4-}$	1.503	−0.190
Keggin-V	$[PW_{12}O_{40}]^{3-}$	1.138	+0.064
	$[AsW_{12}O_{40}]^{3-}$	1.133	–

[a]Scan rate: 10 mV·s^{-1}; working electrode: glassy carbon. [b]Values issued from DFT calculations.

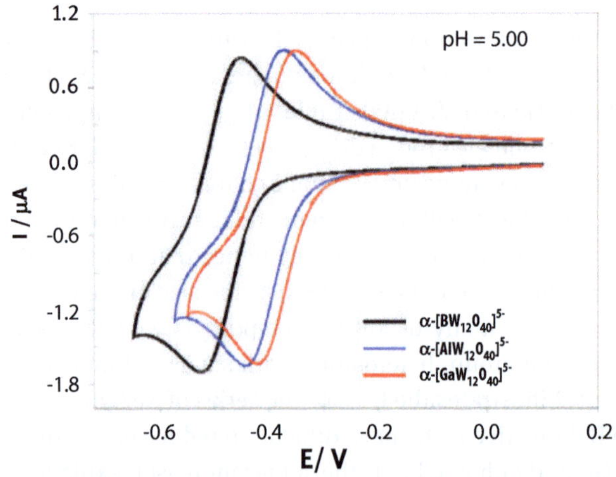

Figure 4.2: Cyclic voltammograms of α-$[BW_{12}O_{40}]^{5-}$ (black line), α-$[AlW_{12}O_{40}]^{5-}$ (blue line) and α-$[GaW_{12}O_{40}]^{5-}$ (red line) at pH 5 (0.4M CH_3COO + CH_3COOH). Polyoxometalate concentration: 0.5 mM; scan rate: 10 mV.s^{-1}; working electrode: glassy carbon; reference electrode: SCE.

To compare theoretical and experimental first reduction processes, we computed the REs shown in Table 4.2, for Keggin-III, IV and V compounds, for the process $[XW_{12}O_{40}]^{q-} + e \rightarrow [XW_{12}O_{40}\ 1e]^{(q+1)-}$.

As we are interested in redox potential differences between species, the RE = −4.10 eV for PW_{12} is taken as the computational reference for the other values. Three sets of REs with the correct trend are obtained, the lowest ones corresponding to the strongest oxidants (XW_{12}, X = P and As) and lowest anion charge (−3). Intermediate

Table 4.2: Computed reduction energies (RE), RE differences (ΔRE), measured apparent potential differences (ΔΦ°) and computed LUMO energies for $XW_{12}O_{40}^{q-}$ compounds in solution.

Family	X	Charge	RE [meV]	ΔRE[a] [meV]	ΔΦ° [a] [meV]	LUMO [b] [eV]
Keggin-III	B	−5	458			−3.83
	Al		373	−85	−81	−3.87
	Ga		337	−121	−104	−3.89
Keggin-IV	Si	−4	191	−72	−37	−4.13
	Ge		119			−4.15
Keggin-V	P	−3	0	−45	−30	−4.38
	As		−45			−4.41

[a]ΔRE and ΔΦ° are relative to the first element of the same group. [b]Lowest Unoccupied Molecular Orbital.

and more positive REs are computed for X = Si and Ge (191 and 119 meV more positive, respectively), with an anion charge of −4 and, finally, the most positive ones (and least oxidizing species) correspond to X = B, Al and Ga (458, 373 and 337 meV above the reference, respectively) since they carry a charge of −5. This trend is simply attributed to the anion charge. The differences encountered within each group are, however, smaller in general, and must be assigned to other factors.

The REs in Table 4.2 can be compared with the experimental half-wave potentials (Φ°) shown in Table 4.1, taking into account equation 1. The ΔRE computed by DFT are, in absolute terms, very similar to those of ΔΦ°, so DFT calculations nicely reproduce the experimental trends. The LUMOs [54] in these fully oxidised compounds have d(W)-like character and are the ones accepting the first electrons upon reduction. In a simplistic fashion, more stable LUMOs give more negative REs and a greater tendency to gain electrons, although this statement is not conclusive. In the present case, the LUMO energies in compounds of the same group are not the only reason for the differences encountered. It can be accepted to be the case for the P/As couple (the energy difference is 30 meV). However, for Si/Ge, their LUMO energies differ by 20 meV only, a much smaller value compared to their mutual ΔRE = 72 meV. Especially for group III compounds (X = B, Al and Ga), where the energy differences of d(W)-like orbitals between the compound are as small as 20–30 meV, we discard this fact as the reason for the large variations in the redox potentials.

One hypothesis is based on the different electrostatic potentials created by the internal XO_4^{q-} units on their surroundings, as the W position, where an incoming electron goes. As previously, we keep to the view that the Keggin tungstates under consideration can be expressed as $[XO_4]^{q-}$@$W_{12}O_{36}$ and focus on the properties of the XO_4^{q-} units in order to explain the observations made on the redox potentials of the studied POMs.

The negative charge that we tentatively assign to each XO_4^{q-} fragment, despite being formally −3, −4 and −5 for different X, can be considered to be somewhat smaller. Actually, a fraction of the electron density is transferred from the internal XO_4 to the $W_{12}O_{36}$ cage [52]. If XO_4^{q-} remained more charged in some cases, we could have an explanation for the different redox potentials measured. However, the computed fragment charges do not fully correlate with the REs. An alternative magnitude that can be computed and mapped is the electrostatic potential, being in addition much more realistic than atomic or fragment charges. A graphical representation of the molecular electrostatic potential (MEP) of XO_4^{q-} shows appreciable differences in Keggin anions of the same group (MEPs for X = P, As and Si, Ge are shown in Figure 4.3). These representations allow us to estimate the electrostatic potential that the W atoms feel in the real Keggin structures since they are mapped over an electron density isosurface coincident with the W positions (*blue* denotes more positive and *red* more negative potentials). Thus, electrons get less destabilised around the more positive potential regions (in blue). In the Keggin-IV and V groups, the differences are small between X, in agreement with the similar redox potentials measured. Larger differences are observed within the Keggin-III group

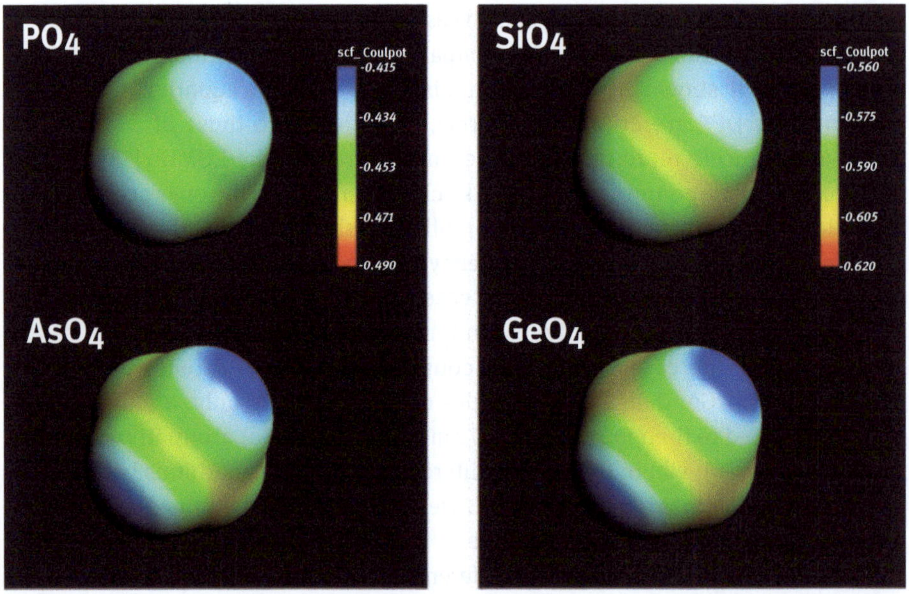

Figure 4.3: Molecular electrostatic potentials for XO_4^{q-} units (X = P, As and Si, Ge) represented over a surface placed exactly at the X-W distance. The apparent differences in the shape and extension of the surface come from the slight geometrical variations from one XO_4 to another. The potential range in each case is shown to the right in atomic units, and it changes from P-As to Si-Ge (red for more negative and blue to less negative potentials). The species displaying more intense blue regions will be reduced at less negative potentials.

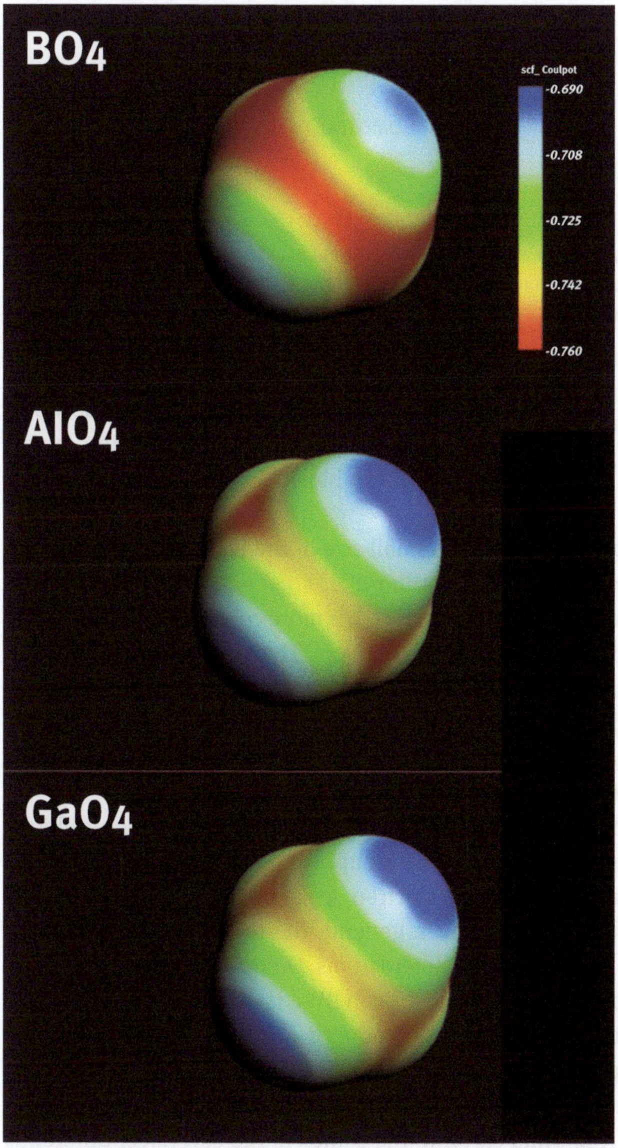

Figure 4.4: Molecular electrostatic potentials for XO_4^{q-} units (X = B, Al and Ga) represented over a surface placed exactly at the X-W distance. The potential range is shown to the right in atomic units.

(see Figure 4.4), where the redox potentials are more different, especially between boron and the other two heteroatoms (Al and Ga). This may be attributed to the larger electronic differences between B (2nd period) and the atoms from the 3rd and 4th periods. In fact, a deeper analysis of the electronic structure of BO_4^{5-} reveals that its highest occupied orbitals are higher than those of XO_4^{5-} of the same group, affecting

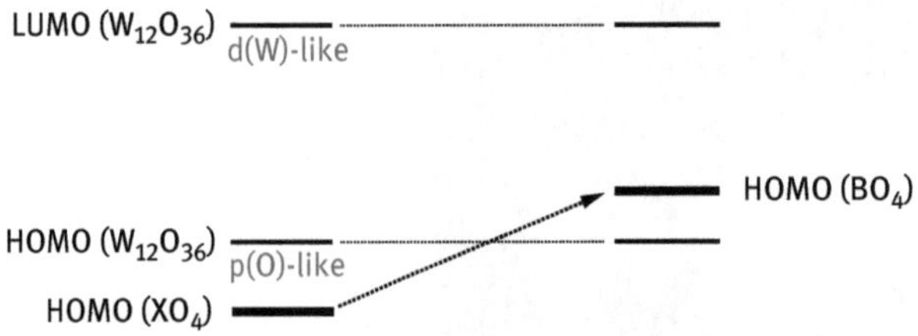

Figure 4.5: Destabilization of the BO_4-like HOMO in BW_{12} with respect to other X heteroatoms, attributed to the combination of high charge and small size of the fragment. Indicated is the character of the main orbitals depicted.

its environment in the Keggin structure. As a matter of fact, the highest occupied molecular orbital (HOMO) in $[BO_4]^{5-}@W_{12}O_{36}$ belongs to the internal anion, a very uncommon feature when dealing with X of the p-block. **Figure 5** represents this situation.

This high orbital energy is linked to the small size of the BO_4^{5-} unit, the smallest of the whole series (DFT computed equilibrium $d(\text{X-O}_{\text{tetra}})$ = 1.545 Å) and its high anionic charge. Similarly, in Keggin IV group, SiO_4^{4-} is more compressed than GeO_4^{4-}, featuring $d(\text{X-O}_{\text{tetra}})$ = 1.653 Å and 1.757 Å, respectively. The MEP obtained for SiO_4^{4-} also shows a slight shift towards more negative potentials compared to GeO_4^{4-}, in agreement with the redox potentials obtained by CV. Also in the latter case, the SiO_4^{4-} orbitals are higher in energy than those of GeO_4^{4-}, although not to the point of being higher than the $W_{12}O_{36}$ oxo-like orbital set.

For X = B at least, for which the electrostatic potential is so much different compared to the other two heteroatoms of the group, we have computed the purely electrostatic repulsion that an incoming electron feels in the LUMO (with d(W) character) without any orbital relaxation. In general, the differences are small, of the order of 70–75 meV for group IV and V heteroatoms. On the other hand, for group III we found that the repulsion of an extra electron in BW_{12} is 240 meV larger than that of AlW_{12} or GaW_{12}. Even if this difference gets reduced to ca. 100 meV after orbital relaxation, it remains large and this could explain the negative shift in the redox wave of X = B vs. the other two heteroatoms of the same group.

In summary, internal XO_4 units carrying the same charge can affect differently the tungstate oxide cage. Within each group of the periodic table, X atoms with lower atomic numbers are also smaller in size, producing a more negative electrostatic potential in the surroundings, and thus a smaller ability of the cage to be electron reduced. The case of BW_{12} is paradigmatic, with the smallest heteroatom of the Keggin-III series, and a very negative reduction potential with respect to the other elements of the same group. Even if the differences in the electrostatic potentials from a qualitative level are modest, they correlate well with the also tiny differences in experimental redox potentials.

4.4 pH-dependent electrochemical behaviour of α_1/α_2-[Fe(H_2O) $P_2W_{17}O_{61}]^{7-}$ isomers

In the second study presented, protonation plays a crucial role. As most POMs, the Wells-Dawson structure can be functionalised or varied by degradation, substitution or complexation [55–57]. The resulting compound has a different behaviour owing to its new structure or composition, and understanding the origin of such changes, as well as finding general trends, are the main goals of experimental and theoretical investigations. An analysis of two isomers of the Wells-Dawson structure, in which one W position has been replaced by Fe, is carried out with electrochemical and computational methods. The α_1 and α_2 isomers of $[Fe(OH_2)P_2W_{17}O_{61}]^{7-}$ (Figure 4.6) differ in the location of the Fe atom, either in the equatorial region (or *belt*) or in one polar region (or *cap*).

The acidity of a solution containing $Fe(OH_2)P_2W_{17}$ was varied and its effects on the redox properties tracked. The most thoroughly studied properties of Wells-Dawson derivatives are related to their redox behaviour. As regards of the Fe-monosubstituted $[Fe^{III}(OH_2)P_2W_{17}O_{61}]^{7-}$ system at neutral pH, the α_1 isomer is reduced before the α_2 form, as expected. However, the α_2 form turns to be more oxidizing as we acidify the solution, reversing the original situation. Electrochemical and DFT inspection can explain that the different coordination of Fe at different pH (-OH or -OH_2 terminal groups) changes the Fe-like orbital that will accept the incoming electron [58]. CVs of α_1 (**1**) and α_2 (**2**) isomers were recorded in several aqueous media (with pH varying from nearly 0 to 8) and in CH_3CN + 0.1 M $LiClO_4$. Below pH 6, $(Fe^{3+})_2$ is easier to reduce than $(Fe^{3+})_1$ i.e. $\Phi_{pc}(Fe^{3+})_2 > \Phi_{pc}(Fe^{3+})_1$. This is an unexpected behaviour since all theoretical

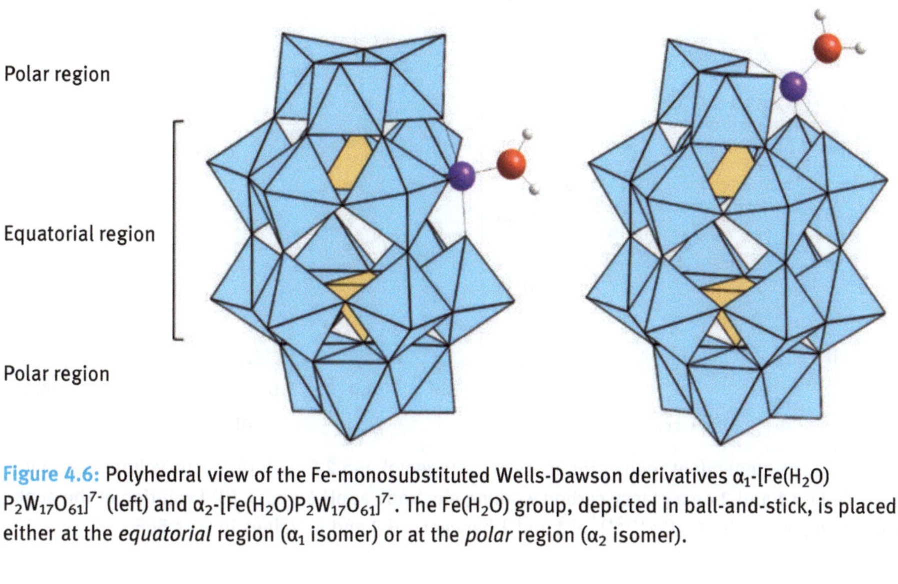

Figure 4.6: Polyhedral view of the Fe-monosubstituted Wells-Dawson derivatives α_1-[Fe(H_2O) $P_2W_{17}O_{61}]^{7-}$ (left) and α_2-[Fe(H_2O)$P_2W_{17}O_{61}]^{7-}$. The Fe(H_2O) group, depicted in ball-and-stick, is placed either at the *equatorial* region (α_1 isomer) or at the *polar* region (α_2 isomer).

and experimental studies performed on this family of compounds (plenary Wells-Dawson structures, $X_2W_{18}O_{62}^{6-}$ X = As or P, and monosubstituted complexes, α_1- and α_2-$[X_2MW_{17}O_{62}]^{q-}$ M = Mo, Tc, V, Re) pointed out that the first electronic exchange preferentially takes place on one of the 12 W atoms located on the equatorial region of the molecule i.e. α_1 position [26, 35, 37, 42, 59–63]. In other words, the α_1 isomer should be always easier to reduce than the corresponding α_2 isomer. However, this accepted and demonstrated rule, in the case of α_1-and α_2-$P_2W_{17}Fe$ forms, was put up-to-default. Indeed, we report here that the influence of the protonation makes a difference in the electrochemical behaviour of both isomers making the reduction of the Fe centre in α_2 position easier than in the case of the α_1 isomer. In contrast, when the protonation effect becomes negligible (for pH ≥ 6 or in organic medium) the normal trend is recovered, i.e. the Fe centre in α_1 position (**1**) is easier to reduce than in the α_2 position (**2**). Figure 4.7 illustrates these facts. Complementing the electrochemical study, DFT calculations help explaining the relative stability and redox potentials of compounds **1** and **2**, and their dependence with protonation. The acidity of the solution revealed determinant in the evolution of the redox properties of both isomers.

Since we are not capable of explicitly imposing a given pH value to our standard DFT calculations, we have generated a number of differently protonated model structures derived from the parent $[P_2W_{17}Fe^{III}O_{62}]^{9-}$ one that are assumed to be dominant at different pH values. Namely, at neutral pH, the deprotonated $[Fe^{III}OP_2W_{17}O_{61}]^{9-}$ structure could be predominant. However, this is not expected regarding the experimental

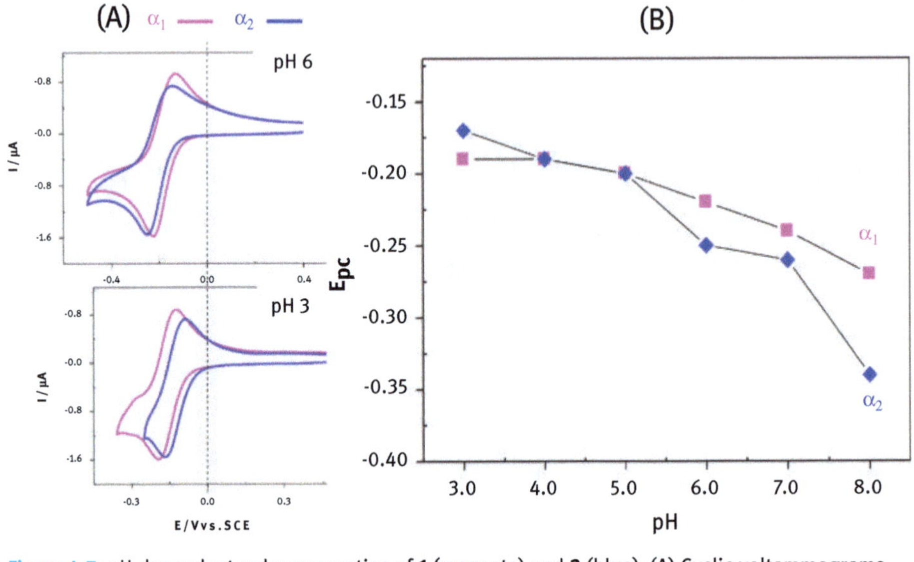

Figure 4.7: pH-dependent redox properties of **1** (magenta) and **2** (blue). (A) Cyclic voltammograms recorded for **1** and **2** on a glassy carbon working electrode at pH 6 and 3. (B) Evolution of the peak reduction potentials, $\Phi_{pc}(Fe^{3+/2+})$, of **1** and **2** between pH 3 and 8. Values in V vs SCE.

evidences that rule out the stability of these molecules [64]. Another likely structure at neutral pH is the mono-protonated one, $[Fe^{III}(OH)P_2W_{17}O_{61}]^{8-}$. The next protonation step will occur when acidity increases to pH 5, obtaining $[Fe^{III}(OH_2)P_2W_{17}O_{61}]^{7-}$. Finally, at even lower pH, two other structures could be formed: one without terminal atom on the Fe site, $[Fe^{III}P_2W_{17}O_{61}]^{7-}$, and one with a water molecule linked to Fe and a protonated bridging oxygen, $[HFe^{III}(OH_2)P_2W_{17}O_{61}]^{6-}$.

Fully optimised structures were obtained for this set of systems either with Fe^{II} or Fe^{III}, and evaluated the reduction free energy [65], ΔG_1 and ΔG_2, for the two isomers. We have also extracted the reduction free energy differences ($-\Delta\Delta G_{2\text{-}1} = -\Delta G_2 + \Delta G_1$) to compare them with the experimental data ($\Delta\Phi$) (see Table 4.3). If the $Fe^{III}(OH)$ species are considered, reduction of **1** is easier than **2** by 17 meV, in good agreement with the experimental half-wave potentials ($\Delta\Phi$ = +30 mV). The tendency is reversed by adding the second proton to the iron-substituted species (simulated moderately acidic pH), when $-\Delta\Delta G_{2\text{-}1}$ = 47 meV. Under conditions of further protonation (pH 1), this phenomenon is more notable and $-\Delta\Delta G_{2\text{-}1}$ = 68 or 90 meV, depending on the model, in favour of **2**. Calculations reproduce the experimental trends.

It is worth pointing out that three different species, $[Fe(OH_2)P_2W_{17}O_{61}]$, $[FeP_2W_{17}O_{61}]$ and $[HFe(OH_2)P_2W_{17}O_{61}]$, are proposed to exist in increasingly acidic solutions. No experimental or theoretical evidence suggest which the predominant species is, or if a mixture of them coexist in solution. Nevertheless, all of them feature the same redox behaviour in good agreement with electrochemical measurements (Table 4.3).

The uncommon feature that isomers **1** and **2** reverse the ordering of their first reduction potentials at pH 6 deserves further insight. From experiments, the first 1e-reduction is assumed to take place at the Fe centre (irrespective of the isomer and the pH), in detriment of the formation of the blue species with one electron delocalised over the W framework ($P_2W_{17}Fe^{3+} + e^- \rightarrow [P_2W_{17}Fe^{3+}\ 1e]$). Since the latter process needs more energy and is not favoured, the delocalised d_{xy}-like molecular orbital of W character appearing at higher energies, also of nonbonding nature, can be ruled out of the competition towards the first incoming electron. The orbitals of the oxidised forms of **1** and **2** ready to accept an extra electron are, in principle, the formally nonbonding d_{xy}(Fe) (perpendicular to the terminal oxygen) and the anti-bonding π^*(Fe-O), which is oriented towards the terminal oxygen. The energy of the

Table 4.3: Computed reduction energies for differently protonated forms of isomers **1** and **2** (ΔG_i, in eV), reduction energy differences ($-\Delta\Delta G_{2\text{-}1}$), and experimental data ($\Delta\Phi$, in V).

	ΔG_1	ΔG_2	$-\Delta\Delta G_{2\text{-}1}$	$\Delta\Phi$
$[Fe(OH)P_2W_{17}O_{61}]^{8-}$	−4.017	−4.000	−0.017	−0.030
$[Fe(OH_2)P_2W_{17}O_{61}]^{7-}$	−4.592	−4.639	+0.047	+0.080
$[FeP_2W_{17}O_{61}]^{7-}$	−4.746	−4.814	+0.068	–
$[HFe(OH_2)P_2W_{17}O_{61}]^{6-}$	−4.853	−4.943	+0.090	

latter orbital is strongly pH-dependent because of its orientation. The Fe-O(terminal) distance can change with protonation following pH variations, and so the π*(Fe-O) energy. In both Fe^{2+} isomers, the d_{xy} orbital is more stable than the π* orbital under conditions of poor protonation ($FeOP_2W_{17}$ and $Fe(OH)P_2W_{17}$ structures) while the inversion occurs for FeP_2W_{17} and $Fe(OH_2)P_2W_{17}$ molecules, assumed to be the predominant species at low pH. This is not so evident for both isomers of the Fe^{3+} form, where the orbital reversal occurs for the α_2 isomer only. This particular behaviour depending on pH is not observed in other metal-substituted Dawson-type tungstodiphosphates, such as $P_2W_{17}M$ with M = V or Mo [59–63].

The protonation state of the system governs the inversion of the order in reduction potentials observed around pH 5 for **1** and **2**. As expected, the computed Fe-O(terminal) distance increases with the number of protons attached to the terminal oxygen. In conditions of no protonation at Fe-O, the computed distance is *d*(Fe-O) ~ 1.66–1.76 Å depending on the isomeric form, with the π*(Fe-O) orbital lying at high energies with respect to the d_{xy}(Fe) one due to its marked antibonding nature. The general evolution of the π*(Fe-O) orbital from neutral (left) to acidic (right) pH is depicted in Figure 4.8, showing the differences in the molecular orbital sequence for the mono- and diprotonated forms of **1** and **2**. For the monoprotonated species, $Fe^{III}(OH)$, the Fe-O distance increases to ~1.87 Å and the π*(Fe-O) is stabilised due to the lower participation of the 2p-O(terminal) orbital, but still remains located above the d_{xy}-Fe orbital. Finally, when the apical group is doubly protonated, $Fe^{III}(OH_2)$, the Fe-O distance becomes very long (2.08 Å) and the π*(Fe-O) orbital turns into a *pure* d_{xz}(Fe) orbital, more stable than the formally nonbonding d_{xy}(Fe) orbital when an extra electron is added.

The more favourable reduction of **2** at pH ≤ 5 compared to **1** can be explained by (i) the dominant role of the $\pi^*(Fe\text{-}OH_2)$ orbital in the reduction process, and (ii) the different orientation of this orbital in either isomeric form with respect to the bridging oxygen atoms surrounding the iron centre. The right hand part of Figure 4.8 shows that the orientation of the π* orbital in **1** coincides with the direction of two Fe-O(bridging) bonds, therefore conferring it a stronger antibonding character than the homologous orbital in **2**. In the latter case, the π* orbital bisects the Fe-O(bridging) bonds, making the 3d(Fe)-2p(O) interaction weaker. Thus, electron reduction takes place in a higher π* orbital in **1** isomer than in **2**, and makes the reduction of the latter compound more favourable at sufficiently acidic pH.

Inspection of the computed atomic spin populations confirms the above statements. The computed change in spin density localised on O(terminal) when going from Fe-OH to $Fe\text{-}OH_2$ is remarkable: 0.36 to 0.06 for both α_1 and α_2 forms of $P_2W_{17}Fe^{III}$. This indicates the decreasing participation of the terminal oxygen in the π*(Fe-O) orbital. For the reduced $P_2W_{17}Fe^{II}$ compounds, the homologous spin density changes from 0.16 to 0.02 on average for both isomers. The smaller spin densities in the case of reduced forms arise from the longer Fe-O(terminal) distances produced by the population of the π*(Fe-O) orbital.

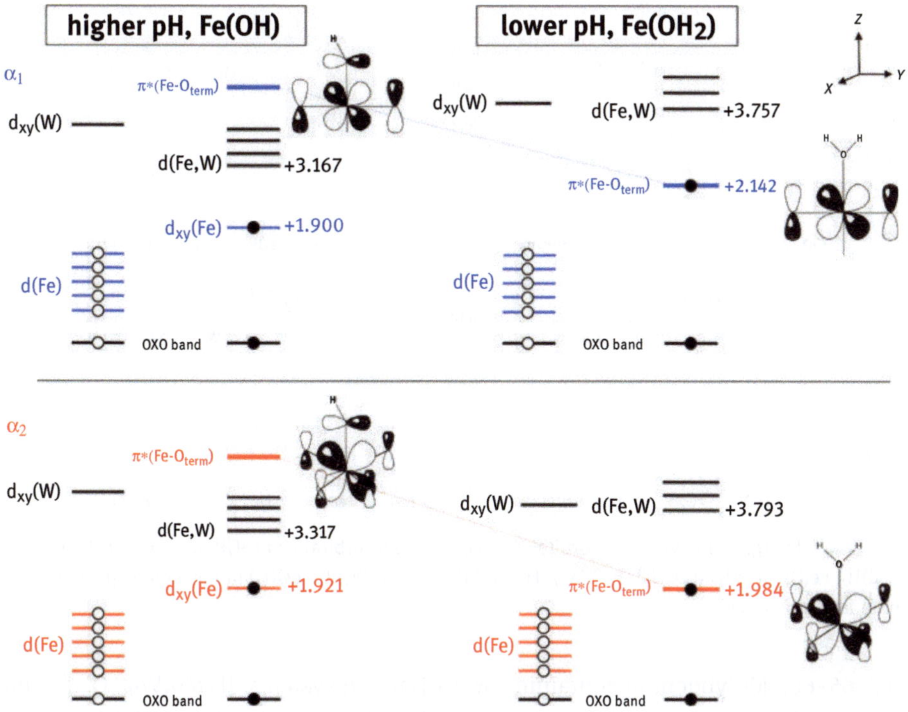

Figure 4.8: Computed frontier orbitals for higher pH, dominating species $Fe^{II}(OH)$, and lower pH conditions, dominating species $Fe^{II}(OH_2)$ of α_1 (**1**, top) and α_2 (**2**, bottom). The character (blue for α_1 and red for α_2 Fe-like orbitals) and relative energies are shown for selected molecular orbitals (in eV vs. the highest orbital of the oxo band). Spin-up and spin-down electrons (empty and filled circles, respectively) are separated in two columns for each compound.

In summary, DFT calculations allowed to interpret the experimental results according to the different molecular orbital energies. It has been shown that protonation on the terminal Fe-O site gradually stabilises the π* orbital with respect to the d_{xy} one, leading to an inversion of the d_{xy} and π* (d_{xz}-Fe) orbital energies when the apical group of iron is water (see Figure 4.9). In both isomers, the d_{xy} orbital is more stable than the π* orbital for $[Fe(OH)P_2W_{17}O_{61}]^{8-}$, assumed to be dominant at neutral pH, while the inversion occurs for $[Fe(OH_2)P_2W_{17}O_{61}]^{7-}$ and $[FeP_2W_{17}O_{61}]^{7-}$, the principal species at low pH.

4.5 Effect of electron (de)localisation and pairing in the redox properties of Wells-Dawson molybdotungstophosphates

From the Wells-Dawson polyoxotungstate, $[P_2W_{18}O_{62}]^{6-}$, controlled stereo-selective, multi-step syntheses allow replacement of one up to six W^{VI} centres by Mo^{VI} or V^{V}

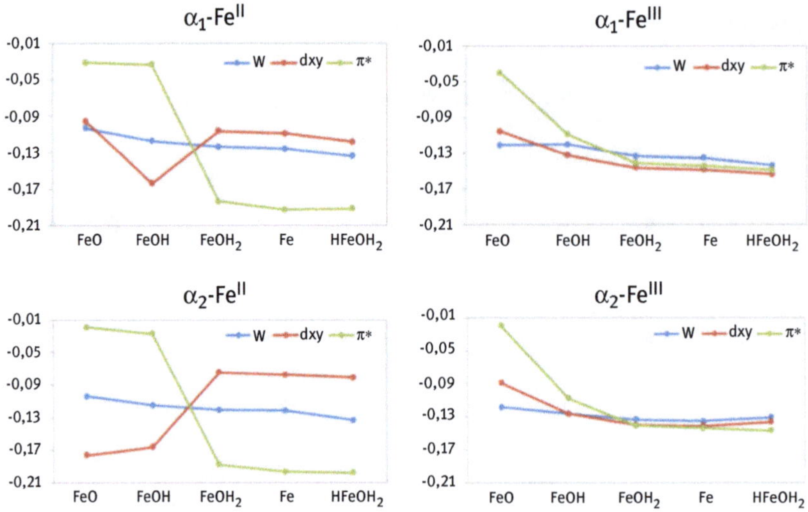

Figure 4.9: Energies (in eV) of d(W) orbital, Fe-like d_{xy} and π* orbitals for $FeOP_2W_{17}$ (FeO), $Fe(OH)P_2W_{17}$ (FeOH), $Fe(OH_2)P_2W_{17}$ ($FeOH_2$), FeP_2W_{17} (Fe) and $HFe(OH_2)P_2W_{17}$ ($HFeOH_2$) molecules respectively of α_1 and α_2 isomers.

[62, 66–68]. Molybdenum-containing Wells-Dawson systems, $[P_2Mo_xW_{18-x}O_{62}]^{6-}$ and in particular the behaviour of electrons transferred to them, is an attractive field. One focus of study is to determine, in a first step, if extra electrons gained by these structures are preferentially transferred into a definite atom or region of the molecule and, in a second step, to check whether the added electrons remain located in a single site or if they delocalise over neighbouring sites or the whole molecule in ordinary conditions, as is the case in the highly symmetrical Keggin compounds.

In this section, the mechanisms governing electron transfer and electron distribution in mixed-metal (W, Mo) Wells-Dawson-type POMs are discussed. The structures selected for the present study are shown in Figure 4.10: α_1- and α_2-$[P_2MoW_{17}O_{62}]^{6-}$, α-$[P_2Mo_3W_{15}O_{62}]^{6-}$ and α-$[P_2Mo_6W_{12}O_{62}]^{6-}$. They contain one Mo^{VI} centre, three equivalent Mo^{VI} centres and six Mo^{VI} centres, equivalent in a 2(cap):4(belt) fashion, respectively.

Figure 4.11(A) shows the CVs of the two isomers, α_1 and α_2-$[P_2W_{17}Mo^{VI}O_{62}]^{6-}$, obtained at pH 3.0. As expected, Mo^{VI} is easier to reduce in equatorial (α_1) than in cap (α_2) position. From DFT calculations it can be deduced that the 1st electron captured by the α_1 isomer partially delocalises over the belt region of the molecule, while it is trapped in one of the caps in the α_2 isomer. In a subsequent electron transfer, expected to be the 1st reduction at the W centres, the α_1 isomer is still easier to reduce than the α_2 isomer (Table 4.4). Interestingly, for this electron transfer, the quantum mechanical calculations for the α_1 and α_2 isomers show that the electron preferentially goes into the metal centres situated in a belt position of the Wells-Dawson structure

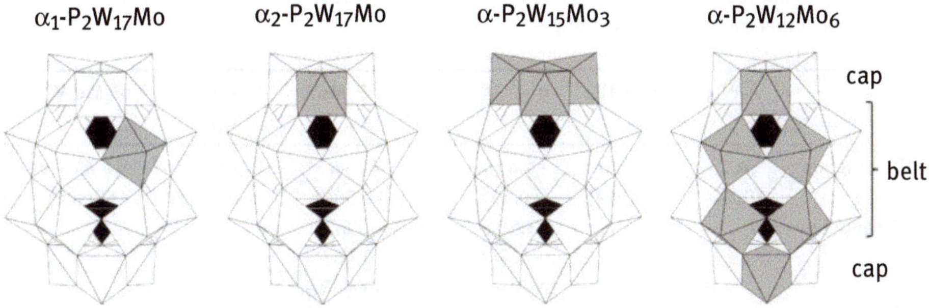

Figure 4.10: Idealised structure of α_1-P_2MoW_{17}, α_2-P_2MoW_{17}, α-$P_2Mo_3W_{15}$ and α-$P_2Mo_6W_{12}$ derivatives. In the fully oxidised form all carry a charge of −6. White and grey octahedra contain W and Mo atoms in the centre, respectively. The two different types of regions are indicated as *cap* (three metal centres each) and *belt* (12 metal centres – two connected six-membered rings). See above for more details.

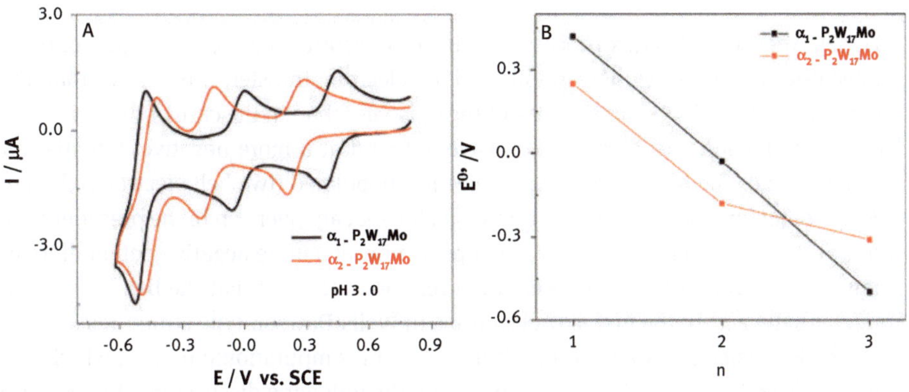

Figure 4.11: (A) CVs of α_1-$[P_2W_{17}Mo^{VI}O_{62}]^{6-}$ (black line) and α_2-$[P_2W_{17}Mo^{VI}O_{62}]^{6-}$ (red line) in 0.5 M Na_2SO_4 + H_2SO_4, pH 3. Polyoxometalate concentration: 0.5 mM; scan rate: 10 $mV.s^{-1}$; working electrode: glassy carbon; reference electrode: SCE. (B) Evolution of the midpoint redox potentials for the 1st three one-electron redox processes (Φ^0_1, Φ^0_2 and Φ^0_3) for α_1-$P_2W_{17}Mo^{VI}$ (black line) and α_2-$P_2W_{17}Mo^{VI}$ (red line).

Table 4.4: Experimental midpoint redox potentials[1] for the 1st three redox processes of α_1- and α_2-$[P_2W_{17}Mo^{VI}O_{62}]^{6-}$. In parentheses, the number of electrons exchanged in each wave.

V vs. SCE	Φ^0_1	Φ^0_2	Φ^0_3
	Mo (1e)	W (1e)	W (2e)
α_1-$[P_2W_{17}Mo^{VI}O_{62}]^{6-}$	0.42	−0.03	−0.50
α_2-$[P_2W_{17}Mo^{VI}O_{62}]^{6-}$	0.25	−0.18	−0.31
$\Delta\Phi(\alpha_1-\alpha_2)$	0.17	0.15	−0.19

[1]Values in V.

Table 4.5: Computed REs and E relative to P_2W_{18} (in parentheses) for the Wells-Dawson compounds discussed in this section. Values in eV.

	P_2W_{18}	α_2-P_2W_{17}	α_2-P_2MoW_{17}	α_1-P_2MoW_{17}	$P_2Mo_3W_{15}$	$P_2Mo_6W_{12}$
1st reduction	−4.234 (0.0)	−2.590 (−1.644)	−4.426 (+0.192)	−4.594 (+0.360)	−4.495 (+0.261)	−4.610 (2e)[a] (+0.376)
2nd reduction			−3.586 (−0.648)	−3.767 (−0.467)		
			α_2-P_2VW_{17}	α_1-P_2VW_{17}		
1st reduction			−4.576 (+0.342)	−4.673 (+0.439)		
2nd reduction			−3.255 (−0.979)	−3.298 (−0.936)		

[a]Two-electron process.

[35, 42]. After this 2nd redox process, and if we concentrate on the belt region of these molecules, which is strongly implicated in electron transfer, we realise that the electron density is higher in the case of the α_1 isomer than in the case of the α_2 isomer. As a consequence, the 3rd reduction wave is found at a more negative potential for isomer α_1 since the belt region is more electron populated (two belt electrons) than in the α_2 isomer (one belt electron) at this stage. Indeed, an inversion in the precedence of the waves occurs, that is, the 3rd wave appears now at a more negative potential for α_1 than for α_2 (Figure 4.11). This observation reinforces the fact that the belt region will preferentially accept the first extra electrons in Wells-Dawson-type structures.

The REs listed in Table 4.5 constitute the main computational results, which will be referenced to P_2W_{18} (RE = −4.234 eV) along the following discussion. Most REs are more negative than −4.234 eV, indicating the presence of stronger oxidant species than P_2W_{18}, in line with the reduction potentials discussed above. In the present section, a theoretical analysis of the distribution of the extra electrons among the metal centres and how this is related with electrochemical measurements is made, making a special emphasis in the different oxidant power of the α_1/α_2 isomers of P_2MoW_{17}.

4.5.1 Calculations on α-P_2W_{18}, α_2-P_2W_{17}, α_2-$P_2Mo_3W_{15}$ and $P_2Mo_6W_{12}$

The plenary α-P_2W_{18} system is an oxidant species as strong as, for instance, the Keggin anion $[PW_{12}O_{40}]^{3-}$, despite carrying a higher negative charge, owing to the fact that the charge −6 is distributed over a larger structure composed of 18 metal centres [69]. For α-P_2W_{18}, the first electron(s) occupy the belt region, which is more electron attracting than the cap regions [35, 42, 59, 60]. Compared to it, the lacunary α_2-P_2W_{17} system is more difficult to reduce, with a RE 1.6 eV less favourable (in non-protonated form) than for α-P_2W_{18}, a fact arising from the large negative charge of –10. However, the electrochemical measurement gives a smaller difference between the reduction waves of these two compounds. In the conditions of measurement α_2-P_2W_{17} is protonated, its total absolute charge being less negative than –10, explaining the theoretically predicted value for α_2-P_2W_{17}. Inspection of the molecular orbital occupied by the first incoming electron shows that it is also delocalised over the equatorial (*belt*) region.

As shown in Table 4.5, α-$P_2Mo_3W_{15}$ and the mono-substituted α_2-P_2MoW_{17} compounds have similar REs, the former being 70 meV more negative. The presence of the Mo_3 unit in one of the caps allows for some degree of electron delocalisation after reduction and, consequently, a more favourable process than the extra electron being more localised in a single Mo^V site. The CV measurements give a difference of 35 mV at pH 3 between the mentioned compounds. The theoretical data show that each Mo in the cap retains the same amount of the extra electron, with some participation of the nearest W neighbours.

In α-$P_2W_{12}Mo_6$, the ellipsoidal Mo_6 framework can favour delocalisation of extra electron(s) even more than in the above-mentioned α-$P_2W_{15}Mo_3$ system. For the DFT calculations we have taken into consideration the experimental fact that the first reduction wave is a 2e process. To obtain computationally a RE (or *E*) comparable with the position of the first reduction CV wave, a 2e-wave, we computed the 2e-reduced and the oxidised forms and therefore obtained –4.610 eV as the value to be compared with the first midpoint potential of 0.465 V vs SCE. The theoretical value is in good agreement with the measurements since it is the most negative RE of the series, slightly more negative than the RE obtained for the 1e-reduction of α_1-$P_2W_{17}Mo$. The more advantageous reduction in the hexamolybdate derivative is a consequence of electron delocalisation observed in the calculations. DFT results also suggest that the 1st electrochemically injected electrons are confined to the four belt-Mo atoms with the participation of some neighbouring belt-W centres. We also computed the hypothetical 1e-reduction process (α-$P_2W_{15}Mo_6^{6-}$ + e^- → α-$P_2W_{15}Mo_6^{7-}$), obtaining atomic spin populations of 0.25 electrons per Mo, and therefore reinforcing the idea that the 1st electron(s) is (are) delocalised over the belt positions only, leaving the two cap Mo centres fully oxidised. These data reveal the importance of delocalisation in the electrochemical properties of POMs.

4.5.2 Calculations on α_1- and α_2-$P_2W_{17}Mo$

First reduction process. The molybdenum mono-substituted Wells-Dawson anions deserve a detailed analysis since they lead to interesting conclusions. Besides the well-known fact that Wells-Dawson compounds containing Mo are more oxidant than the parent species α-P_2W_{18}, the position of Mo within the structure plays a crucial role in the overall oxidising power, not only with respect to the first reduction process but also in the second and third ones. In the cap-substituted α_2 isomer there is some sort of competition for the first incoming electron between the Mo^{VI} atom, in a polar position, and the belt W atoms. Such competition derives from two opposing facts: (i) the empty orbitals of Mo^{VI} have lower energy compared to the W^{VI} ones, and (ii) the empty belt orbitals are lower in energy than the empty cap orbitals. In the end, DFT results show that the cap-$Mo^{VI/V}$ process is 440 mV more favourable than the belt-W reduction for α_2-P_2MoW_{17}. Thus, the 1st extra electron is localised in the cap. The other positional isomer, α_1-P_2MoW_{17}, behaves similarly although a larger degree of electron delocalisation can be observed in the 1e-reduced form based on atomic population analysis. When Mo is in the cap position it retains about 82% of the extra electron, whereas it decreases to 77% when Mo is in the belt site. Since electron delocalisation usually gives extra stabilisation to reduced forms in POMs, the computed 1e-reduction process ($P_2W_{17}Mo^{6-} + e^- \rightarrow P_2W_{17}Mo^{7-}$) is energetically more favourable by ~170 meV for the α_1 form, in excellent agreement with the experimental difference of 170 mV (see Table 4.4). Thus, thermodynamically, the first 1e-reduction process is more favourable for the belt-substituted compound, where the chemical and structural effects add up to favour reduction.

The oxidising power of α_1/α_2-$P_2W_{17}Mo$ must also be compared with that of α-$P_2W_{15}Mo_3$. DFT calculations, in agreement with CV measurements, show that α_1-$P_2W_{17}Mo$ is stronger oxidant than α-$P_2W_{15}Mo_3$ by about 100 mV (see Table 4.5). The advantageous delocalisation in the Mo_3 polar group experienced by the metal electron in the 1e-reduced α-$P_2W_{15}Mo_3$ system cannot be on a par with the extra stabilisation produced in the Mo belt position of α_1-$P_2W_{17}Mo$. The fact that α_2-$P_2W_{17}Mo$ is slightly less oxidising than α-$P_2W_{15}Mo_3$, both being cap-substituted compounds, is easily explained by the enhanced electron delocalisation occurring in the latter compound.

To end up with the discussion on the first reduction processes, we add a comment on the mono-substituted vanadate, $P_2W_{17}V$ since it helps to rationalise the previously discussed facts. The relative shift between the first 1e-wave for α_1- and α_2-$P_2W_{17}V$ is ΔRE = 97 meV (measured ΔE = 89 mV). This small difference compared with P_2MoW_{17} is attributed to the more localised nature of the extra electron in reduced V-containing systems. In other words, V preserves its nature more than Mo when placed in the Wells-Dawson structure and, therefore, its position (cap or belt) is electrochemically less relevant. The computed atomic spin populations for the 1e-reduced α_1 and α_2 tungstovanadates are ~1.0 on the V centre, a value to be compared with 0.82 and 0.77 per Mo atom in the homologous molybdate compounds.

The above discussion allows us to establish a difference of about 90 meV as the energy change of belt vs. cap metal position, which we estimate from the one-electron RE difference for α_1/α_2-$P_2W_{17}V$. Extra RE difference between both isomers, like in Mo-substituted anions, comes from the more delocalised nature of the involved orbitals (which is more pronounced in the belt region). In other words, the ability of an electron to hop from one centre to another, larger in Mo than in V, stabilises the molecular orbital containing that electron and favours reduction. This explains that the RE difference for α_1- and α_2-$P_2W_{17}V$ be smaller than that for α_1- and α_2-$P_2W_{17}Mo$. Therefore, we infer that the extra stabilisation of a belt-localised electron compared to the cap-localised case is intimately related with the different degree of electron delocalisation in the belt region.

We carried out complementary calculations to evaluate further the effect of electron delocalisation upon the reduction potential. We compare two systems: α_1-$P_2W_{17}Mo$ and the hypothetical α-$P_2W_{12}Mo_6$ structure with six neighbouring Mo atoms in a single belt ring (W_3:Mo_6:W_6:W_3). Both molecules are equally charged and contain Mo atoms in the equatorial positions, the difference being the number of Mo atoms. If we consider the 1st reduction as a 1e-process we find a reduction potential difference of 290 mV in favour of α-$P_2W_{12}Mo_6$. Such a difference can only be attributed to the effect of electron delocalisation. A very similar value of ~265 mV was recently computed for the Keggin structure [40]. As a matter of fact, the energies of the LUMOs of the oxidised form for each compound are progressively deeper in energy as the number of implicated Mo atoms increases, namely, the LUMO for α-$P_2W_{12}Mo_6$ is 120 meV below that of α_1-$P_2W_{17}Mo$. If we look at the atomic spin populations of the 1e-reduced forms, we find that in α_1-$P_2W_{17}Mo$ the extra electron is delocalised among the Mo atom and two or three vicinal W atoms. In the case of α-$P_2W_{12}Mo_6$, 80% of the extra electron is delocalised over the Mo_6 ring, and the other 20% among the other W_6 in the belt. The larger the number of metal centres accepting a fraction of the incoming electron, the more favourable the reduction process is. This phenomenon is applicable when comparing α_2-$P_2W_{17}Mo$ and α-$P_2W_{15}Mo_3$, for instance, or α_1-$P_2W_{17}Mo$ and α-$P_2W_{12}Mo_6$.

Second and third reduction processes. At this point, let us discuss the computational results for the 2nd 1e-reduction process in α_1/α_2-$P_2W_{17}Mo$ to complement the CV data. We are especially interested in unravelling the complete CV (1st three reduction waves) of α_1/α_2-$P_2W_{17}Mo$, notably the tricky (at least at first sight) relative positions of the 2nd and 3rd waves. Experimental data cannot reveal if the 2nd metal electron, going to the belt region, is mostly localised (Mo^{IV} character) or partially delocalised ($Mo^VW_{17}1e^-$ character). If we assumed that the 1st 1e-reduction produces Mo^V in either isomer, the 2nd electron must go to the fully oxidised belt-W^{VI} positions, but at a more negative potential due to the molecular charge increment that the 1st reduction entails. But what causes the mutual shift of 100 mV of the 2nd wave for each isomer? The reduction potentials computed for the POM(1e) + $e^- \rightarrow$ POM(2e) process for both isomers predict that shift to be around 150 mV and, thus, we may inspect what is the origin of this phenomenon. We computed the possible solutions for the 2e-reduced systems, namely, the unpaired and the paired electron cases represented in Figure 4.12.

Interestingly, at the level of calculation applied, each of these solutions is the most stable for one of the isomers when M = Mo. In α_2-$P_2W_{17}Mo$, the unpaired situation is the most stable by 70 meV, indicating that the 2nd electron

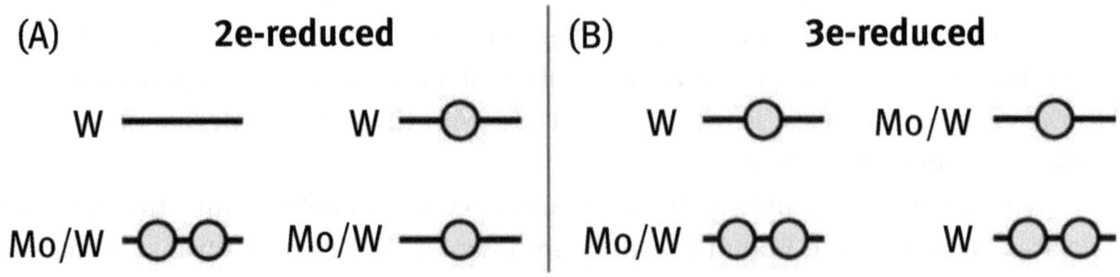

Figure 4.12: Schematic view of the most plausible electron distributions for (A) the 2e-reduced, and (B) the 3e-reduced forms of $P_2W_{17}Mo$. Horizontal lines represent molecular orbitals. When two electrons (circles) occupy the molecular orbital designated Mo/W, some Mo^{IV} character is present, whereas only one electron in the Mo/W-like orbital implies Mo^V.

prefers to delocalise over the W atoms thus avoiding any Mo^{IV} character. On the other hand, the electron-paired solution is 173 meV more stable than the unpaired one in α_1-$P_2W_{17}Mo$. In the two cases (α_1 and α_2), the 2nd electron goes to the belt region but in a different manner and, consequently, with a different energy. The pairing process occurring in α_1 appears as a favourable one, with some Mo^{IV} character as depicted in the scheme, with respect to the non-paired situation in α_2. In α_1, the presence of one electron in the belt Mo^V does not hinder the 2nd one from occupying the same region, but it actually favours it by e–e pairing. In α_2, provided that the 2nd electron is forced to go to the belt region, the 2nd reduction is just favoured by the lower e–e electrostatic repulsion that, from the present data, appears to be a weaker advantage than e–e pairing. The electron pairing argument is reinforced by the well-known and proved fact that the 2e-reduced α-P_2W_{18} species is strongly diamagnetic [45, 70]. The character of an electron can be measured by inspection of the molecular orbital it occupies, and also by atomic populations. Both of them coincide in the more delocalised nature of the belt electrons with or without Mo.

Present calculations show that, after the 2nd 1e-reduction, the α_2-$P_2W_{17}Mo^V1e^-$ situation is the most stable by an energy difference of 70 meV. However, things are different in the α_1 isomer, for which an important Mo^{IV} character is acknowledged. As shown in Table 4.3, the RE difference between the 2nd 1e-processes (α_1 – α_2) agrees with the experimental results and justify them by the different character of the second electron in either isomer in favour of α_1. Thus, the mentioned facts suggest a possible competition between two factors, namely (a) the unfavourable e–e electrostatic repulsion, and (b) the favourable electron pairing. Each isomer is characterised by a dominating factor. In α_1, the 1st electron already occupies a part of the belt region (Mo^V and some W^V character of the vicinal atoms). Although the second electron experiences the repulsive presence of the 1st one, they can pair and thus stabilise the couple (see Figure 4.13, left diagram). On the other hand, the α_2 isomer has the 1st electron trapped in the cap region, the belt region being free of extra electron density

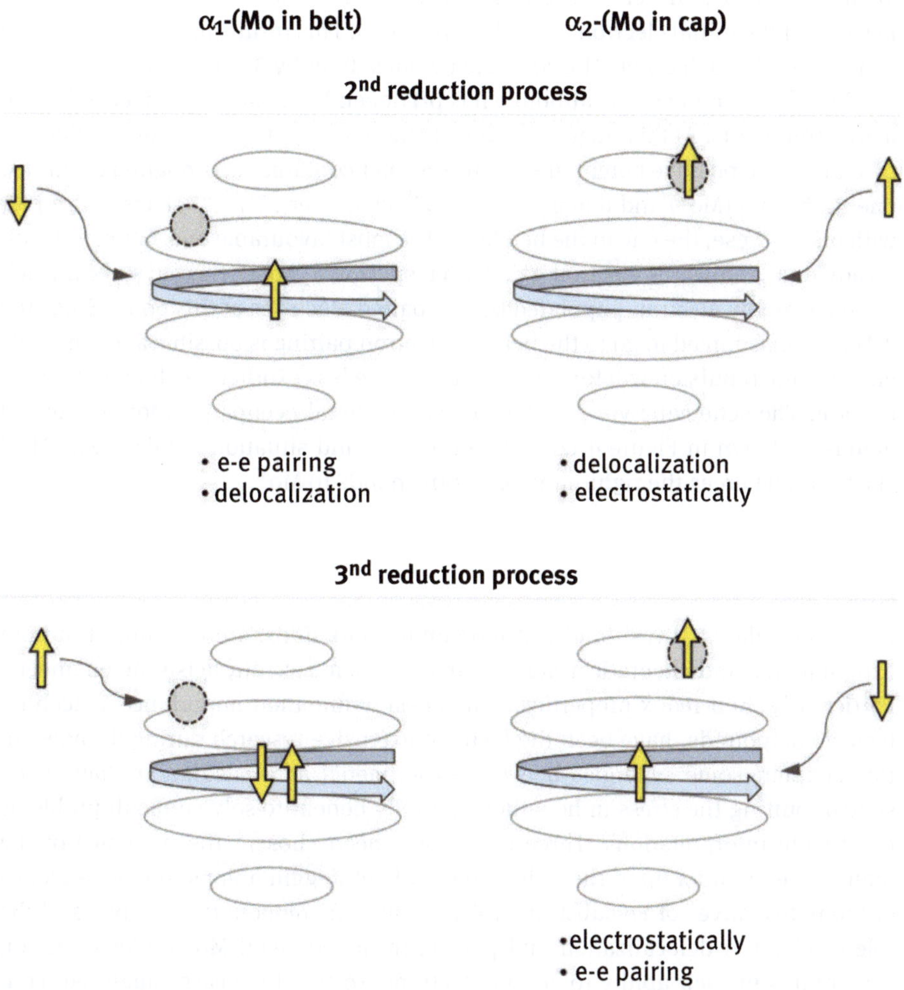

Figure 4.13: Representation of the 2nd and 3rd reduction processes taking place in the α_1 and α_2 isomers of $P_2W_{17}Mo$ (Mo atoms represented by grey circles), and the factors favouring them in each case. The four 3:6:6:3 loops of metal atoms are sketched as thin grey lines. Yellow arrows are electrochemically added electrons, and grey-blue curved arrows represent the delocalised nature of the belt electrons.

prior to the second reduction. This being an electrostatic advantage with respect to the α_1 isomer, electron pairing will not be possible. We may deduce that, as long as the region is sufficiently large for delocalisation, the 1st two electrons will be paired and stabilised. This explanation is schematically depicted in Figure 4.13.

It must be pointed out that the present discussion stands for $P_2W_{17}V$ but giving a different result. Since the 1st 1e-reduction in this vanadotungstate produces a highly localised V^{IV} electron, the 2nd one has hardly a chance of pairing with it (see

Table 4.5). Thus, the electrostatic repulsion will be similar irrespective of the position of the initial electron (cap-V^{IV} or belt-V^{IV}). This results in two 2nd reduction waves close to each other. The computed values differ by 33 mV only.

The 3rd reduction wave, although it is pH dependent, may be justified using the above arguments. At this stage (2e-reduced anions), the situation favours reduction of α_2 at a more positive potential since this isomer contains two unpaired electrons, one in the cap (Mo^V) and one in the belt (W^V character). The 3rd electron can pair with one of these, the one in the belt being the most favourable one. Concerning α_1, no advantages towards electron-gain with respect to α_2 are envisaged since the belt region is highly electron-populated by two paired electrons at this point. Since the 3rd electron is forced to go to the belt, no electron pairing is possible and a notable electrostatic repulsion will force this process to be less exothermic than that for the α_2 form. The schematic view of the molecular orbital occupations for 3e-reduced anions is shown in Figure 4.12), where the left-hand situation implies some Mo^{IV} character, whereas the right-hand one corresponds to Mo^V.

4.6 Summary

Polyoxometalates, a vast family of inorganic compounds, are the subject of both experimental and theoretical studies among chemists, physicists or engineers. Particularly their redox properties, which best define their importance as technological compounds, have been the focus of extensive research during decades. In this chapter, some synergies between experiments and calculations have been shown, putting the stress in how they mutually benefit to solve difficult problems or intricate interpretations. Three cases have been chosen: the influence of the internal heteroatom upon the redox potentials of Keggin anions, the dependence of the redox waves of Fe-substituted Wells-Dawson compounds with pH, and the role of electron delocalization and pairing in mixed-metal Mo/W Wells-Dawson compounds in their ability to accept electrons. In the three cases analysed, considerable merit should be put on the theoretical calculations, which played a crucial role to explain the data collected in the laboratory. DFT demonstrates again its preponderant role over other computational techniques in this class of chemical problems. In the author's opinion, as long as computational power keeps on growing in the near future, the relevance of Computational Chemistry in leading research will become even greater.

References

[1] Pope MT. Heteropoly and isopoly oxometalates. Berlin, Germany: Springer-Verlag, 1983.
[2] Pope MT, Müller A. Polyoxometalate chemistry: an old field with new dimensions in several disciplines. Angew Chem Int Ed 1991;30:34–48.

[3] Merca A, Garai S, Bögge H. An unstable paramagnetic isopolyoxomolybdate intermediate nonhomogeneously reduced at different sites and trapped in a host based on chemical adaptability. Angew Chem Int Ed 2013;52:11765–11769.
[4] Pope MT, Müller A. Polyoxometalates: from platonic solids to anti-retroviral activity. Dordrecht, The Netherlands: Kluwer, 1994.
[5] Pope MT, Müller A. Polyoxometalate chemistry: from topology via self-assembly to applications. Dordrecht, The Netherlands: Kluwer, 2001.
[6] Hill CL, editor. Special issue on polyoxometalates. Chem Rev 1998;98:1–390.
[7] Baker LCW, Glick DC. Present general status of understanding of heteropoly electrolytes and a tracing of some major highlights in the history of their elucidation. Chem Rev 1998;98:3–49.
[8] Izarova NV, Kögerler P. Polyoxometalate-based single-molecule magnets. In: Ruhlman L, Schaming D, editors. Trends in polyoxometalates research. Hauppauge, NY, USA: Nova Science Publishers, Inc., 2015.
[9] Compain JD, Mialane P, Dolbecq A, Mbomekallé IM, Marrot J, Sécheresse F, Iron polyoxometalate single-molecule magnets. Angew Chem Int Ed. 2009;48:3077–3081.
[10] Giusti A, Charron G, Mazerat S. Magnetic bistability of individual single-molecule magnets grafted on single-wall carbon nanotubes. Angew Chem Int Ed 2009;48:4949–4952.
[11] Keita B, Nadjo L. Electrochemistry of polyoxometalates. In: Bard AJ, Stratmann M, editor(s). Encyclopedia of electrochemistry Vol. 7. Weinheim, Germany: Wiley-VCH, 2006:607–700.
[12] Mizuno N, Yamaguchi K, Kamata K. Epoxidation of olefins with hydrogen peroxide catalyzed by polyoxometalates. Coord Chem Rev 2005;249:1944–1956.
[13] Liu S, Möhwald H, Volkmer D, Kurth DG. Polyoxometalate-based electro- and photochromic dual-mode devices. Langmuir 2006;22:1949–1951.
[14] Liu S, Kurth DG, Möhwald H, Volkmer D. A thin-film electrochromic device based on a polyoxometalate cluster. Adv Mater. 2002;14:225–228.
[15] Liu S, Kurth DG, Volkmer D. Polyoxometalates as pH-sensitive probes in self-assembled multilayers. Chem Commun 2002;976–977.
[16] Song YF, Tsunashima R. Recent advances on polyoxometalate-based molecular and composite materials. Chem Soc Rev 2012;41:7384–7402.
[17] Miras HN, Yan J, Long DL, Cronin L. Engineering polyoxometalates with emergent properties. Chem Soc Rev 2012;41:7403–7430.
[18] Yamase T, Pope MT, editors. Polyoxometalate chemistry for nanocomposite design. Dordrecht, The Netherlands: Kluwer, 2002.
[19] Omwoma S, Chen W, Tsunashima R, Song YF. Recent advances on polyoxometalates intercalated layered double hydroxides: from synthetic approaches to functional material applications. Coord Chem Rev 2014;258–9:58–71.
[20] Long DL, Burkholder E, Cronin L. Polyoxometalate clusters, nanostructures and materials: from self assembly to designer materials and devices. Chem Soc Rev 2007;36:105–121.
[21] Rhule JT, Hill CL, Judd DA. Polyoxometalates in medicine. Chem Rev 1998;98:327–358.
[22] López X, Carbó JJ, Bo C, Poblet JM. Structure, properties and reactivity of polyoxometalates: a theoretical perspective. Chem Soc Rev 2012;41:7537–7571.
[23] Souchay P, Hervé G. C R Acad Sci 1965;261:2486–2489.
[24] Pope MT, Varga JGM. Heteropoly blues. I. Reduction stoichiometries and reduction potentials of some 12-tungstates. Inorg Chem 1966;5:1249–1254.
[25] Souchay P, Contant R. C R Acad Sci 1967;265:723–726.
[26] Pope MT, Papaconstantinou E. Heteropoly blues. II. Reduction of 2:18-tungstates. Inorg Chem. 1967;6:1147–1152.
[27] Papaconstantinou E, Pope MT. Heteropoly blues. III. Preparation and stabilities of reduced 18-molybdodiphosphates. Inorg Chem 1967;6:1152–1155.

[28] Massart R, Hervé G. Rev Chim Min 1968;5:501–520.
[29] Varga Jr. GM, Papaconstantinou E, Pope MT. Heteropoly blues. IV. Spectroscopic and magnetic properties of some reduced polytungstates. Inorg Chem 1970;9:662–667.
[30] Papaconstantinou E, Pope MT. Heteropoly blues. V. Electronic spectra of one- to six-electron blues of 18-metallodiphosphate anions. Inorg Chem 1970;9:667–669.
[31] Weinstock IA. Homogeneous-phase electron-transfer reactions of polyoxometalates. Chem Rev 1998;98:113–170.
[32] Sadakane M, Steckhan E. Electrochemical properties of polyoxometalates as electrocatalysts. Chem Rev 1998;98:219–237.
[33] Keita B, Nadjo L. Polyoxometalate-based homogeneous catalysis of electrode reactions: Recent achievements. J Mol Catal A 2007;262:190–215.
[34] Parr GP, Wang Y. Density-functional theory of atoms and molecules. New York, NY, USA: Oxford University Press, 1989.
[35] Keita B, Jean Y, Levy B, Nadjo L, Contant R. Toward a qualitative understanding of the initial electron transfer site in Dawson-type heteropolyanions. New J Chem 2002;26:1314–1319.
[36] Fernández JA, López X, Bo C, de Graaf C, Baerends EJ, Poblet JM. Polyoxometalates with internal cavities: Redox activity, basicity, and cation encapsulation in [X*n*+P5W30O110](15-*n*)- Preyssler complexes, with X = Na+, Ca2+, Y3+, La3+, Ce3+, and Th4+. J Am Chem Soc. 2007;129: 12244–12253.
[37] Yan L, López X, Carbó JJ, Sniatynsky RT, Duncan DD, Poblet JM. On the origin of alternating bond distortions and the emergence of chirality in polyoxometalate anions. J Am Chem Soc 2008;130:8223–8233.
[38] de Graaf C, López X, Ramos JL, Poblet JM. Ab initio study of the antiferromagnetic coupling in the wheel-shaped [Cu20Cl(OH)24(H2O)12(P8W48O184)]25- anion. Phys Chem Chem Phys. 2010;12:2716–2721.
[39] El Moll H, Nohra B, Mialane P, Marrot J, Dupré N, Riflade B, Lanthanide polyoxocationic complexes: Experimental and theoretical stability studies and Lewis acid catalysis. Chem Eur J. 2011;17:14129–14138.
[40] Aparicio PA, López X, Poblet JM. Ability of DFT calculations to correctly describe redox potentials and electron (de)localization in polyoxometalates. J Mol Eng Mater 2014;2:1440004.
[41] Nishiki K, Umehara N, Kadota Y, López X, Poblet JM, Mezui CA, Preparation of α1- and α2-isomers of mono-Ru-substituted Dawson-type phosphotungstates with an aqua ligand and comparison of their redox potentials, catalytic activities, and thermal stabilities with Keggin-type derivatives. Dalton Trans. 2016;45:3715–3726.
[42] López X, Bo C, Poblet JM. Electronic properties of polyoxometalates: electron and proton affinity of mixed-addenda Keggin and Wells-Dawson anions. J Am Chem Soc 2002;124:12574–12582.
[43] Ortega F, Pope MT. Polyoxotungstate anions containing high-valent rhenium. 1. Keggin anion derivatives. Inorg Chem 1984;23:3292–3297.
[44] Alizadeh MH, Harmalker SP, Jeannin Y, Martin-Frère J, Pope MT. A heteropolyanion with fivefold molecular symmetry that contains a nonlabile encapsulated sodium ion. The structure and chemistry of $[NaP_5W_{30}O_{110}]^{14-}$. J Am Chem Soc 1985;107:2662–2669.
[45] Mbomekallé IM, López X, Poblet JM, Sécheresse F, Keita B, Nadjo L. Influence of the heteroatom size on the redox potentials of selected polyoxoanions. Inorg Chem. 2010;49:7001–7006.
[46] Clark CJ, Hall D. Dodecamolybdophosphoric acid *circa* 30-hydrate. Acta Crystallogr 1976; B32:1545–1547.
[47] Day VW, Klemperer WG. Metal oxide chemistry in solution: The early transition metal polyoxoanions. Science 1985;228:533–541.
[48] Müller A. Induced molecule self-organization [10]. Nature 1991;352:115.
[49] Pope MT. Anion guests in heteropolyanions? Nature 1992;355:27.

[50] Weinstock IA, Cowan JJ, Barbuzzi EMG, Zeng H, Hill CL. Equilibria between α and β isomers of Keggin heteropolytungstates. J Am Chem Soc 1999;121:4608–4617.
[51] Jansen SA, Singh DJ, Wang SH. Cluster fragmentation and orbital analysis: A theoretical study of cohesive interactions and electronic properties of polyoxometalates with the Keggin structure. Chem Mater 1994;6:146–155.
[52] Maestre JM, López X, Bo C, Casañ-Pastor N, Poblet JM. Electronic and magnetic properties of α-Keggin anions: a DFT study of $[XM_{12}O_{40}]^{n-}$, (M = W, Mo; X = Al^{III}, Si^{IV}, P^{V}, Fe^{III}, Co^{II}, Co^{III}) and $[SiM_{11}VO_{40}]^{m-}$ (M = Mo and W). J Am Chem Soc 2001;123:3749–3758.
[53] López X, Maestre JM, Bo C, Poblet JM. Electronic properties of polyoxometalates: a DFT study of α/β-$[XM_{12}O_{40}]^{n-}$ relative stability (M = W, Mo and X a main group element). J Am Chem Soc 2001;123:9571–9576.
[54] Lowest Unoccupied Molecular Orbital.
[55] Dolbecq A, Dumas E, Mayer CR, Mialane P. Hybrid organic-inorganic polyoxometalate compounds: from structural diversity to applications. Chem Rev 2010;110:6009–6048.
[56] Mayer CR, Fournier I, Thouvenot R. Bis- and tetrakis(organosilyl) decatungstosilicate, $[\gamma\text{-}SiW_{10}O_{36}(RSi)_2O]^{4-}$ and $[\gamma\text{-}SiW_{10}O_{36}(RSiO)_4]^{4-}$: synthesis and structural determination by multinuclear NMR spectroscopy and matrix-assisted laser desorption/ionization time-of-flight mass spectrometry. Chem Eur J 2000;6:105–110.
[57] Proust A, Matt B, Villanneau R, Guillemot G, Gouzerh P, Izzet G. Functionalization and post-functionalization: a step towards polyoxometalate-based materials. Chem Soc Rev 2012;41:7605–7622.
[58] Vilà N, Aparicio PA, Sécheresse F, Poblet JM, López X, Mbomekallé IM. Electrochemical behavior of α_1/α_2-$[Fe(H_2O)P_2W_{17}O_{61}]^{7-}$ isomers in solution: Experimental and DFT studies. Inorg Chem 2012;51:6129–6138.
[59] Acerete R, Harmalker S, Hammer CF, Pope MT, Baker LCW. Concerning isomerisms and interconversions of 2:18 and 2:17 heteropoly complexes and their derivatives. J Chem Soc, Chem Commun 1979;777–779.
[60] Ciabrini JP, Contant R, Fruchart JM. Heteropolyblues: relationship between metal-oxygen-metal bridges and reduction behaviour of octadeca(molybdotungsto)diphosphate anions. Polyhedron 1983;2:1229–1233.
[61] Harmalker SP, Leparulo MA, Pope MT. Mixed-valence chemistry of adjacent vanadium centers in heteropolytungstate anions. 1. Synthesis and electronic structures of mono-, di-, and trisubstituted derivatives of α-$[P_2W_{18}O_{62}]^{6-}$. J Am Chem Soc. 1983;105:4286–4292.
[62] Abbessi M, Contant R, Thouvenot R, Hervé G. Dawson type heteropolyanions. 1. Multinuclear (^{31}P, ^{51}V, ^{183}W) NMR structural investigations of octadeca(molybdotungstovanado)diphosphates α-1,2,3-$[P_2MM'_2W_{15}O_{62}]^{n-}$ (M, M′ = Mo, V, W): Syntheses of new related compounds. Inorg Chem 1991;30:1695–1702.
[63] Keita B, Mbomekalle IM, Nadjo L, De Oliveira P, Ranjbari A, Contant R. Vanadium-substituted Dawson-type polyoxometalates as versatile electrocatalysts. C R Chimie 2005;8:1057–1066.
[64] Contant R, Abbessi M, Canny J, Belhouari A, Keita B, Nadjo L. Iron-substituted dawson-type tungstodiphosphates: synthesis, characterization, and single or multiple initial electronation due to the substituent nature or position. Inorg Chem 1997;36:4961–4967.
[65] We only consider the electronic contribution to the free energy, $G = E$. The entropic term is assumed to be much smaller, not affecting the values discussed.
[66] Massart R, Contant R, Ciabrini JP, Fruchart JM, Fournier M. ^{31}P NMR studies on molybdic and tungstic heteropolyanions. Correlation between structure and chemical shift. Inorg Chem 1977;16:2916–2921.
[67] Contant R, Ciabrini JP. Stereospecific preparations of new *n*-molybdo-(18-*n*)-tungsto-2-phosphates and related “defect” compounds (n = 2, 4 or 5). J Inorg Nucl Chem 1981;43:1525–1528.

[68] Contant R, Abbessi M, Thouvenot R, Hervé G. Dawson type heteropolyanions. 3. Syntheses and ^{31}P, ^{51}V and ^{183}W NMR structural investigation of octadeca(molybdo-tungsto-vanado)diphosphates related to the $[H_2P_2W_{12}O_{48}]^{12-}$ anion. Inorg Chem 2004;43:3597–3604.
[69] López X, Fernández JA, Poblet JM. Redox properties of polyoxometalates: new insights on the anion charge effect. Dalton Trans 2006;1162–1167.
[70] Kozik M, Casañ-Pastor N, Hammer CF, Baker LCW. Ring currents in wholly inorganic heteropoly blue complexes. Evaluation by a modification of Evans's susceptibility method. J Am Chem Soc 1988;110:7697–7701.

V. Marturano, V. Ambrogi, N. A. G. Bandeira, B. Tylkowski,
M. Giamberini and P. Cerruti

5 Modeling of Azobenzene-Based Compounds

Abstract: Azobenzene is by far the most studied photochromic molecule and its applications range from optical storage to bio-engineering. To exploit the great potential of azobenzene, one must achieve deep understanding of its photochemistry as single molecule in solution AS WELL AS in-chain moiety and pendent group in macromolecular structures. With the advent of computer-aided simulation scientists have been able to match experimental data with computational models. In this chapter, a review on the modeling of azobenzene-containing molecules in different conditions and environments IS provided with a special focus on advanced applications of photo-controllable materials, such as molecular machines and photoactivation of bio-molecules.

Keywords: Azobenzene, photochemistry, modeling, molecular machines, photoswitch

5.1 Introduction

Azobenzene is an aromatic molecule where an azo linkage (–N=N–) joins two phenyl rings. In order to change geometry and electron donating/withdrawing character, different substituents can be bonded to the aromatic moieties, giving rise to a large class of azobenzene-derived compounds. Thanks to its rigid and anisotropic molecular core, azobenzene is an ideal liquid-crystal mesogen, either alone or in macromolecular compounds [1].

Among the numerous spectroscopic and photo-physical properties of azo-materials, their most important feature is their photochemistry. Independently from their substitutions, all azos are able to perform an efficient and reversible photoisomerization, which occurs upon absorption of a photon within the absorption band (Figure 5.1) [2].

Even though the discovery of azo-compounds dates back to the mid-XIX century, the first evidence of the trans-cis isomerization was reported by Hartley in 1937, after observing a lack of reproducibility in absorbance measurements when azobenzene was exposed to light [4]. Since then, azobenzene has been vastly studied and its

This article has previously been published in the journal *Physical Sciences Reviews*. Please cite as: Marturano, V., Ambrogi, V., Bandeira, N. A. G., Tylkowski, B., Giamberini, M., Cerruti, P. Modeling of Azobenzene-Based Compounds. *Physical Sciences Reviews* [Online] **2017**, *2* (11). DOI: 10.1515/psr-2017-0138

https://doi.org/10.1515/9783110482065-005

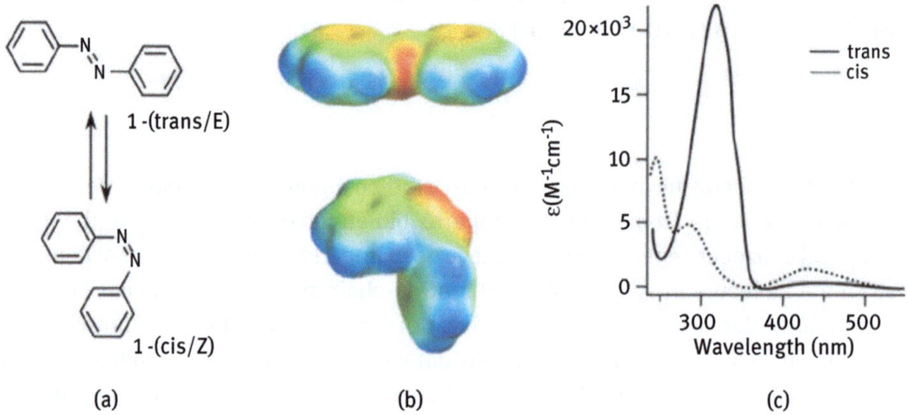

Figure 5.1: (a) Structures of trans and cis isomers of azobenzene. (b) Spacefilling models are colored by electrostatic potential (red—negative to blue—positive). (c) Electronic absorption spectra of the trans and cis isomers of azobenzene dissolved in ethanol [3].

trans-cis photo-isomerization has become by far the most studied photochemical process of all times [5, 6].

In particular, the π-conjugated system gives rise to a strong electronic absorption in the UV-visible range of the spectrum, where the exact spectrum can be tailored via ring-substitution [7]. The spectra of trans- and cis-azobenzenes are distinct but overlapping (see Figure 5.1(c)). Trans-azobenzene shows a weak n–π* band near 440 nm and a strong π–π* transition near 320 nm. Cis-azobenzene has a stronger n–π* band also near 440 nm and shorter wavelength bands at 280 nm and 250 nm. The trans conformation of azobenzene is 10–12 kcal mol^{-1} more stable than the cis isomer so that, in the dark at equilibrium, trans is the dominant isomer (>99.99 %) [8]. The geometry of the trans conformation is almost planar and has a dipole moment near zero. The cis isomer, produced by irradiation with 340 nm light, adopts a bent conformation with its phenyl rings twisted ~55° out of the plane from the azo group (Figure 5.1(b)) and has a dipole moment of 3 Debye. The trans isomer can be regenerated either by darkness – adapting the solution to re-establish the equilibrium or by irradiation at 450 nm. The photo-isomerization events occur with high quantum yields and minimal photo-bleaching. End-to-end distance for each isomer is also substantially different; the distance between the carbons at the para positions of the rings changes by ~3.5 Å from trans to cis isomers. The applicability of azobenzene-based molecules as photo-switches and molecular machines has been possible thanks to the above mentioned geometrical difference and thanks to the fast timescale (few picoseconds) of the isomerization mechanism [9].

The behavior of azobenzene trans-cis isomerization strongly varies on the environment in which the molecules are found; for single-molecule azos, the solvent or the presence of metals can influence the isomerization mechanism, while for macromolecular azo it is important to take into account the position of the azo group (in chain, side chain or cross-linker). To exploit azobenzene's full potential one must understand its environment-depending behavior and one way to accomplish deep insight is to match experimental data with modeling techniques. In this article, a review on the modeling of azobenzene-containing molecules in different conditions and environments will be provided.

5.2 Modeling the trans-cis isomerization of azobenzene and its functionalized derivatives

A great amount of research has been focused on the exact understanding of trans-cis isomerization of azobenzene in various conditions by means of various *ab initio* modeling techniques [10–13]. For example, Armstrong et al. [14] employed MOPAC (Molecular Orbital PACkage), a semiempirical quantum chemistry program based on Dewar and Thiel's NDDO approximation, to determine the normal modes and vibrational energies of azobenzene. One of their most outstanding results is the correct prediction of the increase in frequency in the azo stretch mode upon deuteration of the phenyl rings. The modes show vibrational coupling between the phenyl modes and the azo vibrations as previously hypothesized on the basis of experimental results. It is also interesting to note that Armstrong's calculation attempted to explain the reason why certain in-plane stretching frequencies give rise to relatively intense Raman and resonance Raman scattering. In Raman scattering, in fact, displacements along the N-N and C-N bonds give rise to the strongest scattering as well as the stretches along the azo bond provide the greatest intensity in the Raman scattering [15–17].

As mentioned above, the absorption spectrum of azobenzene shows two distinct bands: a strong π-π^* absorption band peaking at ca. 290 nm (cis-form) and 320 nm (trans-form) and a much weaker n-π^* band with a peak around 440 nm (cis- and trans-form). Spectroscopic measurements proved that the excitation of the trans-isomer in the n-π^* or in the π-π^* band leads to different quantum yields of ≈25% and 12%, respectively, suggesting two different mechanistic pathways for the isomerization. Upon reverse isomerization to cis-azobenzene, the trans-state is reached with a higher quantum yield of ≈50% [18]. Many studies on the photo-isomerization of azobenzene were performed after excitation of the lowest n-π^* transition of the molecule; however for the π-π^* higher energy transition several pathways have been proposed but a unite experimental-theoretical solution is still to be found. Studies on the π-π^* transition of trans-azobenzene in the gas phase after UV excitation with 330 nm were performed by Schultz et al. [19]. Matching the femtosecond time-resolved photoelectron spectroscopy with *ab initio* molecular dynamics they identified two near-degenerate excited states, S_2 ($\pi_{NN}\pi_{NN}^*$) and $S_{3,4}$ ($2 \times \pi_{Ph}\pi_{NN}^*$), and proposed a new model for the isomerization mechanism invoking electron excitation from the aromatic rings. Following this work Satzger et al. [20] also addressed the photo-isomerization of azobenzene after excitation of the π-π* transition by femtosecond time-resolved absorption experiments for both trans-cis and cis-trans transitions and compared experimental data with the different models proposed elsewhere [11, 21]. The authors compared transient absorption data of trans-azobenzene and of cis-azobenzene after π-π* excitation with n-π* excitation data. The experimental results indicate that in both cases (trans-cis) and (cis-trans) the isomerization reaction can be described by a two-step process, where after a short relaxation on the S_2 potential energy surface (decay time 0.13 ps) the main isomerization motion takes place on the S_1 potential energy surface. The difference as compared to n-π* excitation is that the isomerization reaction starts on a different point on the S_1 potential energy surface (outside the Franck-Condon region). This causes different relative intensities of the decay associated spectra. Nevertheless, the spectral shapes and decay times of the spectral components are similar to the ones obtained after n-π* excitation.

It is clear so far that the modeling of azobenzene photo-isomerization, its excited-state dynamics and nonadiabatic crossings between different electronic states has been a demanding subject of study for physicists and chemists even in the gas state [22–25]. Interestingly, one of the most challenging unexplored field is the understanding the molecular mechanism of the trans–cis conversion in a crowded environment, i.e. in a condensed or liquid phase. The incredible complexity and great importance of this topic were addressed by Tiberio et al. [26]. Even though almost all applications featuring an azo photo-isomerization take place in solution or solid-state (e.g. polymers), the clear majority of the theoretical information present in literature involves the isomerization in gas phase, where the conformational change is not hindered by the environment and where we can expect that the mechanism can be different. The authors underlined how the main challenge in the study involving solvent environment effects on the trans–cis isomerization can be attributed to the need for adequate modeling parameters to take into account both basic photochemistry with atomistic approach to the host–guest system. The approach adopted by Tiberio et al. consists in the study of n-π* [27] azobenzene transitions from the ground to the excited state and back during the time evolution in various low-molar-mass organic solvents (n-hexane, toluene, ethanol, anisole, ethylene glycol) by means of nonequilibrium molecular dynamics (MD) simulations. The simulations confirmed the results obtained by Ciminelli et al. [28] indicating that the dominant isomerization mechanism in solution is a mixed torsional–inversion mechanism, while in vacuum the isomerization follows predominantly a torsional one. Indeed, the pure inversion seems to occur only after the decay into the S_0 state, while in the excited state this pathway appears unlikely, as a high-energy barrier must be overcome. The analysis on azobenzene solvent-depending behavior showed how higher solvent viscosity seemed to increase the pure inversion contribution. Moreover, increasing viscosity clearly affects the rotational dynamics of AB, notably resulting in faster dynamics for the trans-isomer.

Azobenzene back isomerization, characterized by a relatively low activation energy (about 95 kJ/mol, i.e. less than 1 eV/molecule [29]), typically takes several hours or days can be accelerated by the presence of gold nanoparticles down to several minutes [30–33].

Titov et al. [34] attempted to clarify, via experimental and theoretical investigation, the mentioned catalytic effect of gold nanoparticles on thermal cis-trans isomerization of azobenzene molecules (Figure 5.2). One of the explanations for the catalytic activity of gold nanoparticles is related to electron transfer between the adsorbed molecule and the gold particle surface [35, 36]. Titov et al. performed density functional theory calculations of the activation energy barriers in order to investigate the effect on the isomerization rate of electron attachment and detachment from various azobenzene-containing molecules in aqueous solution. The results interestingly show that activation barriers are greatly lowered for azobenzene-containing molecules. Moreover, applying Eyring's transition state theory, they calculated absolute rates for thermal cis-trans isomerization to be about 1000 times larger for azobenzene-containing molecules upon electron attachment and withdrawal. Such large numbers are not in quantitative agreement with the obtained experimental values; however, that was also not expected because of differences of the experimental conditions and the idealized gas-phase type theoretical models adopted here.

As vastly reported in the literature photochemical and thermal isomerizations of azobenzene are strongly affected not only by intermolecular interactions but also by intramolecular ones. For example, bulky substituents at the

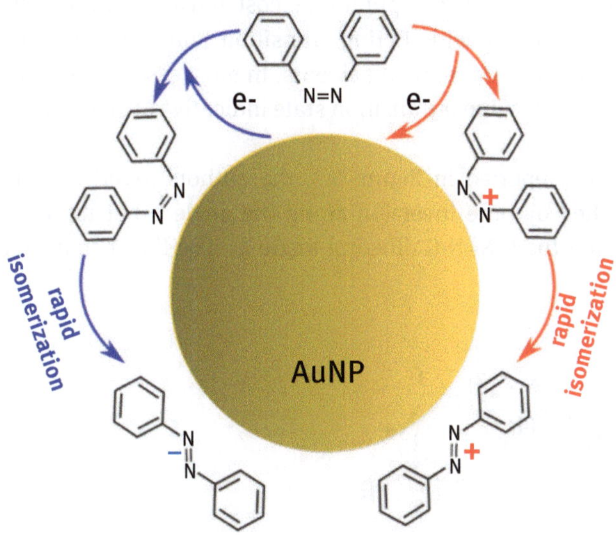

Figure 5.2: Gold nanoparticle catalysis of thermal cis-trans isomerization in azobenzene-containing molecules upon electron attachment and withdrawn.

ortho-positions alter the photochemical isomerization [37, 38]. The effect of ring strain in macrocyclic azobenzene dimers has been studied previously in literature [39, 40]. Interestingly, Norikane et al. [41] were able to give a fresh insight on previously unraveled relationship between structure and photochemical/thermal isomerization of azobenzene moieties in azobenzene macrocyclic dimers with comparison to noncyclic azobenzene dimers. In both cases, upon irradiation with UV light, each (trans/trans) isomer exhibited a stepwise photo-isomerization to give the corresponding (trans/cis) isomer, followed by isomerization to yield the (cis/cis) isomer. Considering the quantum yields for the isomerization, it is clear that the photochemical isomerization of macrocyclic azo was altered by its cyclic structure. Ring strain, in fact, could cause shifts in the absorption spectra of the azobenzene chromophore and perturbations in molecular motion in the excited state. Ab initio calculations, performed using Gaussian 98 program package with HF/6-31G** basis, were employed to estimate the ring strain of each isomer in their ground electronic states. In accordance with experimental data, it was found that the ring strain in the macrocyclic azobenzene dimer increases the activation energy and activation entropy for the thermal cis-trans isomerization of the azobenzene moiety. This knowledge of the correlation between structure and reactivity of azobenzene gives significant information for controlling the isomerization to establish photo-responsive functional materials that utilize the isomerization of azobenzene.

One of the most extensive theoretical study on azo-derivatives cis-trans thermal isomerization and the effect of substituents and solvents has been reported by Dokic et al. [42]. About 90 azobenzene-derived molecules have been studied in that work. For all of them, cis and trans isomers have been calculated and transition states have been investigated. Balancing computational cost and accuracy, B3LYP/6-31G* model was chosen, together with Eyring transition state theory and a polarizable dielectric continuum model to treat solvents. In most cases the isomerization reaction proceeds through a linear transition state indicative of an inversion mechanism.

Following the schematics reported in Figure 5.3, the authors found that the transition state is not reached by pure inversion along the angle α but rather by simultaneous rotation around the C-N=N-C dihedral angle ω. The linear transition

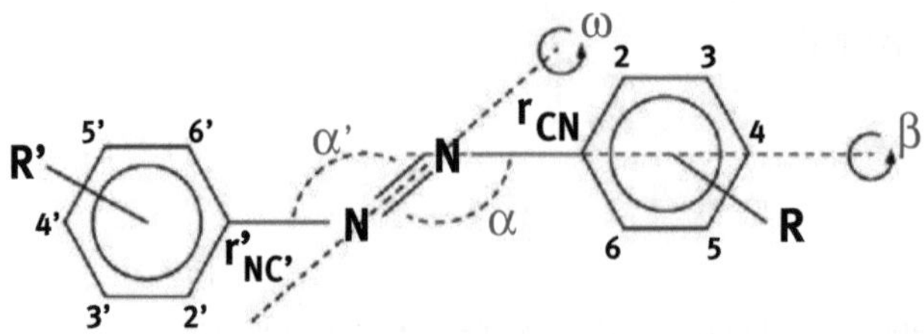

Figure 5.3: Schematics for the labeling procedure of the atoms in azobenzene derivatives.

state can be stabilized very efficiently with acceptors in ortho and para positions. In the case of double and triple substitution, substituents on one ring have an additive effect on activation energies and kinetics, while for substitution on both rings no simple additivity rule is found. For push–pull azobenzenes the reaction mechanism depends on the solvent, changing from inversion to rotation in polar environments. On a semiquantitative level, the environment increases the rates according to a Kirkwood scenario in which the molecules are treated as dipoles in a polarizable continuum. Bulky azobenzenes, which were also investigated experimentally by Dokic [42], are obtained by adding tert-butyl groups. It is found that generally these bulky compounds behave similar to conventional azobenzenes. In this case the kinetics obtained by theory are in good agreement with experimental data, at least as far as activation energies are concerned. Also trends observed by systematic variation of substituents and the polarity of the solvent are nicely reproduced and can be rationalized.

5.3 Modeling molecules with the azobenzene unit in their constitutive structure

As discussed above, azobenzene molecules are almost never found on their own, but most applications revolve around azobenzene-based molecules or polymers [1, 43–46]. Understanding the behavior of the azo moieties in complex structures is of the key importance for obtaining high-quality performance of the final application. This section reports some of the main studies on molecular modeling and theoretical studies of azobenzene-based materials, including polymers, inorganic materials and non-covalent interactions.

Photo-irradiation of polymers containing azobenzene either in the backbone or as a pendant group can cause reversible changes in physical properties such as surface wettability, viscosity and solubility [47, 48]. These types of specialty polymers containing photochromic groups which induce reversible property changes are termed photo-responsive polymers [49].

Siewierski et al. [50] reported on the preparation of azobenzene-based monolayers on silicon. Exposure to 354 nm light slightly decreased the water contact angles of the azobenzene monolayers. In contrast to modeling studies which predicted ≈3 Å changes in the position of terminal groups upon trans-to-cis isomerization and a consequent decrease in film thickness [51], an unexplained experimental observation is the increase in film thickness (as measured by X-ray reflectivity) upon photoirradiation. Zhang et al. [52] described a new series of hydrogen-bonding species in a novel hydrazide-modified p-metoxyazobenzene derivatives, called D3, D6 and D10, synthesized according to a procedure reported elsewhere [53]. The hydrazide moiety can be viewed as a hybrid structure, composed of a primary amide function and a terminal amino group, which has mutual electronic and steric influence on the hydrogen-bonding formation between hydrazide groups. The exhibited thermal behavior of these compounds may also relate to the π–π stacking of p-methoxy azobenzene or phenyl groups. The intermolecular hydrogen bonding and molecular alignment patterns were evidenced from experimental data obtained via both temperature-dependent infrared (IR) spectroscopy and *ab initio* computational models [54]. Temperature-dependent IR spectra effectively illustrate the reversible re-organization of the molecules, and 2D correlation IR spectral analysis for the amide and amino groups reveals two different hydrogen-bonding patterns: parallel arrangement with ordered alignment in D10 and D6, and less-ordered anti-parallel alignment in D3. Length and odd/even number of the alkyl spacer in azo-containing macromolecules play an important role in the molecular alignments and the hydrogen-bonding formations, as well as in the liquid crystal properties of Dn.

Molecular dynamics (MD) simulation has been employed by Heinz et al. [55] to examine the molecular-level orientation and reorganization of model azobenzene derivatives in the interlayer space of layered silicates upon laser excitation. The modeling method was compared to experimental data, such as X-ray diffraction and UV/vis absorption measurements. The synthesis and characterization of hybrid structures of this kind have long been reported in literature, always featuring the great limit posed by the evaluation and control on the location and the orientation of the intercalated azo dyes [56, 57] as shown in Figure 5.4.

MD results show uniform reversible changes in basal plane spacing of montmorillonite up to 2.8 Å (14 %) upon trans-cis isomerization of attached azo-containing ionic surfactants. Experimentally, Okada et al. have shown nonuniform reversible optical switching of the gallery height for semiflexible surfactants up to 10 Å (41 %) in

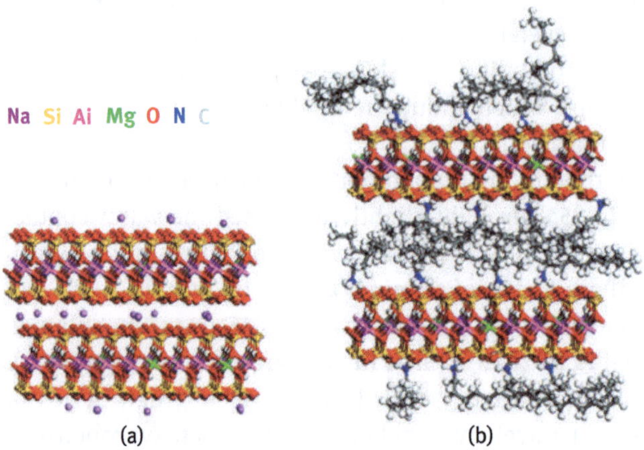

Figure 5.4: Molecular structure of (a) sodium montmorillonite and (b) n-dodecylammonium montmorillonite. The gallery spacing of the layered structure depends on the grafting density and on the molecular volume of the cationic surfactant.

the presence of phenol and uniform reversible optical switching of 0.9 Å (4 %) without co-intercalates [56]. The absence of changes in gallery spacing for azobenzene derivatives with attached flexible hydrocarbon chains at low packing density without co-intercalates are well explained by simulation.

Barrett et al. [58] presented their study on azo-based block copolymers. In their work, a pump/relax procedure was applied to analyze the thermal cis-trans isomerization behavior of the azo groups. All samples show a fast isomerization process due to strained cis isomers, trapped below the glass transition temperature of the film, followed by a slower isomerization process. More recently, Toshchevikov et al. [59] have proposed a microscopic theory of light-induced deformation in amorphous side-chain azobenzene polymers taking the internal structure of azobenzene macromolecules explicitly into account. When the system is irradiated with polarized laser light, a mechanical stress, caused by reorientation of macromolecules due to interactions of chromophores in side chains with the light, is applied to the system. All experimental results were in optimal agreement with quantum chemical calculations obtained using density functional theory [60]. Depending on the architectures of oligomers, a sample can be either stretched or uniaxially compressed along the polarization direction of the laser light. The light-induced stress can be larger than the yield stress at characteristic light intensities used in experiments. This result explains the possibility of irreversible sample deformation under homogeneous illumination and, hence, the possibility of the inscription of surface relief gratings. The authors showed that the irreversible elongation of a sample decreases with increasing temperature, with critical temperature (corresponding to the point when the irreversible elongation disappears) independent of the glass-transition temperature, T_g. These results demonstrate the great potential of the proposed microscopic orientation approach for describing the photoelastic properties of different azobenzene polymers [61].

Polyimides containing azobenzene side groups have been synthesized by polycondensation reaction by Sava et al. [62]. The authors present photochromic studies proving that the azobenzene groups are able to isomerize even when incorporated in a rigid chain system. However, the cis-isomer maximum conversion degree reported in this work is lower than the flexible chain polymers. For example, in solutions the maximum cis-isomer content can reach 70%, in solid state only maximum values of 39% can be obtained. Taking into consideration the ratio between the trans–cis isomerization and the cis–trans relaxation processes, this class of materials is recommended both for nano-manipulation and surface relief grating applications also because the polymers present a good thermostability and the degradation process starting for all the products above 300°C and have a glass transition temperature in the range of 185–230°C.

Azobenzene can also be found in a bridged form or used as cross-linker, and relative computer models can be found in literature. For example, Jiang et al. [63] reported their study on density-functional-based molecular dynamics simulations applied to isomerization of azobenzene and bridged azobenzene induced by n-π^*

excitation. The quantum yield was found to be about 25% for azobenzene, in agreement with experiment [64], and about 45% for bridged azobenzene (B-Azo), which is again consistent with experiment [65]. In those simulations that did produce isomerization, the range of calculated S_1 lifetimes was 330–480 fs for azobenzene and 22–32 fs for bridged azobenzene. In the simulation shown in detail, the lifetime of the S_1 excited state in trans-azobenzene was found to be about 400 fs, with the whole trans–cis isomerization process completed in approximately 500 fs. The S_1 lifetime in trans-B-Azo was found to be only about 30 fs, an order of magnitude shorter, just as in the earlier work by Böckmann et al. [66]. However, the full trans–cis isomerization for B-Azo requires about 430 fs, which is comparable to the timescale for azobenzene. The reason for this is that the twisting of the phenyl rings in B-Azo around the CN bonds is hindered by the presence of the bridge structure.

5.4 Achieving photo-controllable materials

This section is dedicated to the application of molecular simulation in azobenzene-based materials and technology, from biomolecules control, to molecular machines and other applications.

5.4.1 Biomolecules

A great amount of interest has been dedicated to the study of optical control of peptide/protein conformation which could be a powerful biochemical tool for tuning protein properties in diverse systems [67, 68]. Powerful chromophores, such as azobenzene, can theoretically be used for reversible conformational control of proteins [69, 70].

Kumita et al. [71] reported on the reversible means of controlling helix stability that involved the incorporation of a photo-isomerizable azobenzene cross-linking reagent into an engineered peptide system obtaining a promising result in need of improvement. The same group published an upgraded version of cross-linked peptides, designed applying a molecular modeling strategy, that significantly extend the versatility of this approach to the photocontrol of protein structure [72].

As shown in Figure 5.5, the cis form of the cross-linker is predicted to be compatible with an α-helical conformation of the peptide, whereas the trans form of the cross-linker is too extended for such a spacing and its related to the unfold peptide in solution. The importance of this work relies on the fact that the activities of proteins with key helical domains involved in function, such as leucine zippers [73] and helix-loop-helix transcription factors [74], and certain proteins involved in signal transduction [75], might be reversibly photo-controlled using azo-based reagents. Such photo-controlled proteins could be useful for probing the importance of timing in biochemical networks.

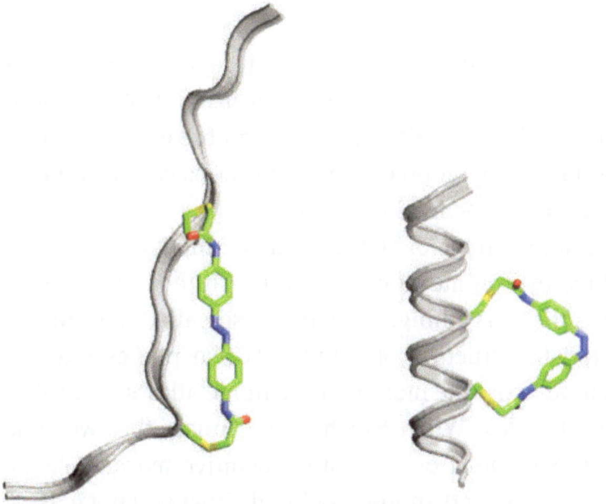

Figure 5.5: Energy-optimized molecular models of peptides attached to the azobenzene cross-linker in its trans (left) and cis (right) form as reported by Flint et al. (2002).

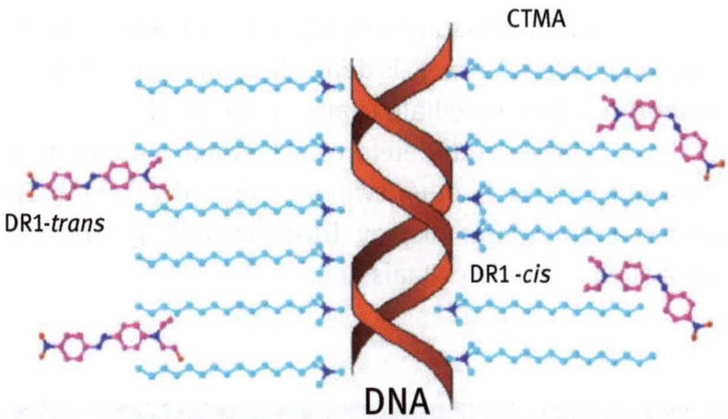

Figure 5.6: Illustration of the semi-intercalation of DR1 in a DNA–CTMA biopolymer showing the origin of memory effect upon light excitation.

Mitus et al. [76] formulated the concept of dye semi-intercalation employing Monte Carlo simulation and backing up experimental data by Sou et al. [77]. Featuring a complementary mechanism between full intercalation [78] and guest–host behavior [79], the semi-intercalation theory describes the scenario where the dye molecule is partially immobilized by the surfactant chains imposing steric hindrance to its photo-isomerization, as depicted in Figure 5.6 for DNA–cetyltrimethylammonium (CTMA) complexes.

The use of azobenzene polymers in the photo-control of biological systems is a vast and very significant topic, properly reviewed by Goulet-Hanssens et al. [80].

5.4.2 Molecular machines

Molecular machines are nanometer-sized factories and self-repairing structures where complex and responsive processes operate under exquisite control and translational and rotational movement is directed with precision and fueled by chemical and light energy [81]. Molecular modeling on precision applications can be a breakthrough method to obtain fine tuning of the final properties [82].

One of the most interesting applications of computer-aided modeling to molecular machines has been provided by Kohlmeyer et al. [83]. The authors described a new class of molecular machines that has a hingelike motion, so-called light-driven molecular hinges. The characteristic structure of a hinge is two planes that are connected at their edges through an axis; the motion of the hinge allows a transformation between open and closed states. When two hinges connect the two rigid planes, efficient motion of the hinges must occur in a cooperative manner; i.e., a configuration in which one hinge is open and another is closed either will not exist or will be very unstable. The molecular motion in above described system is induced by photochemical trans-cis isomerization of azobenzene units; cooperative effects and ring strain restrict the molecular motion [84]. The motion (close and open) (see Figure 5.7) can be operated by alternate irradiation with UV and visible light. The trans/trans and cis/cis isomers are thermally stable at 40°C, and the photochemical closure reaction (from trans/trans to cis/cis i\somer) is dependent on the intensity of the light used because of the short-lived intermediate (trans/cis isomer).

The change in molecular shape is large, and therefore such systems might be of use in various fields, e.g., as storage devices or other switching devices. The highly-regulated motion of the hinge has a potential to display drastic switching behavior, especially for anisotropic media such as liquid crystals.

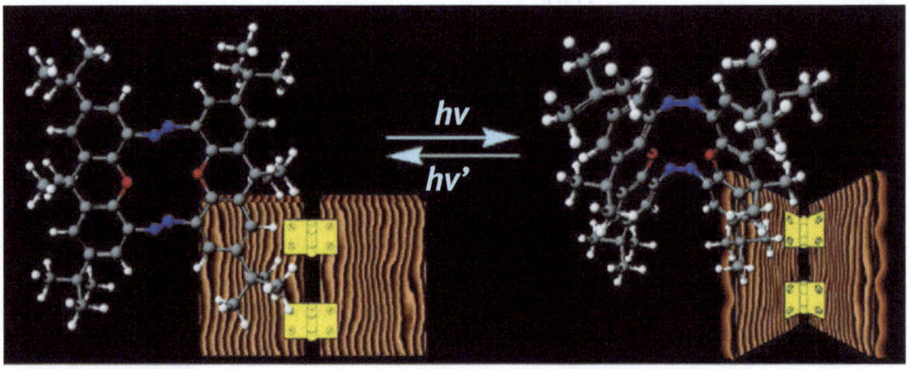

Figure 5.7: Molecular structures of trans-trans (left) and cis-cis (right) isomers of a light-driven azobenzene-based molecular hinge, optimized by AM1 calculations (top), and graphical representations of these two isomers (bottom).

5.5 Conclusions

This article provides an insight on the modeling techniques applied to azobenzene molecules in gas or condensed state and to azobenzene moieties in larger molecules.

The mechanism for cis-trans isomerization is a complex subject that still garners debate in the literature. The most recent accounts using dynamic (time dependent) simulations favor a mixture of the twisting and rotation reaction coordinates, two pathways which were previously assumed to be independent. The interconversion process is also influenced by the solvent medium and steric effects which might preclude one particular switching mode.

Despite the development of numerous model chemistries to simulate the mechanism associated with light-induced isomerization, it is of key importance to find innovative approaches able to precisely explain and support all the experimental observations. Understanding the urgency for new statistical approaches that correctly reproduce light-driven phenomena in azobenzene-containing molecules could represent a breakthrough step for the improvement of all new azobenzene-based commodity and specialty applications.

References

[1] In: Zhao Y, Ikeda T, editor(s). Smart light-responsive materials: azobenzene-containing polymers and liquid crystals. Hoboken, NJ: John Wiley & Sons, 2009

[2] Forber CL, Kelusky EC, Bunce NJ, Zerner MC. Electronic spectra of cis-and trans-azobenzenes: consequences of ortho substitution. J Am Chem Soc. 1985;107(21):5884–90.

[3] Beharry AA, Woolley GA. Azobenzene photoswitches for biomolecules. Chem Soc Rev. 2011; 40(8):4422–37.

[4] Hartley GS. The cis-form of azobenzene. Nature. 1937;140:281.

[5] Griffiths J. II. Photochemistry of azobenzene and its derivatives. Chem Soc Rev. 1972;1(4):481–93.

[6] Kumar GS, Neckers DC. Photochemistry of azobenzene-containing polymers. Chem Rev. 1989;89(8):1915–25.

[7] Barrett CJ, Mamiya JI, Yager KG, Ikeda T. Photo-mechanical effects in azobenzene-containing soft materials. Soft Matter. 2007;3(10):1249–61.

[8] Dias AR, Da Piedade MM, Simoes JM, Simoni JA, Teixeira C, Diogo HP, et al. Enthalpies of formation of cis-azobenzene and trans-azobenzene. J Chem Thermodyn. 1992;24(4):439–47.

[9] Knoll H. Handbook of organic photochemistry and photobiology (ed. Horspool W. M., Lenci F.). Boca Raton, Florida: CRC Press. Chap. 89 2004

[10] Rau H. Spectroscopic properties of organic azo compounds. Angewandte Chemie International Edition English. 1973;12(3):224–35.

[11] Monti S, Orlandi G, Palmieri P. Features of the photochemically active state surfaces of azobenzene. Chem Phys. 1982;71(1):87–99.

[12] Cimiraglia R, Hofmann HJ. Rotation and inversion states in thermal E/Z isomerization of aromatic azo compounds. Chem Phys Lett. 1994;217(4):430–35.

[13] Biswas N, Umapathy S. Density functional calculations of structures, vibrational frequencies, and normal modes of trans-and cis-azobenzene. J Phys Chem. 1997;101(30):5555–66.

[14] Armstrong DR, Clarkson J, Smith WE. Vibrational analysis of trans-azobenzene. J Phys Chem. 1995;99(51):17825–31.

[15] Lin-Vien D, Colthup NB, Fateley WG, Grasselli JG. The handbook of infrared and Raman characteristic frequencies of organic molecules. San Diego, California: Elsevier, 1991
[16] Biancalana A, Campani E, Di Domenico G, Gorini G, Iacoponi A, Masetti G. Direct observation of a 220 cm− 1 structure in the lowest π–π* absorption band in the vapour phase of trans-azobenzene. Spectrochimica Acta A: Mol Biomol Spectrosc. 1999;55(14):2883–87.
[17] Weaver MJ, Corrigan DS, Gao P, Gosztola D, Leung LW. Some applications of surface Raman and infrared spectroscopies to mechanistic electrochemistry involving adsorbed species. J Electron Spectros Relat Phenomena. 1987;45:291–302.
[18] In: Dürr H, Bouas-Laurent H, editor(s). Photochromism: molecules and systems. Amsterdam, The Netherlands: Elsevier, 2003
[19] Schultz T, Quenneville J, Levine B, Toniolo A, Martínez TJ, Lochbrunner S, et al. Mechanism and dynamics of azobenzene photoisomerization. J Am Chem Soc. 2003;125(27):8098–99.
[20] Satzger H, Root C, Braun M. Excited-state dynamics of trans-and cis-azobenzene after UV excitation in the ππ* band. J Phys Chem. 2004;108(30):6265–71.
[21] Lednev IK, Ye TQ, Hester RE, Moore JN. Femtosecond time-resolved UV– visible absorption spectroscopy of trans-azobenzene in solution. J Phys Chem. 1996;100(32):13338–41.
[22] Bernardi F, Olivucci M, Robb MA. Potential energy surface crossings in organic photochemistry. Chem Soc Rev. 1996;25(5):321–28.
[23] Yarkony DR. Diabolical conical intersections. Rev Mod Phys. 1996;68(4):985.
[24] Jasper AW, Zhu C, Nangia S, Truhlar DG. Introductory lecture: nonadiabatic effects in chemical dynamics. Faraday Discuss. 2004;127:1–22.
[25] Garavelli M. Computational organic photochemistry: strategy, achievements and perspectives. Theoretical Chemistry Accounts: Theory, Computation, and Modeling (Theoretica Chimica Acta). 2006;116(1):87–105.
[26] Tiberio G, Muccioli L, Berardi R, Zannoni C. How does the trans–cis photoisomerization of azobenzene take place in organic solvents? Chem Phys Chem. 2010;11(5):1018–28.
[27] Chang CW, Lu YC, Wang TT, Diau EWG. Photoisomerization dynamics of azobenzene in solution with S1 excitation: a femtosecond fluorescence anisotropy study. J Am Chem Soc. 2004;126 (32):10109–18.
[28] Ciminelli C, Granucci G, Persico M. The photoisomerization mechanism of azobenzene: A semiclassical simulation of nonadiabatic dynamics. Chemistry–A Eur J. 2004;10(9):2327–41.
[29] Bandara HD, Burdette SC. Photoisomerization in different classes of azobenzene. Chem Soc Rev. 2012;41(5):1809–25.
[30] Yoon JH, Yoon S. Photoisomerization of azobenzene derivatives confined in gold nanoparticle aggregates. Phys Chem Chem Phys. 2011;13(28):12900–05.
[31] Nachtigall O, Kördel C, Urner LH, Haag R. Photoresponsive switches at surfaces based on supramolecular functionalization with azobenzene–oligoglycerol conjugates. Angewandte Chemie International Edition. 2014;53(36):9669–73.
[32] Hallett-Tapley GL, D'Alfonso C, Pacioni NL, McTiernan CD, González-Béjar M, Lanzalunga O, et al. Gold nanoparticle catalysis of the cis–trans isomerization of azobenzene. Chem Commun. 2013;49(86):10073–75.
[33] Simoncelli S, Aramendía PF. Mechanistic insight into the Z–E isomerization catalysis of azobenzenes mediated by bare and core–shell gold nanoparticles. Catalysis Sci Technol. 2015;5 (4):2110–16.
[34] Titov E, Lysyakova L, Lomadze N, Kabashin AV, Saalfrank P, Santer S. Thermal Cis-to-Trans Isomerization of Azobenzene-Containing Molecules Enhanced by Gold Nanoparticles: An Experimental and Theoretical Study. J Phys Chem C. 2015;119(30):17369–77.
[35] Ricci D, Bongiorno A, Pacchioni G, Landman U. Bonding trends and dimensionality crossover of gold nanoclusters on metal-supported MgO thin films. Phys Rev Lett. 2006;97(3):036106.

[36] Van Bokhoven JA, Louis C, Miller JT, Tromp M, Safonova OV, Glatzel P. Activation of oxygen on gold/alumina catalysts: in situ high-energy-resolution fluorescence and time-resolved X-ray spectroscopy. Angewandte Chemie. 2006;118(28):4767–70.
[37] Nishimura N, Sueyoshi T, Yamanaka H, Imai E, Yamamoto S, Hasegawa S. Thermal cis-to-trans isomerization of substituted azobenzenes II. Substituent and solvent effects. Bull Chem Soc Jpn. 1976;49(5):1381–87.
[38] Samanta S, McCormick TM, Schmidt SK, Seferos DS, Woolley GA. Robust visible light photoswitching with ortho-thiol substituted azobenzenes. Chem Commun. 2013;49(87):10314–16.
[39] Lu YC, Diau EWG, Rau H. Femtosecond fluorescence dynamics of rotation-restricted azobenzenophanes: new evidence on the mechanism of trans→ cis photoisomerization of azobenzene. J Phys Chem. 2005;109(10):2090–99.
[40] Luboch E, Bilewicz R, Kowalczyk M, Wagner-Wysiecka E, Biernat JF. Azo macrocyclic compounds. Adv Supramolecular Chem. 2003;9:73–163.
[41] Norikane Y, Tamaoki N. Photochemical and thermal cis/trans isomerization of cyclic and noncyclic azobenzene dimers: effect of a cyclic structure on isomerization. European J Org Chem. 2006;2006(5):1296–302.
[42] Dokic J, Gothe M, Wirth J, Peters MV, Schwarz J, Hecht S, et al. Quantum chemical investigation of thermal cis-to-trans isomerization of azobenzene derivatives: substituent effects, solvent effects, and comparison to experimental data. J Phys Chem. 2009;113(24):6763–73.
[43] Yesodha SK, Pillai CKS, Tsutsumi N. Stable polymeric materials for nonlinear optics: a review based on azobenzene systems. Prog Polym Sci. 2004;29(1):45–74.
[44] Nakano H, Takahashi T, Kadota T, Shirota Y. Formation of a surface relief grating using a novel azobenzene-based photochromic amorphous molecular material. Advanced Mater. 2002;14 (16):1157–60.
[45] Zhao YL, Stoddart JF. Azobenzene-based light-responsive hydrogel system. Langmuir. 2009;25 (15):8442–46.
[46] Juan ML, Plain J, Bachelot R, Royer P, Gray SK, Wiederrecht GP. Multiscale model for photoinduced molecular motion in azo polymers. ACS Nano. 2009;3(6):1573–79.
[47] Jiang W, Wang G, He Y, Wang X, An Y, Song Y, et al. Photo-switched wettability on an electrostatic self-assembly azobenzene monolayer. Chem Commun. 2005;28:3550–52.
[48] Delorme N, Bardeau JF, Bulou A, Poncin-Epaillard F. Azobenzene-containing monolayer with photoswitchable wettability. Langmuir. 2005;21(26):12278–82.
[49] Lee MJ, Jung DH, Han YK. Photo-responsive polymers and their applications to optical memory. Molecular Crystals Liquid Crystals. 2006;444(1):41–50.
[50] Siewierski LM, Brittain WJ, Petrash S, Foster MD. Photoresponsive monolayers containing in-chain azobenzene. Langmuir. 1996;12(24):5838–44.
[51] Xing L, Mattice WL. Atomistic simulations of self-assembled monolayers that contain azobenzene. Langmuir. 1996;12(12):3024–30.
[52] Zhang HH, Wu Y, Bai BL, Li M. Classification of the hydrogen-bonding species in a series of novel hydrazide based azobenzene derivatives investigated by two-dimensional correlation infrared spectroscopy and molecular modeling. Spectrochimica Acta A: Mol Biomol Spectrosc. 2006;63(1):117–25.
[53] Bai B, Wang H, Xin H, Zhang F, Long B, Zhang X, et al. Hydrazide-based organogels and liquid crystals with columnar order. New J Chem. 2007;31(3):401–08.
[54] Jonas U, Shah K, Norvez S, Charych DH. Reversible color switching and unusual solution polymerization of hydrazide-modified diacetylene lipids. J Am Chem Soc. 1999;121(19): 4580–88.
[55] Heinz H, Vaia RA, Koerner H, Farmer BL. Photoisomerization of azobenzene grafted to layered silicates: simulation and experimental challenges. Chem Mater. 2008;20(20):6444–56.

[56] Okada T, Watanabe Y, Ogawa M. Photoregulation of the intercalation behavior of phenol for azobenzene–clay intercalation compounds. J Mater Chem. 2005;15(9):987–92.
[57] Ogawa M, Ishii T, Miyamoto N, Kuroda K. Intercalation of a cationic azobenzene into montmorillonite. Appl Clay Sci. 2003;22(4):179–85.
[58] Barrett C, Natansohn A, Rochon P. Thermal cis-trans isomerization rates of azobenzenes bound in the side chain of some copolymers and blends. Macromolecules. 1994;27(17):4781–86.
[59] Toshchevikov V, Saphiannikova M, Heinrich G. Microscopic theory of light-induced deformation in amorphous side-chain azobenzene polymers. J Phys Chem B. 2009;113(15):5032–45.
[60] Parr RG, Weitao Y. Density-functional theory of atoms and molecules Vol. 16. Oxford, UK: Oxford University Press, 1994
[61] Gleim W, Finkelmann H. Thermoelastic and photoelastic properties of crosslinked liquid-crystalline side chain polymers. Macromol Chem Phys. 1987;188(6):1489–500.
[62] Sava I, Resmerita AM, Lisa G, Damian V, Hurduc N. Synthesis and photochromic behavior of new polyimides containing azobenzene side groups. Polymer. 2008;49(6):1475–82.
[63] Jiang CW, Xie RH, Li FL, Allen RE. Comparative studies of the trans– cis photoisomerizations of azobenzene and a bridged azobenzene. J Phys Chem. 2010;115(3):244–49.
[64] Bortolus P, Monti S. Cis-trans photoisomerization of azobenzene. Solvent and triplet donors effects. J Phys Chem. 1979;83(6):648–52.
[65] Siewertsen R, Neumann H, Buchheim-Stehn B, Herges R, NäTher C, Renth F, et al. Highly efficient reversible Z– E photoisomerization of a bridged azobenzene with visible light through resolved S1 (nπ*) absorption bands. J Am Chem Soc. 2009;131(43):15594–95.
[66] Böckmann M, Doltsinis NL, Marx D. Unraveling a chemically enhanced photoswitch: bridged azobenzene. Angewandte Chemie International Edition. 2010;49(19):3382–84.
[67] Adams SR, Tsien RY. Controlling cell chemistry with caged compounds. Annu Rev Physiol. 1993;55(1):755–84.
[68] Curley K, Lawrence DS. Light-activated proteins. Curr Opin Chem Biol. 1999;3(1):84–88.
[69] Liu D, Karanicolas J, Yu C, Zhang Z, Woolley GA. Site-specific incorporation of photoisomerizable azobenzene groups into ribonuclease S. Bioorg Med Chem Lett. 1997;7(20):2677–80.
[70] Willner I, Rubin S. Control of the structure and functions of biomaterials by light. Angewandte Chemie International Edition English. 1996;35(4):367–85.
[71] Kumita JR, Smart OS, Woolley GA. Photo-control of helix content in a short peptide. Proc Natl Acad Sci. 2000;97(8):3803–08.
[72] Flint DG, Kumita JR, Smart OS, Woolley GA. Using an azobenzene cross-linker to either increase or decrease peptide helix content upon trans-to-cis photoisomerization. Chem Biol. 2002;9 (3):391–97.
[73] Kumita JR, Flint DG, Woolley GA, Smart OS. Achieving photo-control of protein conformation and activity: producing a photo-controlled leucine zipper. Faraday Discuss. 2003;122:89–103.
[74] Löwik DW, Leunissen EHP, Van Den Heuvel M, Hansen MB, Van Hest JC. Stimulus responsive peptide based materials. Chem Soc Rev. 2010;39(9):3394–412.
[75] Drepper T, Krauss U, Zu Berstenhorst SM, Pietruszka J, Jaeger KE. Lights on and action! Controlling microbial gene expression by light. Appl Microbiol Biotechnol. 2011;90(1): 23–40.
[76] Pawlik G, Mitus AC, Mysliwiec J, Miniewicz A, Grote JG. Photochromic dye semi-intercalation into DNA-based polymeric matrix: Computer modeling and experiment. Chem Phys Lett. 2010;484(4):321–23.
[77] You H, Spaeth H, Linhard VNL, Steckl AJ. Role of surfactants in the interaction of dye molecules in natural DNA polymers. Langmuir. 2009;25(19):11698–702.
[78] Kawabe Y, Wang L, Horinouchi S, Ogata N. Amplified Spontaneous Emission from Fluorescent-Dye-Doped DNA–Surfactant Complex Films. Advanced Mater. 2000;12(17):1281–83.

[79] Dumont M, El Osman A. On spontaneous and photoinduced orientational mobility of dye molecules in polymers. Chem Phys. 1999;245(1):437–62.
[80] Goulet-Hanssens A, Barrett CJ. Photo-control of biological systems with azobenzene polymers. J Polymer Sci A: Polymer Chem. 2013;51(14):3058–70.
[81] Browne WR, Feringa BL. Making molecular machines work. Nat Nanotechnol. 2006;1(1):25–35.
[82] Duchstein P, Neiss C, Görling A, Zahn D. Molecular mechanics modeling of azobenzene-based photoswitches. J Mol Model. 2012;18(6):2479–82.
[83] Kohlmeyer RR, Chen J. Wavelength-Selective, IR Light-Driven Hinges Based on Liquid Crystalline Elastomer Composites. Angewandte Chemie International Edition. 2013;52 (35):9234–37.
[84] Norikane Y, Tamaoki N. Light-Driven Molecular Hinge: A New Molecular Machine Showing a Light-Intensity-Dependent Photoresponse that Utilizes the Trans– Cis Isomerization of Azobenzene. Org Lett. 2004;6(15):2595–98.

Iwona Gulaczyk

6 Gas-phase high-resolution molecular spectroscopy for LAV molecules

Abstract: High-resolution infrared (IR) spectroscopy is essential to the analysis of molecular rotation-vibration spectra. The high-resolution spectra deliver much information about structure and dynamic of molecules, but often they are very complex. For nonrigid molecules the complexity arises from transition tunneling splittings. Methylamine is a classic example of a nonrigid molecule in which two large amplitude motions, inversion and torsion, occur simultaneously. It has six equivalent potential minima, for which an effective vibration-inversion-torsion-rotation Hamiltonian has been developed. In the chapter assignment and analysis of several spectral regions of methylamine have been briefly presented explaining the assigning techniques and theoretical treatment of experimental lines.

Keywords: IR high resolution spectroscopy, large amplitude motions, effective Hamiltonian, methylamine

6.1 Introduction

Spectroscopy in general provides a wide range of information on the structure, potential energy surface and internal dynamics of molecules. The most accurate data is obtained from high-resolution rovibrational spectra. The main goal of rotational and vibration-rotational spectroscopy is spectrum interpretation, i. e., pointing out the states between which a transition occurs. In order to extract information from rovibrational spectra, a detailed and scrupulous assignment and analysis is needed. The difference between rotational and rovibrational spectroscopy is as follows. In rotational spectroscopy, one can observe transitions between rotational energy levels within the same vibrational level, whereas in vibration-rotation spectroscopy observed transitions are between piles of rotational energy levels belonging to two different vibrational levels. As vibrational transitions can be observed in the liquid or solid state of the sample, the rotational ones only in the gas phase. The assignment and analysis of such spectra are carried on using theoretical models describing molecular motions in the context of quantum mechanics. After the spectral

This article has previously been published in the journal *Physical Sciences Reviews*. Please cite as: Gulaczyk, I. Gas-phase high-resolution molecular spectroscopy for LAV molecules. *Physical Sciences Reviews* [Online] **2017**, *2* (11). DOI: 10.1515/psr-2017-0139

https://doi.org/10.1515/9783110482065-006

assignment has been done, one can determine the shape of multidimensional potential energy surface and molecule geometry. It is not straightforward though, even for rigid molecules whose potential energy surface consists of one deep minimum only. An equilibrium configuration is a configuration that corresponds to such a minimum. For molecules with several minima of similar in energy, it is a challenge to determine the molecular geometry. It is a case for non-rigid molecules performing large amplitude motions which take a molecule from one configuration to another.

Methylamine, which has been of spectroscopic interest for many years, is considered as a nonrigid (floppy) molecule with two large amplitude motions. Thus, in order to describe spectroscopy and dynamics of methylamine, its equilibrium structure is not enough, since the molecule undergoes two tunneling motions, the inversion (umbrella motion) of the amino group and internal rotation of the methyl group with respect to the rest of the molecule. The high-resolution spectra of such nonrigid molecules like methylamine are often very complicated because of transition tunneling splittings but at the same time very rich in information about their barrier heights or arrangement of their functional groups.

To be able to understand spectra of floppy molecules, it is essential to use group theory. The group theory is used among others to assign symmetry to energy levels, to develop selection rules for transitions or to calculate the overall shape of tunneling splitting. The size of tunneling splittings is dependent on the barrier height of a given large amplitude motion, which may be calculated from ab initio methods. Having the barrier heights calculated, one can predict the inversion–torsion–rotation levels and finally the spectrum of a molecule. There are two group theories used in high-resolution molecular spectroscopy, i. e. the point group theory and the permutation-inversion group theory. The point group symmetry is used for so-called rigid molecules like water or carbon dioxide, i. e., the molecules with one equilibrium configuration only. The point group theory is widely described in the spectroscopic text books [1–4]. As for the permutation-inversion (PI) group, it is used to describe a symmetry of nonrigid molecules like methanol, methylamine, hydrazine, 2-methylmalonaldehyde since such molecules have more than one equilibrium configuration and the point group symmetry cannot be applied any longer. In the PI groups the symmetry elements consist of the permutations of identical nuclei with or without inversion [4]. Once internal rotation or inversion is feasible in a molecule, the point group fails and the PI theory has to be applied. For instance, methylamine molecule, CH_3NH_2, belongs to the point group symmetry C_s (in its rigid equilibrium geometry) and to the G_{12} permutation-inversion symmetry group.

In this review, the studies of different high-resolution FT spectral regions of a methylamine molecule will be presented. The molecule performs two large amplitude motions which reflect strongly in its rovibrational spectra. The published results will be summarized to describe the current state of knowledge on the high-resolution IR spectra of methylamine.

6.2 Methylamine molecule

The methylamine molecule (CH_3NH_2) is a seven-atom organic molecule and the simplest primary alkylamine in chemistry. Its CH_3 group is connected to NH_2 group by C-N bond. Methylamine performs two large amplitude motions, the CH_3-torsional motion (internal rotation) and the NH_2-wagging motion (inversion), which makes the CH_3NH_2 a prototype for complex non-rigid molecules. A schematic view of the methylamine molecule is presented in Figure 6.1. The coupling between internal rotation of methyl group and inversion of two hydrogen atoms in the amine group gives rise to the tunneling splitting. Since the barriers in the potential functions are of intermediate height, the observed splittings are relatively large even in the ground state and the assignment and fitting of the spectrum is a very tedious task.

Both large amplitude motions are hindered by potential barriers of 23.2 kJ/mol for the wagging, and 8.6 kJ/mol for the torsion [5]. Figure 6.2 presents six equivalent minima in the potential energy surface of methylamine. The minima are separated by inversion and torsional barriers. The barriers to inversion and to torsion are low enough to allow a tunneling between minima, which leads to splittings of energy levels. In each such minimum, a rotation-inversion-torsion function is localized.

Methylamine as a non-rigid molecule has a very high symmetry and is described by permutation-inversion symmetry group G_{12} which is isomorphic to the C_{6v} point group [7]. Each energy level is labeled by the value of the usual quantum numbers J and $K = K_a$ and by a torsion-wagging-rotation symmetry species Γ corresponding to an irreducible representation of the G_{12} group. The group G_{12} consists of six components: A_1, A_2, B_1, B_2, E_1, E_2. Each K = 0 rovibrational transition splits into four (A, B, E_1, E_2) lines. For $0 < K \leq 4$ the lines of A and B symmetry are further split due to asymmetry (A_1, A_2, B_1, B_2). The symmetry labels of doubly degenerate levels E_1 and E_2 have additional ±1 labels to distinguish between two levels with the same J and K_a values, so finally for $K \neq 0$ there are A_1, A_2, B_1, B_2, E_{1+1}, E_{1-1}, E_{2+1} and E_{2-1} symmetry species with respective intensities, 1:1:3:3:3:3:1:1. For the K = 0 levels, the +1 and −1 labels

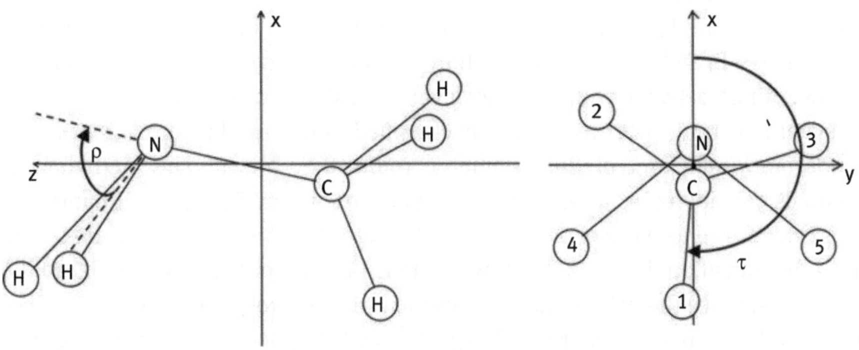

Figure 6.1: A schematic view of the methylamine molecule with inversion ρ and torsion τ coordinates.

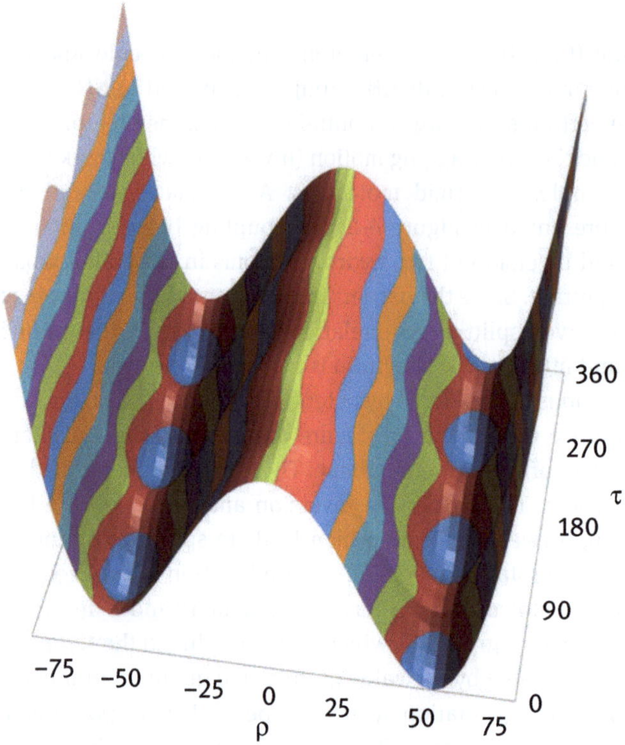

Figure 6.2: Potential energy surface of methylamine where ρ and τ are inversion and torsional angles, respectively, given in [deg] (taken from paper [6]).

have no meaning since there is only one E_1 and E_2 level. Therefore, all K = 0 E levels are designated as +1 levels. When asymmetry splitting disappears for K > 5, the (A_1, A_2) and (B_1, B_2) components overlap and their intensities are doubled. The torsion barrier tunneling causes a splitting between states of A or B symmetry and E_1 or E_2 symmetry. The tunneling through the inversion barrier splits states of A and B symmetry or E_1 and E_2. Thus, the following transitions are only allowed: $A_1 \leftrightarrow A_2$, $B_1 \leftrightarrow B_2$, $E_1 \leftrightarrow E_1$ and $E_2 \leftrightarrow E_2$ [7], and they are called the symmetry selection rules, which determine between which energy levels' transitions can occur.

Asymmetric rotor spectra are usually classified as near-prolate or near-oblate. It depends on which limit of the symmetric rotors they are at. The methylamine molecule is considered as a near prolate asymmetric rotor. The spectral bands are further categorized as parallel or perpendicular type. This distinction is made on the basis of the transition dipole moment which may be parallel or perpendicular to the near-top axis. Since the methylamine molecule is a near prolate asymmetric rotor, a type A band is a parallel band and type B and C bands are considered as two perpendicular bands.

The methylamine molecule possesses 15 normal vibrational modes. It comes out from the 3N-6 formula which calculates the degrees of freedom of the non-linear molecule (for linear molecules the formula is 3N-5). The experimental harmonic vibration frequencies (in cm^{-1}) are presented in Table 6.1.

In this review in the order of appearance four vibrations in methylamine will be described, i. e. the wagging, CN stretching, torsional and CH_3 stretching vibrations. The wagging vibration represented by symbol ν_9, with its experimental frequency of 780 cm^{-1} has the A′ symmetry, which means that this vibration is symmetric with respect to reflection in the symmetry plane, which passes through the C-N axis and bisects the NH_2 angle (Figure 6.1). The C-N stretch vibration, ν_8, with the frequency of 1044 cm^{-1} has also the A′ symmetry. The lowest vibration in frequency, the torsional vibration, ν_{15}, being antisymmetric has A″ symmetry and appears at 264 cm^{-1}. And finally, the CH_3 stretch vibration is also antisymmetric with respect to reflection in the symmetry plane (A″ symmetry) with the experimental frequency of 2985 cm^{-1}.

6.3 Measurements of spectra

For chemists, the most desirable aspects of research are molecular structures, vibrational energies and excited electronic states of the molecules of interest. To obtain these information, both rotational and rovibrational spectra are needed. High-resolution gas phase IR spectra provide information on vibrational levels and rotational

Table 6.1: Experimental vibrations in cm^{-1} of 15 modes of methylamine (on the basis of Table 4 from paper [8]).

Mode[a]	Exp. [9, 10]
A′	
NH_2 wag.	**780**
CN stretch	**1044**
CH_3 rock	1130
CH_3 s-deform.	1430
CH_3 d-deform.	1474
NH_2 scis.	1623
CH_3 s-stretch	2820
CH_3 d-stretch	2962
NH_2 s-stretch	3360
A″	
Torsion	**264**
CH_3 rock	972
NH_2 twist.	1335
CH_3 d-deform.	1485
CH_3 d-stretch	**2985**
NH_2 a-stretch	3424

[a]The symmetry refers to the point group C_s of methylamine.

structures as well. Moreover, using this technique, it is also possible to study molecules with zero permanent dipole moment, whereas rotational spectra (microwave and millimeterwave) require a permanent dipole moment. The IR spectral region extends from 100 to 4000 cm^{-1}.

The molecular spectra of methylamine which are going to be described in the next chapters are the high-resolution spectra recorded at the University of Oulu in Finland using Bruker IFS-120HR Fourier transform spectrometer. The resolution due to MOPD (Maximum Optical Path Difference) was 0.00125 cm^{-1} and the relative wavenumber precision was almost one order of magnitude better than the respective resolution, due to careful calibration of the spectra. The pressure was 0.036 Torr and the path length of 3.2 m in the optimized White cell.

6.4 Assignments and analysis of rovibrational spectra with Loomis-Wood for Windows program

The high-resolution rovibrational spectra of polyatomic molecules are often very complex. A clue to the spectra assignment is to put labels with proper quantum numbers on a given transition, and this is the initial step of any analysis. To perform that, some special techniques need to be applied. One such technique is checking the assignment by ground or excited state combination differences.

Figure 6.3 illustrates the rovibrational transitions, where a transition between two vibrational levels v″ (lower) and v′ (upper) is ruled by the formula $\Delta v = \pm 1$. Between these two vibrational levels rotational transitions occur with the selection rule $\Delta J = \pm 1$, where J is the angular momentum quantum number (J″ and J′ denote lower and upper states, respectively). The group of lines with $\Delta J = +1$ correspond to the R branch with increasing wavenumber and with $\Delta J = -1$ to P branch with decreasing wavenumber. Each transition is labeled R(J) or P(J), where the value J corresponds to J″. To derive information about a series of lower and upper states, between which transitions occur, differences of spectral line frequencies between transitions with a common upper state, like in Figure 6.3 series R(0) and P(2), depend on properties of the lower state. Similarly, for transitions with the same lower state, like in Figure 6.3 series R(2) and P(2), differences between these transitions are dependent on properties of the upper states only. Thus, the rotational energy gap in the vibrational lower state can be obtained by lower state combination differences (LSCDs), $\tilde{\nu}[R(0)]$-$\tilde{\nu}[P(2)]$, or similarly the energy interval in the excited state by upper state combination differences (USCDs), for instance, $\tilde{\nu}[R(2)]$-$\tilde{\nu}[P(2)]$. It is worth mentioning that by lower state one can understand any lower state, but in most cases it is a ground state (usually well determined) and then the name of LSCDs becomes GSCDs (ground state combination differences). If there is an assignment of a hot band, then obviously the lower state is not the ground state but the appropriate lower state (it

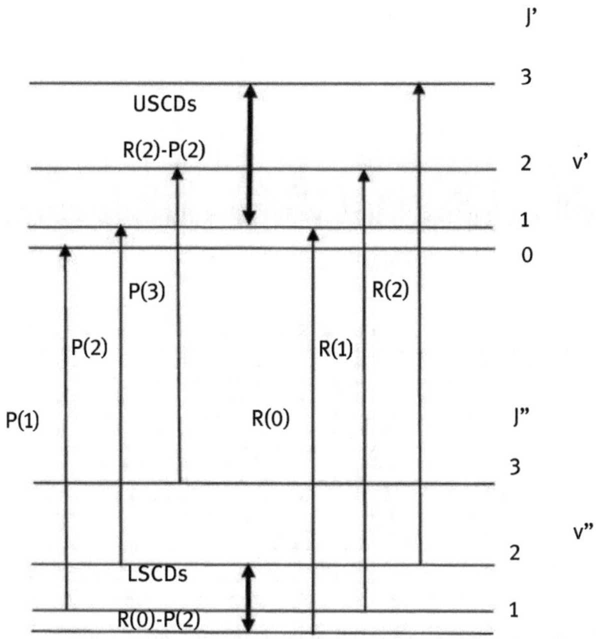

Figure 6.3: Schematic view of P and R branches along with combination differences.

will be discussed in more details in one of the chapters below, where the assignment and analysis of the first excited state of methylamine are described).

Among many computer programs for assigning the spectra, the Loomis-Wood for Windows (LLW) package was built just on the basis of combination differences [11]. This is the graphical software, which is extensively used in the investigations and is very helpful, especially in troublesome assignments. The program is dedicated to assigning Fourier Transform Infrared rotation-vibration spectra. The LWW package can be also used for molecules with large amplitude motions. The LWW is based on the idea of the Loomis-Wood (LW) algorithm which was presented for the first time in 1928 [12]. The algorithm of the LW uses the periodicity of a pattern of lines from the spectrum. The lines are arranged in such a way that the periodicity is converted into recognizable spectral pattern in the LW diagram. In other words this symbolic representation of sequences of transitions in LW diagrams translates the numerical information from the wavenumber listings of branches into a graphical representation. Figure 6.4 shows such a LW diagram from LWW package, where lines (peaks) of a given series (for instance $^{R}R_5$ of A_2 symmetry of the first torsional band of methylamine) shown in a fragment of the spectrum above are presented in the LW diagram by small triangles arranged one by one. The spacing between adjacent R branch lines and P branch lines is close to 2B, where B is the rotational constant [11].

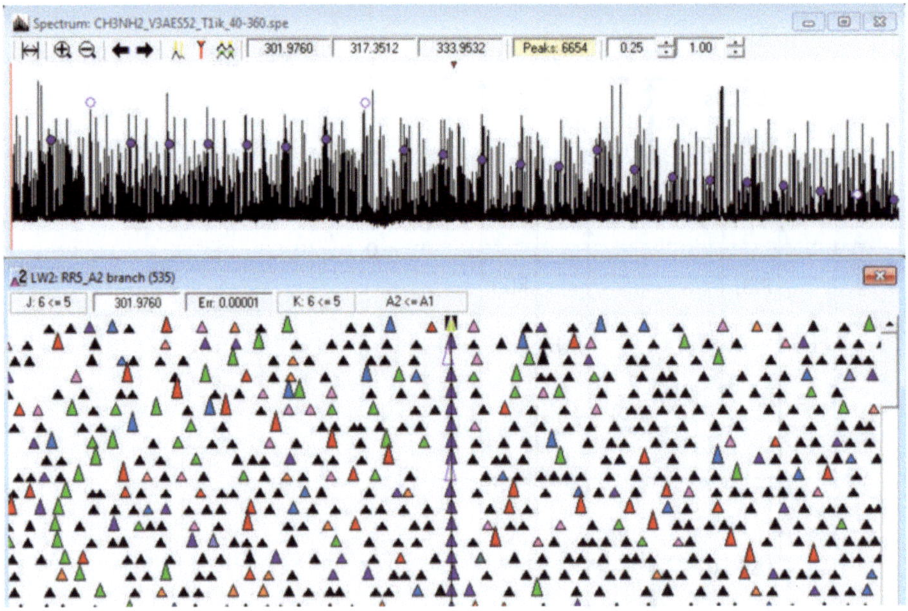

Figure 6.4: LW diagram from LWW software presenting a series of lines and the corresponding spectral lines above.

To start the assignment with LWW, it is necessary to have an energy file of rotational sublevels. Usually, the starting rovibrational energies are of poor quality, but it can be improved after the first assignments are correctly done. The assignments in the LWW are confirmed by Lower State Combination Differences (LSCDs), which are the built-in part of the program. This is a big advantage while assigning the spectral lines since after a line has been indicated, it is possible to check at once the correctness of the assignment by LSCDs. The program allows simultaneous display of several LW diagrams that are mutually connected by LSCD, i. e. a couple of series of branches that belong to the same "family of branches", which means the series sharing the common upper energy level. These series appear as visually recognizable patterns of the same shape in the LW diagram. In Figure 6.5 three LW diagrams of the series belonging to such a "family" are presented, i. e. series $^{R}Q_5$, $^{R}R_5$ and $^{P}P_7$ of A_2 symmetry species with the same upper state of $K = 6$. If it happens that one of the series has a different shape, the assignment is wrong and should be corrected. In other words, the label attached is not the right one. The lower state of the band under assignment (not necessarily the vibrational ground state) should be determined with accuracy allowing for the LSCD checking. Along with the LW diagrams, one has the access to the plot of the IR spectrum with labels of predicted or assigned lines in the so-called spectrum window. This plot is directly linked with its peaklist (table of wavenumbers and intensities). The LWW program allows a simple calculation of a

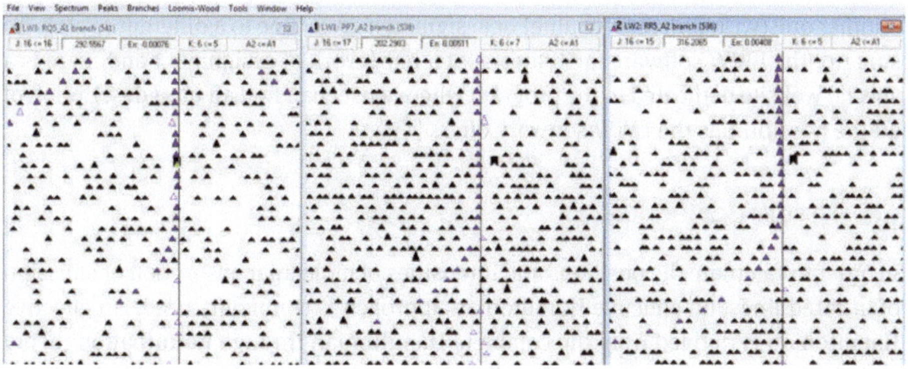

Figure 6.5: LW diagrams showing the "family of branches" assigned and confirmed by the LSCDs.

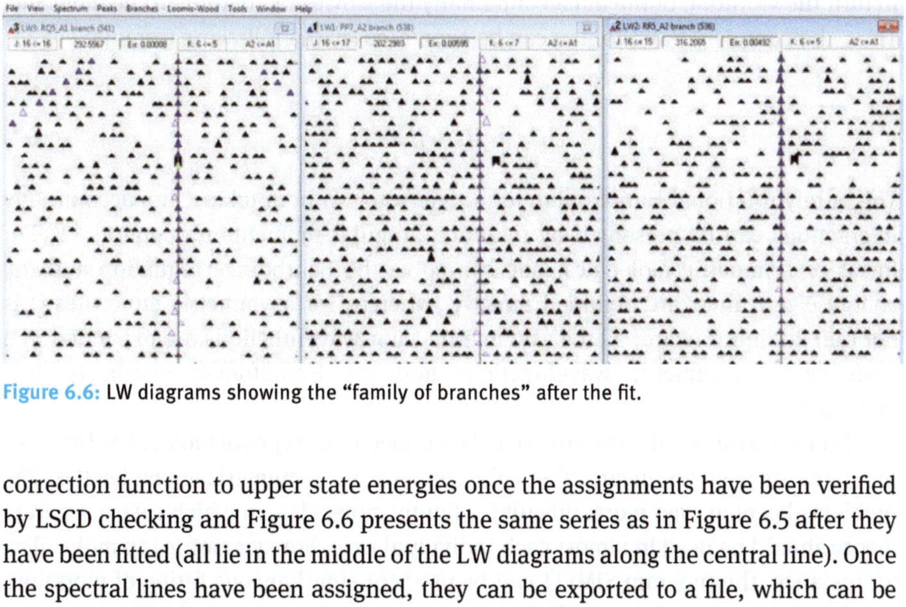

Figure 6.6: LW diagrams showing the "family of branches" after the fit.

correction function to upper state energies once the assignments have been verified by LSCD checking and Figure 6.6 presents the same series as in Figure 6.5 after they have been fitted (all lie in the middle of the LW diagrams along the central line). Once the spectral lines have been assigned, they can be exported to a file, which can be used in external fitting procedures with complex Hamiltonian models. The software gives a possibility of choosing the series for exporting or any other purpose. The rovibrational energies refined by least-squares fitting external program can be imported back into the LWW program where they replace previous energies. The assignment can be carried on this way until the entire range of the spectrum is satisfactory assigned.

The LWW software has a lot of other advantages and useful tools like a possibility to merge two spectra in different spectral ranges, to generate the whole set of branches to be assigned or to plot energy files which help find the line crossing. There are also many functions which enable a LWW user to perform very useful actions when assigning spectra like moving the assignment of a present series to another one (with different K value) or assigning a given line to a different J value. To sum up, the LWW software facilitates a lot a hard work of assignment and there is a special website dedicated to the program where a user can find an utter description of all the possibilities the LWW software offers [13].

6.5 Theoretical model

In order to calculate the positions and intensities of molecular absorption or emission of infrared lines, an effective Hamiltonian approach is commonly used. An effective Hamiltonian is derived from the complete Hamiltonian through perturbation theory and almost in all cases effective Hamiltonians are constructed by considering Born-Oppenheimer approximation, which separates electronic, vibrational and rotational motion. Among all these motions, the electronic one is believed to be the fastest, then in turn the vibration of the nuclei and finally the slowest is rotation of the molecule. Thus, the complete Hamiltonian can be expressed by the following formula:

$$H_{evr} = H_e + H_v + H_r \tag{6.1}$$

The purely rotational Hamiltonian for a rigid rotor without taking into account any interactions can be presented by rotational angular momentum operators [2]. For linear or symmetric molecules, one can solve the Schrödinger equation and the energy eigenvalues are obtained directly, whereas for asymmetric molecules it is not that straightforward, as the asymmetric rotor wavefunctions are expanded in a basis set of symmetric wavefunctions and the Hamiltonian matrix is then diagonalized.

There is no single theory which can be applied to all types of molecules. Different programs are used for linear, spherical, symmetric or asymmetric top molecules. The problems become even more difficult for floppy molecules, in which case a separate theory should be used in almost each individual case. For symmetric top molecules, for instance, the program SIMFIT can be used for simultaneous fitting of vibration-rotation and rotational spectra [14].

For the analysis of asymmetric rotor spectra, the powerful package SPFIT/SPCAT written by H. M. Pickett [15] is commonly applied. These programs were originally aimed at cataloguing rotational spectra of many different molecules. This was the reason for a very general way of setting up the molecular Hamiltonian and efficient factorization of its energy matrix. Thus, the input of the SPFIT program consists of a set of assigned experimental transitions or energy levels and a file with molecular parameters. A least square procedure is applied to the assigned data. From the obtained set of parameters the SPCAT program calculates back the transition frequencies and intensities.

It is not possible to use the standard fitting programs for molecules performing large amplitude vibrations. Thus, it is necessary to create individual approaches dedicated to the specific group of LAV molecules or sometimes even to a single molecule. Among others there are some programs written particularly for LAV molecules: XIAM (internal rotation program for up to three symmetric internal rotors and up to one quadrupole nucleus [16]), ERHAM (Effective Rotational HAMiltonian program for molecules with up to two periodic large-amplitude motions [17]), BELGI (The BELGian Internal Rotor Program [18]), RAM36 (the Rho Axis Method for 3-and 6-fold barriers [19]). More detailed description of each of the program is available at the link cited in [20].

As the object of this article is a methylamine molecule and its spectra, mainly the quantum theory concerning this molecule will be under consideration. Thus, in most studies concerning methylamine spectra, the phenomenological Hamiltonian based on the group-theoretical high-barrier tunneling formalism was used [7]. This formalism is based on the assumption that a complete set of vibration-rotation states, which are localized in one minimum of the potential surface, interacts with the equivalent complete set of vibration-rotation states localized in neighboring minima. The minima in the potential surface in Figure 6.2. correspond to the six configurations (frameworks) shown in Figure 6.7. The coordinate ρ stands for the inversion angle, whereas the τ coordinate is the internal-rotation angle. These six minima are tied together by two types of tunneling path. Both types entail some torsion (internal rotation), whereas only one involves inversion. For example, when

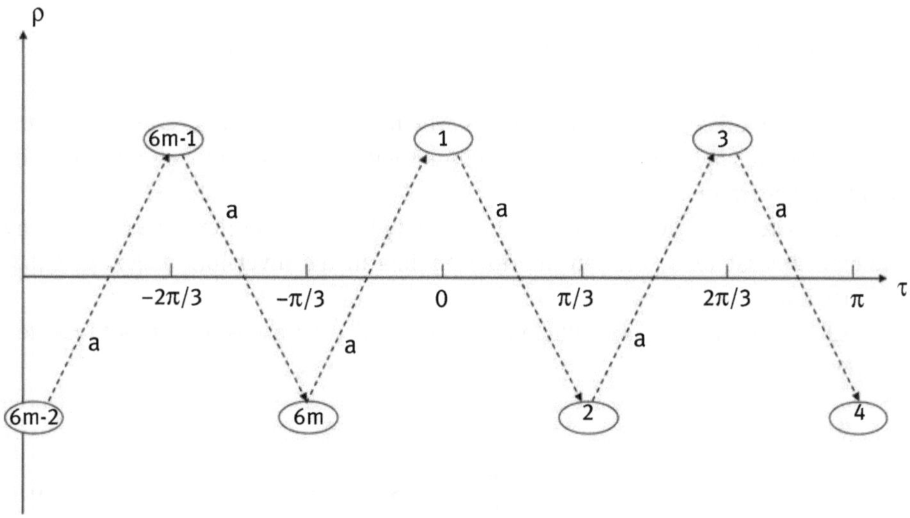

Figure 6.7: Schematic view of the six equivalent configurations in methylamine (frameworks) and the path between subsequent configurations under the symmetry operation $a = (123)(45)$.

the torsion of the CH_3 group occurs only (angle τ changes by 120°) and the inversion does not take place (coordinate ρ does not change), this case is illustrated by tunneling motion from position 1 to 3 in Figure 6.7. If inversion takes place (change from positive value of ρ to negative), then the internal rotation is involved as well, since coordinate τ changes by 60° and this situation is shown by tunneling motion from framework 1 to 2 in Figure 6.7. The system of 6 minima is repeated *m* times to minimize the torsion-rotation coupling in the molecule-fixed axis system. This is achieved by counterrotation of the molecular axis system by -ρτ when the methyl rotor is rotated by +τ [7].

The formalism proposed for methylamine by Ohashi and Hougen [7] treats each analyzed state as an isolated one, which is split into several sublevels of A_1, A_2, B_1, B_2, E_1 and E_2 symmetry and the Hamiltonian operator is defined as follows:

$$\begin{aligned}\hat{H} = {} & h_v + h_J J^2 + h_K J_z^2 + (f_+ J_+^2 + f_- J_-^2) + q J_z + \\ & + \frac{1}{2}[s_+(J_+ J_z + J_z J_+) + s_-(J_- J_z + J_z J_-)] + \\ & + (r_+ J_+ + r_- J_-) + \left(f_+^{(2)} J_+^4 + f_-^{(2)} J_-^4\right) + \\ & \textit{higher order terms}\end{aligned} \tag{6.2}$$

where *higher order term* represents expansion of terms in J(J + 1) and/or K^2 as it is in a standard expression for the asymmetric top molecule (ordinary centrifugal distortion terms). For instance, the term h_1 has the following form:

$$h_1 = \nu_0 + BJ(J+1) + (A-B)K^2 - D_J J^2(J+1)^2 - D_{JK} J(J+1)K^2 - D_K K^4 + \dots \tag{6.3}$$

The Hamiltonian contains effective terms which include ΔK = 0, ±1, ±2, ±4 operators. The coefficients *h*, *f*, *s*, *r*, $f^{(2)}$ depend on vibrational coordinates and are multiplied by rotational operators.

Moreover, each term in eq. (6.2) is expanded in a Fourier series describing different interactions between multiple minima in the inversion-torsion potential surface (Figure 6.2).

The *c* = *h*, *f*, *r* and *s* coefficients are expended as a cosine series presented below

$$c = c_1 + \sum_{n=2}^{3m} 2c_n \cos(n-1)\tau \tag{6.4}$$

and the *q* coefficient as a sine series

$$q = \sum_{n=1}^{3m} 2i\, q_n \sin(n-1)\tau \tag{6.5}$$

where n is the ordinal number of a molecular configuration interacting with the configuration number 1, and τ the angle depending on the type of interaction and symmetry species.

$$\tau = \left(\frac{2\pi}{6}\right) k\rho + \theta(\Gamma\,) \tag{6.6}$$

In eq. (6.4) and eq. (6.5), the n-th term represents a tunneling from framework 1 to framework n. The subscript $n = 1$ corresponds to nontunneling motion (τ angle is 0 in Figure 6.7). The even subscripts $n = 2,4,6\ldots$ describe tunneling involving the inversion of the NH_2 group whereas the odd ones $n = 3,5,7\ldots$ refer to tunneling involving mainly the internal rotation of the CH_3 group. The respective parameters used in the fitting model (molecular constants describing the observed spectrum of methylamine) are defined as h_n, f_n, q_n etc.

The experimental data are the frequencies of the transitions between the energy levels of a studied molecule and the role of the theoretical model is to reproduce the observed frequencies. As a result, the spectroscopic parameters with information on the energy structure of the system are set. The Ohashi–Hougen Hamiltonian operator was applied successfully in fitting the ground vibrational state of methylamine [21] and its first excited torsional state [6, 22, 23]. For other states of methylamine, i. e. the wagging or C-N stretching states, the results were not as good. In these cases it was not possible to fit the line within experimental accuracy. The main reason for the failure is the coupling between the wagging, ν_9, and the third excited torsional, $3\nu_{15}$, states or between C-N stretching and the fourth excited torsional ($4\nu_{15}$) states, which will be in more detailed described in the subsequent chapters. Thus, in order to obtain a better fit result, it is necessary to take into account at least two states, which interact with each other and a global fit should be conducted then. One can think also of another reason for which the tunneling model gives worse fitting results. For higher rovibrational states the inversion-torsion splittings get higher than for the ground state or the first excited torsional state, i. e. the higher are the rovibrational states, the smaller are the barriers between the equilibrium minima and the inversion torsion functions are no longer well localized in each minimum, but get dispersed. The inversion torsion functions can be presented by the probability density, and Figure 6.8 shows the situation described above.

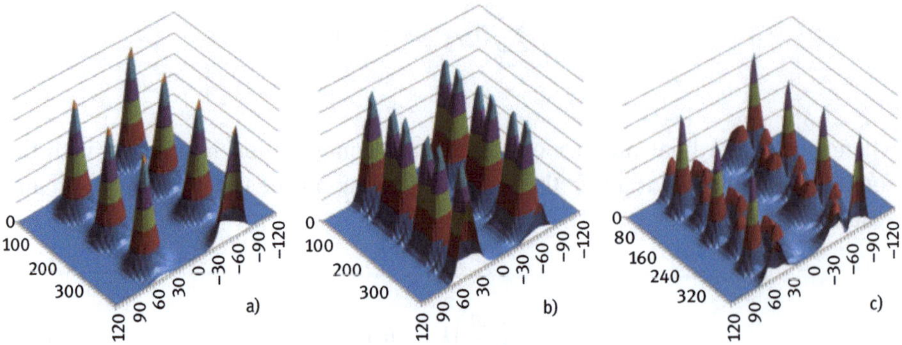

Figure 6.8: The probability density of the inversion torsion functions for $J = K = 0$ rotational states of A_1 symmetry in (a) the ground vibrational state, and (b) first and (c) second excited torsional states of methylamine.

6.6 Assignment and analysis of the wagging state, ν_9

The rotational structure of the ground state of methylamine has been extensively studied since 1957 [24]. An analysis of tunneling-rotational levels of the first excited inversion (ν_9 wagging) state of methylamine was needed for understanding of a complex dynamics of the molecule. The first high-resolution study of the ν_9 wagging band was performed in 1992 [25]. Since the transition moment of the wagging vibration is neither perpendicular nor parallel to axis of the top, it consists of both components. The wagging band has a hybrid structure which is further modified through splitting due to inversion and internal rotation of the methylamine molecule.

The spectrum of the ν_9 wagging band spreads from 640 to 960 cm^{-1} (Figure 6.9). The wagging band consists of perpendicular and parallel components. The first identified series were the ^RR series belonging to perpendicular component of the band on the high frequency side of the spectrum along with their weaker ^PP counterparts on the low frequency side [25]. At the band centre, ^QQ branches spread over some region of the spectrum, which is the effect of the torsion and inversion. As for the parallel component of the wagging band, the ^QR and ^QP branches are generated and depending on the K values, some of them are even stronger in intensity than the perpendicular components but to assign these series was not that straightforward since they have their beginnings at the very center of the band, where the lines are overlapped and really dense.

In the first attempt to assign and analyze the wagging band of methylamine, the best fit that could be achieved was of standard deviation of 0.0095 cm^{-1} for almost 5000 lines and $K \le 10$ and $J \le 15$ [25]. Many line series have been found, identified and fit, even those of A and E_2 symmetry (weaker components). Still some unanswered questions remained, even for stronger components in intensities of B and E_1

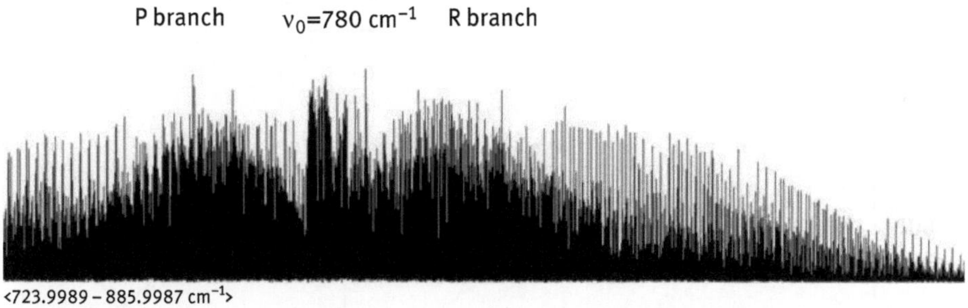

Figure 6.9: A part of the wagging band, ν_9, of methylamine in the FTIR spectrum.

symmetry. Some series were missing. In an independent approach [26, 27] about 8800 transitions were assigned but still some transitions could not be identified.

It was believed that some levels of the wagging state were in a strong resonance with some "dark" states. By the "dark" state one can understand the state which resonates with so called "bright" state (in this case it is the ν_9 state). Perturbations are the main complication in the analysis of asymmetric rotor bands and the mostly encountered perturbations are caused by Coriolis interaction and/or Fermi-type resonance. Each perturbation expresses the mixing of a "bright" rotational level of the vibrationally excited state with rotational levels of nearly located "dark" vibrational states. When a perturbation mixing takes place, the "dark" rotational levels borrow intensities from the bright state, hence they become observable transitions in the spectrum. Perturbations cause frequency shifts of the observed transitions. The candidates for "dark" states in the wagging region seemed to be the third excited torsional state (especially the upper part of the torsional splitting) and the fourth torsional state (the lower part of the torsional splitting), which might interact with the wagging state (Figure 6.10).

Having this insight into possible interactions, the wagging band of methylamine was put through the analysis in a new high-resolution spectrum [28]. In general, the assignment of the band was extended to about 13,000 transitions up to $J = 40$ in most series and many of the missing lines in previous studies were identified. The only two series that still were not found were of the A ($K = 0$) and E_{2+1} ($K = 3$). The reason for which these two could not be assigned is the perturbation. The resulting standard deviation was 0.018 cm^{-1}, which is not within the experimental accuracy and again the perturbation is blamed for not perfect result of the fit. Figure 6.11 shows one of the series that was luckily found and assigned to $K = 0$ of B symmetry, the $^{P}Q_1(B)$ branch. One can notice that the branch changes its way twice, which is also reflected in its counterparts' shapes, the braches P and R. That made it very difficult to find these series. The series of $K = 0$ of B symmetry are shifted of 4 cm^{-1} because of the strongest perturbation taking place in the wagging band. Going back to the indicated

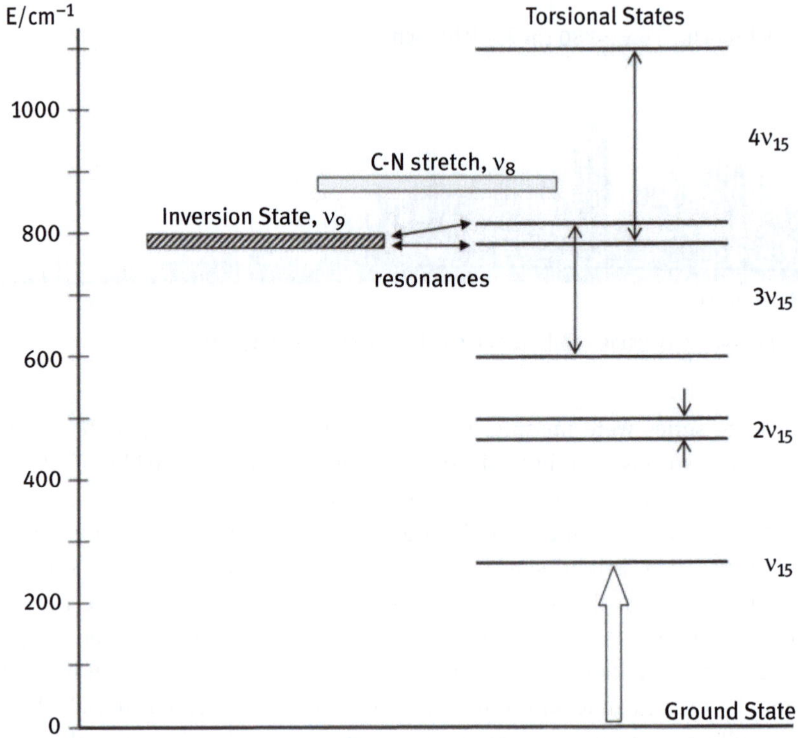

Figure 6.10: Energies of the inversion-torsional states of CH_3NH_2 (presented in paper [6]).

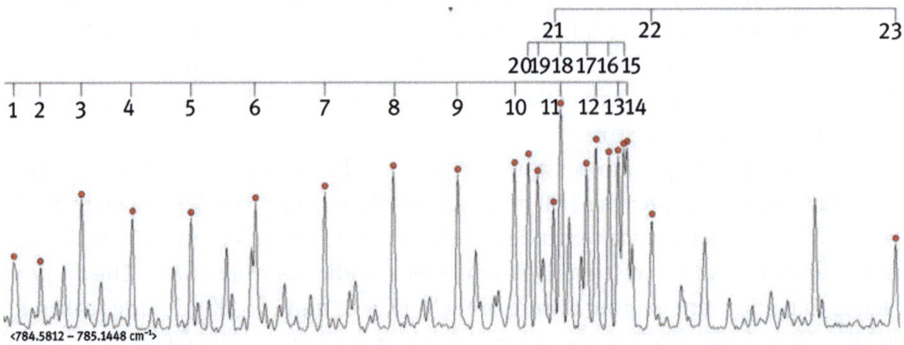

Figure 6.11: Part of the methylamine spectrum showing the structure of the $^{P}Q_1(B)$ branch. The labels show the values of J′(taken from paper [28]).

perturbers ("dark" states), the $3\nu_{15}$ can be involved only in local Coriolis interactions since it has a different symmetry with respect to the reflection in the symmetry plane of the NH_2 group. As for the $4\nu_{15}$ state, it can cause both the Coriolis and the Fermi types of interactions with the wagging state, because it has the same symmetry as the wagging state [29].

To overcome the difficulties with the fitting of all the assigned and perturbed data, the theoretical model should consist of two or even three interacting states. The effective Hamiltonian for inversion-torsion coupling in methylamine has been already created and will be implemented soon. Such a Hamiltonian treats explicitly the coupling between two or three states. The approach is based on the idea of the group-theoretical Hamiltonian applied simultaneously to two mutually perturbing states.

Concluding, the inversion band has been successfully reanalyzed and completed [28].

6.7 Assignment and analysis of the C-N stretching band of methylamine, ν_8

The analysis of the C-N stretching band was the next step of a systematic study on the high-resolution IR spectrum of methylamine, CH_3NH_2. The C-N stretching band spreads in the spectrum from 960 to 1200 cm^{-1}.

The C-N stretching vibration is a parallel A-type band with the change of dipole moment along the axis of least moment of inertia. Thus, the K-selection rule is $\Delta K = 0$ and the J-selection rules are $\Delta J = -1$, 0 or +1 and they correspond to IR transitions labeled as P, Q or R, respectively. In the spectrum, the C-N stretching band turns up as a strongly overlapped structure with lines of relatively weak intensities, which makes the assignment rather hard (Figure 6.12). The C-N stretch band is centered at about 1044 cm^{-1} with dense J-multiplets in the P and R branches and a tight Q branch. It was not straightforward to assign K values to transitions properly, and before the unambiguous results were obtained with help of the GSCDs, a lot of attempts of assignment had to be made.

Despite many difficulties with correct labeling of the transitions, over 3500 lines with a resolution of 0.001 cm^{-1} for K from 0 to 12 have been assigned [30, 31]. All allowed transitions for B symmetry species were well assigned up to $K' = 12$ (K'

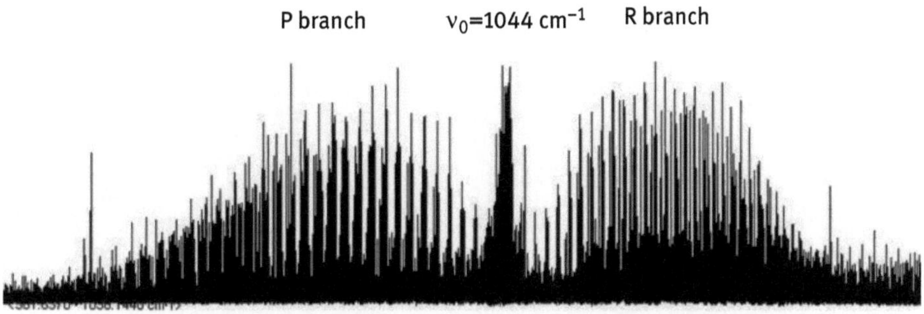

Figure 6.12: A part of the C-N stretch band, ν_8, of methylamine in the FTIR spectrum.

denotes K value of the higher state), since the B lines are of the highest intensities. The assignment of E_1 transitions was more troublesome comparing to the B symmetry species, although the lines of both symmetries are more or less of similar intensities. Just to remind, the relative intensities of individual lines presented by statistical weights are the following: $A_1{:}A_2{:}B_1{:}B_2{:}E_{1+1}{:}E_{1\text{-}1}{:}E_{2+1}{:}E_{2\text{-}1}$ = 1:1:3:3:3:3:1:1. Finally the transitions up to K′ = 11 and 9 for E_{1+1} and $E_{1\text{-}1}$ symmetries were assigned, respectively. These difficulties with identification could be caused by many irregularities appearing in the spectrum and shown in Figure 6.13 and Figure 6.14 (taken from paper [31]) and by some resonances in the C-N stretch region.

As for irregularities mentioned, in Figure 6.13 one can see a part of the C-N stretching band with the J = 4←5 manifold of the ^{Q}P branch consisting of lines for K′ = 1 values and different symmetry species. It can be noticed that due to accidental overlapping some lines have higher intensities than predicted from the statistical weights. Moreover, in Figure 6.14 a structure of the ^{Q}P branch for J = 8 and B

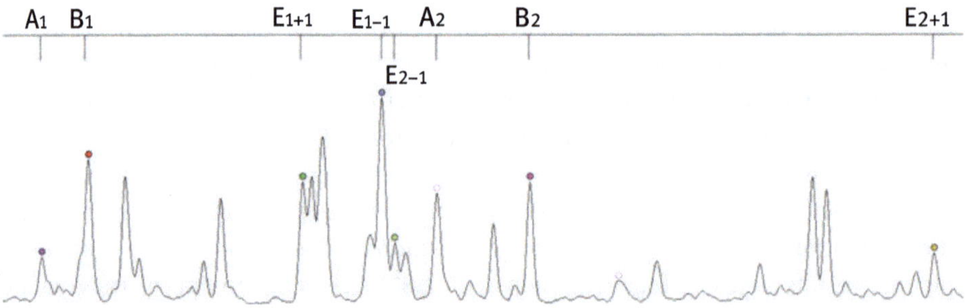

<1037.1353 – 1037.5173 cm^{-1}>

Figure 6.13: Part of the C-N stretching band in the – range showing the J = 4←5 manifold of the ^{Q}P branch consisting of lines for K′ = 1 values and different symmetry species.

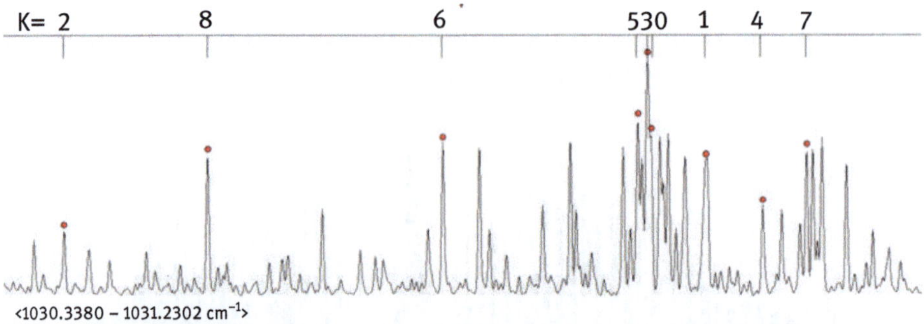

<1030.3380 – 1031.2302 cm^{-1}>

Figure 6.14: Part of the ν_8 band in the – range showing the structure of the ^{Q}P branch for J = 8 and B symmetry for different K values.

symmetry for different K values is presented where there is no simple correlation between the position of the line and the value of K.

Apart from identification of the lines of B and E_1 symmetries, also the low-intensity transitions to which A and E_2 symmetry species transitions belong were observed and successfully assigned. For A symmetry, all series up to K′ = 9 have been identified whereas for E_{2+1} and E_{2-1} symmetry up to K′ = 8. There is only lack of the K′ = 5 series for the E_{2-1} symmetry. All assignments have been confirmed by the GSCD and are consistent with the experimental laser lines [32–34], and are available in the paper on the analysis of the C-N stretching band [31].

Not only the transitions belonging to the ν_8 band were identified but also the transitions, which were assigned to K′ = 4 series of E_{2-1} symmetry species of a "dark" state. To check possible resonances in the excited states of methylamine the inversion-torsion-rotation energy levels were calculated for selected J = K values from the two-dimensional potential surface [5].

In Figure 6.15 the energies of the inversion-torsional states are presented: the first excited torsional state-ν_{15}, the second excited torsional state-$2\nu_{15}$, the third excited torsional state-$3\nu_{15}$, the fourth excited torsional state-$4\nu_{15}$, the excited inversion state-ν_9, the combination state-$\nu_9+\nu_{15}$ and the C-N stretching state-ν_8. The energies were calculated for values up to J = K = 5. For the C-N stretching state, its energies were calculated using the effective Hamiltonian and all experimental data. Figure 6.15 shows a diagram with energy ladders for all indicated states for K = 4 where the nearest neighboring states to the C-N stretching state are the combination state $\nu_9+\nu_{15}$ and the $4\nu_{15}$ state, thus these states were taken into consideration as "dark" states.

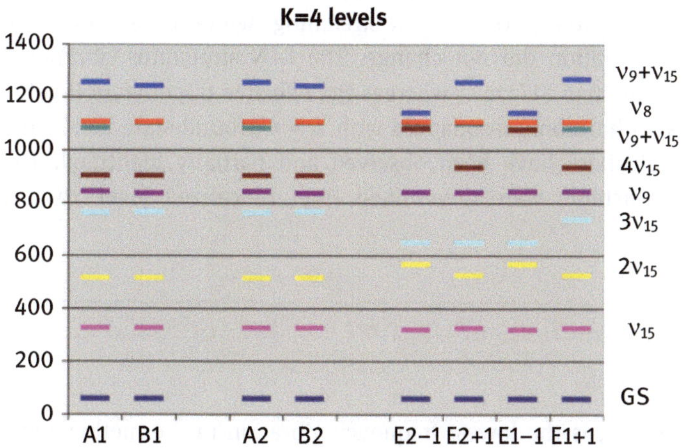

Figure 6.15: Calculated energy levels schemes of CH_3NH_2 for J = K = 4 for different symmetry species (GS – ground state; ν_{15}, $2\nu_{15}$, $3\nu_{15}$, $4\nu_{15}$ – excited torsional states; ν_9 – excited wagging; ν_8 – C-N stretching state; $\nu_9+\nu_{15}$, $\nu_9 + 2\nu_{15}$ – combination states).

The perturbation observed in the C-N stretching for K′ = 4 for the $E_{2\text{-}1}$ symmetry seemed to be a Fermi-type resonance since all lines of the series are shifted of about 0.4 cm^{-1}. It was proved that a Fermi-type resonance is allowed only between states of the same type, for A′-A′ or A″-A″ states [29]. The C-N stretching state, ν_8, is an A′ type state, the inversion-torsion combination state, $\nu_9+\nu_{15}$, is an A″ type state, and the fourth excited torsional state, $4\nu_{15}$, is an A′ type state. Concluding, the state responsible for the perturbation in K′ = 4 series of $E_{2\text{-}1}$ symmetry, is the $4\nu_{15}$. Between the ν_8 and the $\nu_9+\nu_{15}$ states only Coriolis coupling is possible. Many other perturbations are also observed, some Fermi-type resonances as well as local Coriolis interactions.

Taking into account all the experimental data and the resonances that occur in the C-N stretch spectrum, the fit was performed using the Hougen–Ohashi single state model with the final result of 0.04 cm^{-1} [31]. The band centre was determined at 1044.8134 cm^{-1}. The group theoretical Hamiltonian is dedicated to fit the data of one state only. As the observed vibrational state is in resonance with other vibrational states, the Hougen–Ohashi Hamiltonian gives the results for the C-N stretching state, ν_8, which differ from the experimental data.

In any case, it was possible to assign the transitions in the C-N stretching region of the spectrum for the first time and fit all the identified data in spite of the difficulties caused by weak line intensities, perturbations and strong lines overlapping.

At the same time, the C–N stretching infrared fundamental of CH_3NH_2 has been investigated by R. M. Lees et al. [35] using high-resolution laser sideband and Fourier transform synchrotron spectroscopy. In that work, many sub-bands have been assigned for *K* values ranging also up to 12 for the stronger B and E_1 inversion species for the $v_t = 0$ torsional state, along with many of the weaker sub-bands of the A and E_2 species. Both ground and C–N stretch origins were fitted to a phenomenological Fourier series model. It was noticed that the amplitude of the torsional energy oscillation increased substantially for the C–N stretching, while the amplitude of the inversion energy oscillation did not change. The C–N stretching vibrational energy was determined at 1044.817 cm^{-1}, whereas the effective upper state B-value at 0.7318 cm^{-1}. Several anharmonic resonances with $v_t = 4$ ground-state levels (the fourth excited torsional state) have been observed and partially identified. The interaction coupling constants were determined from *J*-localized level-crossing resonances.

6.8 Assignment and analysis of the first excited torsional state, ν_{15}

The torsional vibration, ν_{15}, is the lowest frequency vibration in the methylamine molecule. In comparison to the assignments of the wagging band and C-N stretching, the assignment of the first torsional band is different since it is not much perturbed. As it is shown in Figure 6.10, the first excited torsional state is well isolated from other

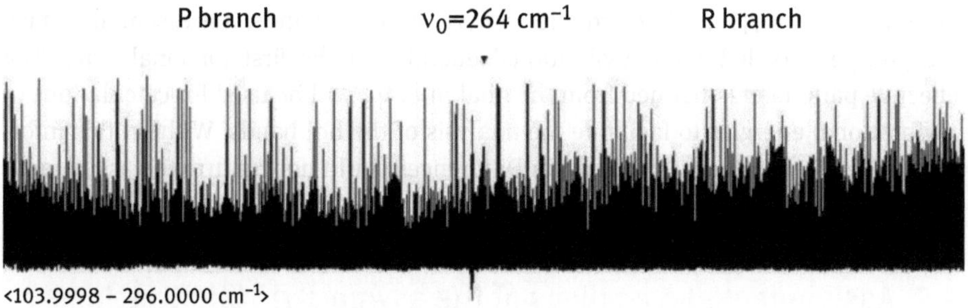

Figure 6.16: A part of the first torsional band of methylamine, ν_{15}, in the FTIR spectrum.

vibrational states, i. e. the second, third and fourth excited torsional state ($2\nu_{15}$, $3\nu_{15}$, $4\nu_{15}$), the inversion state, ν_{15} or the C-N stretch, ν_8.

The ν_{15} spectrum of CH_3NH_2 shown in Figure 6.16 ranges from 100 to 360 cm^{-1}. The center of the band is determined at 264 cm^{-1}. The band is a B-type band, since the changing dipole moment lies in the axis of intermediate moment of inertia (b axis) and thus the symmetry rules are with $\Delta K = \pm1$ and $\Delta J = 0, \pm1$. The respective IR transitions are labeled as P, Q or R. The type B bands do not have a strong central Q branch. The Q lines spread over and overlap the P and the R branches. The high-frequency side (the right part) of the spectrum is dominated by strong, well resolved ^rQ branches and ^rR branch series. The low-frequency side (the left part) of the band consists of a group of weak lines, mainly of the ^pP an ^pQ series. Near the band center, the Q branches spread over a spectrum region and many weak lines (of A and E_2 symmetry) are covered by stronger lines, which makes the identification process hard.

The analysis of the first torsional band started in 1988 [22], then a year later was reinvestigated [23], and much improved in 2016 [6]. Since some of the series of the first torsional band were already identified, the reassignment began with the previously assigned lines, which most of them were the intense ones of B and E_1 symmetry. After the lines have been reassigned, the energy file needed for the LWW was recalculated and the next step of identification started, to find completely new lines. Although the lines, especially the weak ones, were hidden in a dense spectrum, it was possible to assign almost 12,000 transitions and finally fit them with the standard deviation $\sigma = 0.00079$ cm^{-1} using the single state effective Hamiltonian with the set of 88 molecular parameters [6]. It should be mentioned at this point, that apart from the IR data assigned in the first torsional band, also the data from previously measured MW and IR pure rotational transitions were taken into account in the fitting [22, 23]. All the assignments were confirmed by GSCDs. The single state model worked properly, mostly because of the fact that the ν_{15} state lies in the isolated region and is not expected to be in the interaction with other states.

In the region of the first torsional band, several hot bands are observed, $\nu_{15}\rightarrow 2\nu_{15}$, $\nu_{15}\rightarrow 3\nu_{15}$, $\nu_{15}\rightarrow 4\nu_{15}$ [36, 37]. In order to analyze these hot bands, it is essential to have the precise knowledge on rovibrational energies of the first torsional state. The effective parameters obtained from the final fit of ν_{15} will be used for calculations of rovibrational energies to facilitate the analysis of the hot bands. Without this information, the Lower State Combination Differences could not confirm the assignment of the hot bands originating from the first torsional state.

6.9 Assignment and analysis of the asymmetric CH-stretching state, ν_{11}

As far as assignment of higher frequency vibrational bands of LAV molecules is concerned, it is still a challenge. In that region (near 3000 cm^{-1}) the density of vibrational states is so high that there are many interactions between them, resonances, which hinder identification and assignment. For methylamine, only one of the high frequency bands was studied, i. e. the asymmetric CH stretching band, ν_{11} (in Table 6.1 it is denoted as CH_3 d-stretch with its experimental vibrational frequency of 2985 cm^{-1}) [38].

The high-resolution infrared spectrum of methylamine in the CH stretching fundamental band region was recorded using slit-jet direct absorption spectroscopy with the resolution of 0.0025 cm^{-1} [38]. The region covered by the ν_{11} band ranges from 2965 to 3005 cm^{-1}. The transition dipole moment of the ν_{11} band moves along the b-rotational axis, thus the rules for the b-type transitions are as follows: $\Delta K = \pm 1$ and $\Delta J = 0, \pm 1$. The spectrum in this region was difficult to assign since apart from the complexity caused by torsion and inversion tunneling, like in lower frequency vibrations (NH_2 wagging, CN stretching and the torsion), it is strongly perturbed by "dark" states. In the spectrum, there are multiplets of 2 or 3 mixed bright and dark states [38]. Since many perturbations occur in that region, one does not observe the complete series of lines, only fragments of the branches are visible. With help of the deperturbation theory and LSCDs [38], it was possible to assign over 600 lines in the ν_{11} band. It was found out that in the CH stretching band the torsion-inversion levels appeared in a different order than in the ground state (in the ground state the order of the energy levels tunneling multiplet follows the order: $A_1<B_1<E_1<E_2$ [39]), which is also the case for other vibrational states, but in the ν_{11} state, the levels order does not correspond to the observed one. It is so, because in so perturbed region probably not all the dark states were identified and taken into account.

6.10 Summary

High-resolution IR rovibrational spectra of floppy molecules like methylamine are still, despite the advances in the experimental techniques, very complex to analyze.

Each rovibrational analysis requires individual approach, the assignments and studies of LAV molecules spectra cannot be unified. It was seen in preceding chapters describing analyses of the methylamine rovibrational spectra that each spectral region was different and involved diverse problems. It is very important for complex studies to do the assignments in a systematic way. For instance, with a well-determined ground state of methylamine [21, 39], it was possible to perform further analyses (for instance for NH_2 wagging [25, 28], CN stretching [30, 31, 35]). After the successful studies on the first excited torsional state [6, 22, 23], one can gradually move toward the next spectral regions, for instance, the studies of the torsional overtones like the second excited torsional state, $2\nu_{15}$ [36, 37, 40] then in turn the third excited torsional state, $3\nu_{15}$, etc. As the next step the analyses of hot bands appearing in these regions should be performed and finally it would be worthwhile going back to resonances in the region of about 1000 cm^{-1} where the previous fit results were not perfect in spite of some resonances [28, 31].

References

[1] Herzberg G. Molecular spectra and molecular structure: electronic spectra and electronic structure of polyatomic molecules Vol. 3. Malabar, FL: Krieger Publishing Company, 1991a.

[2] Herzberg G. Molecular spectra and molecular structure: infrared and raman of polyatomic molecules, vol. 2. Malabar, FL: Krieger Publishing Company, 1991b.

[3] Hollas JM. Modern spectroscopy. England: John Willey & Sons Ltd, 2004.

[4] Bunker PR, Jensen P. Molecular symmetry and spectroscopy. Ottawa: NRC Research Press, 1998.

[5] Kręglewski M. Vibration - Inversion - Torsion - Rotation Energy Levels and Vibrational Resonances in the Methyl Amine Molecule. In: Laane J, editor(s). Structures and conformations of non-rigid molecules Vol.410. Berlin: Kluwer Academic Publishers, 1993:29–43. 978-94-010-4920-7.

[6] Gulaczyk I, Kręglewski M, Horneman VM. J Mol Spectrosc 2016. DOI: 10.1016/j.jms.2016.12.007.

[7] Ohashi N, Hougen JT. J Mol Spectrosc 1987;121:474–501.

[8] Pelegrini M, Roberto-Neto O, Machando FB. Chem Phys Lett 2005;414:495–99.

[9] Gray AP, Lord RC. J Chem Phys. 1957;26:690.

[10] Hamada Y, Tanaka N, Sugawara Y, Hirakawa AY, Tsuboi M, Kato S, J Mol Spectrosc. 1982;96:313.

[11] Łodyga W, Kręglewski M, Pracna P, Urban Š. J Mol Spectrosc 2007;243:182.

[12] Loomis FW, Wood RW. Phys Rev. 1928;32:223.

[13] Łodyga Wiesław. DOI:http://lww.amu.edu.pl

[14] Pracna P, Sarka K, Demaison J, Cosléou J, Harlemont F, Khelkhal M, et al. J Mol Spectrosc 1997;184:93–105.

[15] Pickett HM. J Mol Spectrosc 1991;148:371–77.

[16] Hartwig H, Dreizler H. Z Naturforsch 1996;51a:923–32.

[17] Groner P. J Mol Spectrosc 2012;278:52–67.

[18] Hougen JT, Kleiner I, Godefroid M. J Mol Spectrosc 1994;163:559–86.

[19] Ilyushin VV, Kisiel Z, Pszczółkowski L, Mäder H, Hougen JT. A new torsion-rotation fitting program for molecules with a six-fold barrier: application to the microwave spectrum of toluene. J Mol Spectrosc 2010;259:26–38.

[20] Kisiel Zbigniew. DOI: http://www.ifpan.edu.pl/~kisiel/asym/asym.htm http://www.ifpan.edu.pl/~kisiel/introt/introt.htm.
[21] Ohashi N, Takagi K, Hougen JT, Olson WB, Lafferty WJ. J Mol Spectrosc 1987;126:443–59.
[22] Ohashi N, Takagi K, Hougen JT, Olson WB, Lafferty WJ. J Mol Spectrosc 1988;132:242–60.
[23] Ohashi N, Tsunekawa S, Takagi K, Hougen JT. J Mol Spectrosc 1989;137:33–46.
[24] Nishikawa T. J Phys Soc Jpn 1957;12:668–80.
[25] Kręglewski M, Winther F. J Mol Spectrosc 1992;156:261–91.
[26] Sztraka L, Alanko S, Koivusaari M. Acta Chim Hung 1993;130:887–900.
[27] Sztraka L, Alanko S, Koivusaari M. J Mol Struct 1997;410–411:391–95.
[28] Gulaczyk I, Łodyga W, Kręglewski M, Horneman VM. Mol Phys 2010;108:2389–94.
[29] Gulaczyk I, Kręglewski M. J Mol Spectrosc 2009;256:86–90.
[30] Gulaczyk Iwona. The C-N stretching band of methylamine. 21st International HRMS Conference, Poznań , 2010.
[31] Gulaczyk I, Kręglewski M, Horneman VM. J Mol Spectrosc 2011;270:70.
[32] Dyubko SF, Svich VA, Fesenko LD. JETP Lett 1972;16:418–19.
[33] Plant TK, Coleman PB, DeTemple TA. New optically pumped far-infrared lasers. IEEE J Quantum Electron 1973;QE-9:962–63.
[34] Lees RM, Sun ZD, Li-Hong X. Int J Infrared Millim Waves 2008;29:148–56.
[35] Lees RM, Sun ZD, Billinghurst BE. J Chem Phys 2011;135:104306.
[36] Ohashi N, Shimada H, Olson WB, Kawaguchi K. J Mol Spectrosc 1992;152:298–306.
[37] Gulaczyk I. unpublished data work in progress.
[38] Dawadi MB, Lindsay CM, Chirokolava A, Perry DS, Xu LH. J Chem Phys 2013;138:104305.
[39] Illyushin VV, Alekseev EA, Chou YC, Hsu YC, Hougen JT, Lovas FJ, et al. J Mol Spectrosc 2008;251:56.
[40] Oda M, Ohashi N. J Mol Spectrosc 1989;138:246–50.

Renata Jastrzab, Małgorzata T. Kaczmarek, Bartosz Tylkowski and Akira Odani

7 Computer analysis of potentiometric data of complexes formation in the solution

Abstract: The determination of equilibrium constants is an important process for many branches of chemistry. In this review we provide the readers with a discussion on computer methods which have been applied for elaboration of potentiometric experimental data generated during complexes formation in solution. The review describes both: general basis of modeling tools and examples of the use of calculated stability constants.

Keywords: Potentiometric method, SUPERQUAD, complexes formation

7.1 Introduction

Potentiometric method is a technique of the electrode potential determination, which is proportional to free concentration of the component in solution. Potentiometry has been used to the systems in which ligands bind metal ions and H^+. Measurement of concentration of the H^+ or metal ion is a useful tool for calculation of stability constants generally in the aqueous solution [1, 2, 4]. The simple measurement of $[H^+]$ by a glass pH electrode is highly accurate and the determinations of pK_a could be carried out on the same instrument. The pH titration method is the most important method used for determination of the stability constants. It should be mentioned that the potentiometric method cannot be applied in the following systems: (I) unprotonated ligands such as crown ethers and (II) organic media (except MeOH, EtOH, and aqueous organic solvents). This limitation is due to the absence of proton release in eq. (7.1).

$$M + HL \overset{K}{\rightleftarrows} ML + H \tag{7.1}$$

Special achievements in reaction equilibrium research carried out in aqueous solutions of complex compounds were contributed by two Danish scientists in the forties and fifties of the twentieth century, Niels Bjerrum and Janik Bjerrum. They

This article has previously been published in the journal *Physical Sciences Reviews*. Please cite as: Jastrzab, R., Kaczmarek, M. T., Tylkowski, B., Odani, A. Computer analysis of potentiometric data of complexes formation in the solution. *Physical Sciences Reviews* [Online] **2018**, *3*. DOI: 10.1515/psr-2017-0140

https://doi.org/10.1515/9783110482065-007

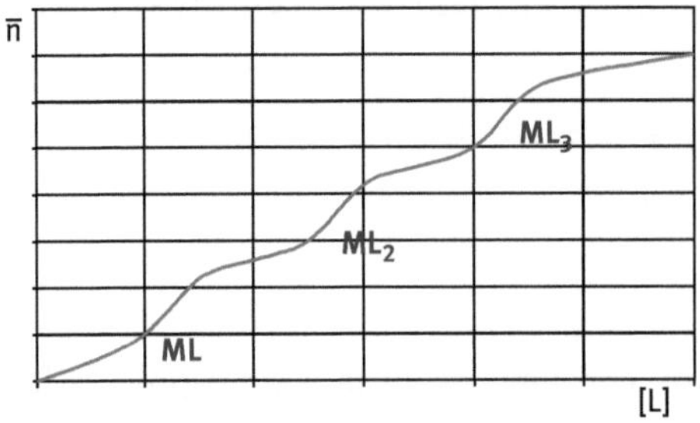

Figure 7.1: Formation function of Bjerrum [5].

introduced the concept of the function formation of the monocore complexes ($\overline{n}$) for systems obeying two laws: (i) the law of mass action and (ii) the mass balance equation (Figure 7.1).

$$\overline{n} = \frac{c_L - [L]}{c_M}$$

where c_L total ligand concentration
c_M – total metal concentration
$[L]$ – concentration of free ligand

According to their theory, complexes are formed in the way presented by the chemical reaction equations, e. g. formation of a simple type ML_n.

$$M + L \overset{K_1}{\rightleftarrows} ML$$

$$ML + L \overset{K_2}{\rightleftarrows} ML_2$$

$$ML_2 + L \rightleftarrows ML_3$$

$$ML_{n-1} + L \overset{K_n}{\rightleftarrows} ML_n$$

For each of the mentioned above reactions, gradual stability constants of the forming complexes can be determined [5].

$$K_1 = \frac{[ML]}{[M][L]},\ K_2 = \frac{[ML_2]}{[ML][L]},\ K_3 = \frac{[ML_3]}{[ML_2][L]},\ \ldots\ldots.\ K_n = \frac{[ML_n]}{[ML_{n-1}][L]};$$

The stability constant in the computational methods is generally the overall stability constant β, which is usually denoted by:

$$pM + qL + rH^+ \overset{\beta_{pqr}}{\rightleftarrows} M_pL_qH_r$$

$$\beta_{pqr} = \frac{[M_pL_qH_r]}{[M]_p[L]_q[H]_r}$$

A value of β_n is an overall stability constant of the nth complex, i. e. a product of the individual $\beta_{n=}K_1 \bullet K_2 \bullet ... \bullet K_n$.

$$K_1 = \beta_1 = \frac{[ML]}{[M][L]}$$

$$K_2 = \frac{\beta_2}{\beta_1} = \frac{[ML_2]}{[ML][L]} \quad (\log K_2 = \log b_2 - \log b_1)$$

$$K_n = \frac{\beta_n}{\beta_{n-1}} = \frac{[ML_n]}{[ML_{n-1}][L]} \quad (\log K_n = \log b_n - \log b_{n-1})$$

The limitation of the Bjerrum method corresponding to consequent formation of complexes is the possibility of determination of the values of stability constants for analogical series of compounds only, e. g. ML, ML_2, ML_n or M(HL), $M(HL)_2$, $M(HL)_n$, etc.

7.2 Computer programs applied to stability constants calculations

Dynamic development in research on thermodynamic determination of stability constant of complex formation reaction occurred in the sixties, along with development of computation techniques. First computer programs, which used the nonlinear least squares method, were mainly applied in potentiometric studies concerning reaction with proton release.

$$M^+ + HL \rightleftarrows ML + H^+$$

In the reactions of this kind, usually organic ligands are involved, in which at the released proton position, a metal ion is attached. In such processes, the change of H^+ concentration is regarded as the measure of ligand's complexity.

$$K = \frac{[ML][H^+]}{[M][HL]}$$

where K is the stability constant (equilibrium constant, formation constant, binding constant, ionization constant, or association constant). The same law is applied to the ligand-proton systems:

$$L + H \overset{pK_a}{\rightleftarrows} HL$$

$$pK_a = \frac{[HL]}{[L][H]}$$

where L is the deprotonated ligand and pK_a is $1/K_a$ (K_a = dissociation constant) [6].

The first computer softwares were based on quadric and cubic interpolation methods. One of such programs applied in analytical laboratories was GAUSS, which was launched in the sixties by D.D. Perrin and I.G. Sayca. That program was used for stability constants calculations of solutions with one type of metal ion and one type of ligand [7, 8]. At the same time, another computer program appeared in laboratories, named LETAGROUP and written by L.S. Sillen. The calculation method implemented in this program allowed for wider use of procedures for systems with more than one ligand, which enabled constant stability calculations for one metal, two ligands systems, mixed complexes. A restriction of that method narrowed it to one cycle and determination of one unknown parameter only, since the quadric interpolation is a one-dimensional search. At the first stage of the computation method there was doubt whether the stability constants obtained by the classical graph method (linear method) were same as the values by the computer (non-linear method), and hence it lasted for a while both calculations were employed.

In the late sixties, I.G. Sayce created a program SCOGS based on the FORTRAN coding. The program applied the non-linear least squares method where volumes of added titrant to a system were treated as values for comparison. The program allowed for stability constant calculations of M_2L_2H complex forms appearing in a solution next to each other. In the seventies, the SCOGS became a basic tool for iterative processes in analytical laboratories [9–13]. Other programs were developed by the following FORTRAN source codes: PKD1 [14], PSEQUAD, [15], PKAS [16] and DALSFEK [17].

In the mid-seventies, P. Gans, A. Sabatini and A. Vacca published data on a new program – the MINIQUAD [18]. In contrast to programs used previously, this program allowed calculation of the values for all the proposed model complex forms (single core complexes, multi-core complexes, protonated complexes, simple complexes and hydrocomplexes) [19, 20]. In 1985 the MINIQUAD was improved and published as the SUPERQUAD [21], and in 1996 the latest version of the program appeared, known as the HYPERQUAD. This version was extended with the ability to analyze the spectrophotometric results [22]. The last three versions of the program, especially the SUPERQUAD, are currently the most commonly used ones in laboratories in the world as programs designated to calculate stability constants of complexes formed in solution.

7.3 General basis of computer methods

Computer methods used to calculate stability constants are ruled by a fundamentally different strategy of experimental results analysis than the strategy in the classical methods. In numerical methods, once the experiment has been carried out, the approximately estimated model is introduced (i. e., the set of complex forms with approximated values of stability constants). Under the proposed model, which uses an iterative method, the program changes the values of the logβ as long as the titration curve, generated theoretically and taking into account the calculated values of the stability constants, and the experimental curve converge. As input one need experimental values such as analytical concentration of the metal, analytical concentration of ligands, the volume of the test sample, the concentration of titrant. Importantly the initial value of the stability constant is requested and the impossible values (for example, $K_1 < K_2$) result no-convergence of the calculation.

Mathematically, all discussed above computer programs use non-linear least squares method.

$$S = \sum_{i=1}^{n} w_i (f_i^{obs.} - f_i^{cal.})^2$$

where $f_i^{obs.}$ – observed value
$f_i^{cal.}$ – calculated value
w_i – statistical weight
n – number of measurements

Starting values of the logβ are improved by running an approximation process until the sum of squares S is minimized, i. e., the rests (the rest is a difference between the observed and calculated value). In case of a system with one parameter being calculated, a result of the function is the S_{min} value (Figure 7.2).

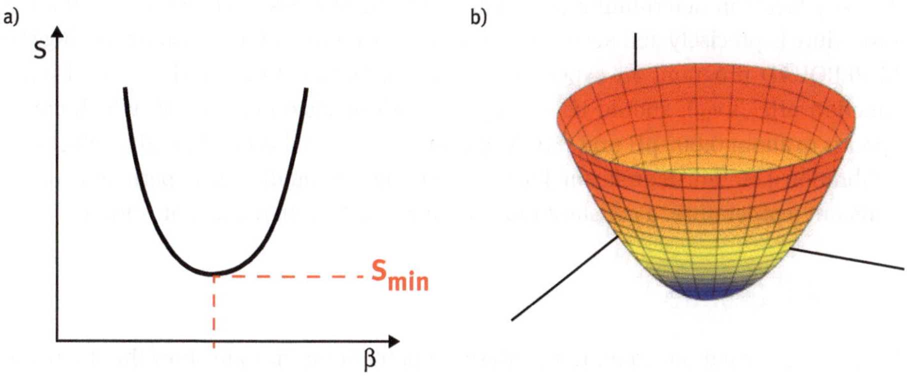

Figure 7.2: Solutions of the least squares function a) simple function, b) hyperspace.

In practice, systems with one unknown value are not observed. The objects of studies are usually systems with many parameters e. g. two metals and two ligands (i. e., bioinorganic systems).

$$a(M_1) + b(M_2) + c(L_1) + d(L_2) + eH^+ \rightleftarrows (M_1)_a(M_2)_b(L_1)_c(L_2)_dH_e$$

In this type of systems constants of ligand protonation are usually determined, as well as stability constants of simple complexes, protonated forms, hydrocomplexes and forms of heteroligands. Protonated constants and general stability constants of $[(M_1)_a(M_2)_b(L_1)_c(L_2)_dH_e]$ complexes generated in such a system are expressed by the following formula:

$$\beta = \frac{[(M_1)_a(M_2)_b(L_1)_c(L_2)_dH_e]}{[M_1]^a[M_2]^b[L_1]^c[L_2]^d[H]^e}$$

In such a case, a number of possible solutions creates multidimensional function and solution of this function is a minimum of so called hyperspace, which corresponds to the values of $\log\beta_1$, $\log\beta_2$, $\log\beta_3$ ...... $\log\beta_n$ (Figure 7.2). When the minimum value corresponding to the sum of deviations between experimental and theoretical values (value S) is too big (significantly differs from the assumed evaluation criteria in the program, where the criteria is determined by the experiment used), the assumed model can be rejected entirely and then it is described as so-called divergence of refinement process. In that situation, it is necessary to change a proposed earlier model.

Practically, refinement processes usually start with the possibly simplest models containing a major species and in the subsequent stages they are expanded in further forms of complexes.

7.4 Factors affecting convergence process

The key factor in determining the quality and the success of computer calculation procedure is precisely and accurately conducted experiment. For instance, using the SUPERQUAD program, an experiment – potentiometric titration described by the function $pH_n = f(v_{tit})$, should be carried out with precision of 0.1 mV, which corresponds to about 0.002 pH unit [21]. A systematic error is also eliminated by electrode calibration [23, 24]. Significant factors affecting the quality of experiment are: a constant temperature, a constant ionic strength and an inert gas atmosphere.

7.5 Model verification

Most of the carried-out computer analyses of potentiometric data show that there are not many problems with selection of statistically best model [13, 18, 19, 25–27]. It happens, however, that because of computer analysis, a convergence of many different models is obtained and then a problem occurs of selecting the statistically best model. If only one verification criterion is applied for a selection of the correct set of forms, it would lead to the incorrect results assessment, and finally to the selection of a wrong model. To choose the statistically best model, a multi-step verification procedure is proposed [28].

7.5.1 The convergence criterion

The aim of the positive ending of computer refinement process is to achieve convergence (convergence conditions are established for the experiment). If the input values of complex stability constants β are wrong, there is a risk of rejection of a proper model (combination of the species) on this stage of verification. In order to avoid this kind of mistake, it is advised to use as input results for similar ligands and metal ions. In some cases, when there are no literature data on investigated ligands, neither on their derivatives, one can determine the initial values of the stability constants for simple complexes using other calculation methods such as the Bjerrum method [5] or the calculation at one point/one stability constant.

7.5.2 Comparison of standard deviations

To determine the standard deviation of the sample, it is necessary to compute the variance in the first stage according to the following formula

$$s_x^2 = \frac{\sum_{i=1}^{n} (x_x - x)^2}{n - 1}$$

where x_i denotes the ith observation in the sample, x – arithmetic mean of the sample, n – number of observations.

The expression n-1 appearing in the denominator specifies the number of degrees of freedom, i. e., the number of independent information, which one can use to calculate the parameter. A standard deviation of the sample (s_x) is a positive square root of the variance.

$$s_x = \sqrt{s_x^2}$$

While the statistically best model is selected, analysis is subject to two types of sigma standard deviations: standard deviation of the titrant (SDT, e. g. in the SCOGS program) and standard deviation of stability constant (ESDC, e. g. in the HYPERQUAD program).

According to D.D. Perrin, standard deviation of the titrant (SDT) is a measure of compliance of the verified model with the experimental data. Standard deviation of the stability constant (ESDC) is a measure of precision of the calculated values of the stability constants β [29]. If the standard deviation of the titrant exceeds 0.01 in the verified model, it indicates a high probability of the existence of a statistically better model in the analyzed system and further testing is necessary.

7.5.3 Comparison of titration curves

Comparison test of titration curves is a comparison of theoretical curve (curve 2, Figure 7.3), which corresponds to the analyzed model, and experimental curve of titration (curve 1, Figure 7.3).

Differences between compared curves can suggest that coordination connections in the system which were not considered in the analysis (e. g. differences between compared curves for low values of pH point out that protonated forms were created, and differences in a high range of pH values indicate possibility of appearing hydrocomplexes, Figure 7.3). Once the protonated complexes and hydrocomplexes have been taken into account in the proposed model, the convergence of experimental and theoretical curves was achieved (Figure 7.3). It is clear from our research that the suggested model is considered as a statistically good model, if the differences

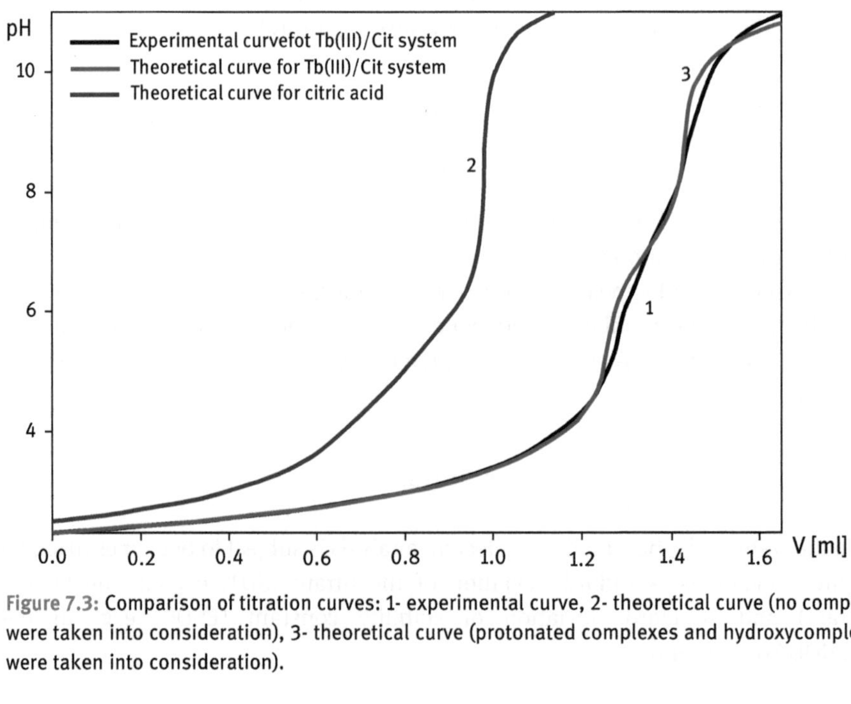

Figure 7.3: Comparison of titration curves: 1- experimental curve, 2- theoretical curve (no complexes were taken into consideration), 3- theoretical curve (protonated complexes and hydroxycomplexes were taken into consideration).

between the compared curves, especially in the pH range of formation of protonated complexes, do not exceed 0.002 of pH unit. The application of the comparison criterion for titration curves requires an extra computer program for generating theoretical titration curves, e. g. HySS [30].

7.5.4 Hamilton test (R)

Parameter R obtained in Hamilton's test is a measure of compliance between experimental and theoretical data.

$$R = \left[\frac{\sum_{i=1}^{n} w_i \left[f_i^{(obs)} - f_i^{(calc)} \right]^2}{\sum_{i=1}^{n} w_i \left[f_i^{(obs)} \right]^2} \right]^{1/2}$$

w_i is the weight of the i-th observation, where the weight is the value of 0–1.

It is assumed that the model with the lower value of R coefficient is considered as a better one [20]. It is possible to obtain a better agreement between two models by introducing some additional parameters. Testing relations of coefficients R (Hamilton's test), the values R_A and R_B (for models with lower and higher number of parameters, respectively) are compared by calculating the value $R = R_A/R_B$ in crystal structural analysis. The parameters R are gathered in tables [30], where the value $R_{b,n\text{-}m,\alpha}$ is found, where b is a hypothesis extension (a number of extra parameters in the extended model), (n-m) is a degree of freedom number, and α is a significance level. If $R > R_{b,n\text{-}m,\alpha}$, the tested model with the value R_A can be rejected with probability of an error smaller than α.

7.5.5 Chi square test (χ^2)

The χ^2 test is dedicated to hypothesis testing in which the population of rests has a given (in this case normal distribution) type of distribution.

$$\chi_2 = \sum_{i=1}^{r} \frac{(n_i - np_i)^2}{np_i}$$

where r- number of classes, n_i- number of rests in the i-th class, p_i- theoretical probability of belonging to the i-th class.

The limitation of the compliance test χ^2 is the requirement of a large sample, since the result is divided into different classes. For each class, from hypothetical distribution, theoretical values are calculated. Then they are compared with empirical values using the appropriated statistic χ^2. If the differences between empirical values and theoretical ones are too significant, the hypothesis that the population is distributed according to the assumed theoretical distribution has to be rejected [31].

7.5.6 Comparison of results from different computer programs

In some cases, while choosing the statistically best model, one can compare results obtained from two different computer programs. If as a result of the application of two different programs for the tested system several models converge, then one should select the model, which was obtained in both programs and is statistically consistent (Student's test). For example, as a result of the analysis of the tested system under the computer programs SCOGS and HYPERQUAD the values of stability constants for a given complex were obtained for each of the programs, respectively, β^s and β^H. In the next stage the following values were calculated $\Delta = \log\beta^s - \log\beta^H$ and $\delta = \delta_s^2 + \delta_H^2$ (δ_s, δ_H – standard deviations for $\log\beta^s$ and $\log\beta^H$). If $\Delta/\delta < 1.96$, then the differences between the results obtained from two programs are insignificant. If $1.96 < \Delta/\delta < 2.52$, then they are probably significant, and for $\Delta/\delta > 3.3$, the differences are highly significant. Practically, if the value Δ/δ is lower than 1.96 for several models and differs, then the model with the lowest Δ/δ value is considered as the best one.

7.6 Examples of the use of calculated stability constants

As a result of a computer analysis of data obtained from potentiometric titration of the studied system, a set of stability constants β_i is obtained as well as a number of coordination connections in the examined system. The values of stability constants allow the computation of reaction equilibrium constants in the system. Analysis of the distribution curves determines the pH values at which the maximum concentration of the individual complexes in the system is reached. This enables to determine the optimal conditions for spectral studies (UV, NMR, IR, Raman, EPR). The results obtained in this way and combined with the analysis of the equilibrium studies allow to determine the structure of the complexes in solution for example, that is, determination of the atoms in the ligands, which are involved in the interaction.

Example 1. Determination of solution structure of the complex in aqueous solution

Computer analysis of potentiometric data was applied to the system of Cu/ADP/Ser-P and it was found that in this system the protonated complex $Cu(ADP)H_3(Ser\text{-}P)$ is formed with stability constant of $\log\beta113 = 42.54$. On the basis of an analysis of the distribution curves a pH range was determined for occurrence of the complex and a formation reaction of that connection was proposed

$$\mathrm{Cu(ADP)} + \mathrm{H_3(Ser - P)} \rightleftarrows \mathrm{Cu(ADP)H_3(Ser - P)}$$

and also equilibrium constant for that reaction was calculated

$$K_e = \frac{[Cu(ADP)H_3(Ser-P)]}{[Cu(ADP)][H_3(Ser-P)]}$$

In addition, the results of the conducted spectral studies VIS and EPR show that coordination environment is formed with chromophore oxygen {O}3{N}. The NMR results suggest that also atoms of nitrogen phosphoserine and donor nitrogen atoms of the ring purine of nucleotide are involved in the interaction. By analyzing these results, it was found that Cu(ADP)H_3(Ser-P) is a molecular complex, in which a metalation center is a phosphate group of nucleotide, and completely protonated polyamine interacts noncovalently with donor atoms of the nitrogen of the ADP purine ring, which is located outside the inner coordination sphere of the central atom [32] (Figure 7.4).

Example 2. Determination of the solution structure of the complex in non-aqueous solution

Due to the insolubility of organic compounds, e. g Schiff base ligands in water the potentiometry studies can be performed in mixtures of organic solvents: dimethyl sulfoxide [33], methanol [34], ethanol [35] dioxane [36] with water.

The formation of complexes between *N,N'*-bis(salicylidene)-4-methyl-1,3-phenylenediamine (H_2L) ligand (Figure 7.5) and zinc ion or lanthanide ions (Ln = La^{3+}, Eu^{3+}, Gd^{3+}, Tb^{3+}, Ho^{3+} and Lu^{3+}) was studied by pH potentiometry in non-aqueous solution (DMSO:water 30:70 (v/v) mixture; $pK_{w} = 14.501$).

The protonation constants of the *N,N'*-bis(salicylidene)-4-methyl-1,3-phenylenediamine as well as the stability constants of the metal complexes were determined using HYPERQUAD software and that distribution of respective forms was obtained using HALTFALL program. For *N,N'*-bis(salicylidene)-4-methyl-1,3-phenylenediamine (H_2L) three protonation constants were determined from computer analysis of the titration data: $pK_1 = 8.54(7)$, $pK_2 = 6.36(7)$ (protonation of the phenolic units) and $pK_3 = 4.52(4)$ $pK_4 < 2$ (protonation of an imine nitrogen) [33, 38]. A result of the computer analysis of the potentiometric titration data for Zn^{2+}/H_2L showed formation complex with mono-deprotonated (HL) ligand Zn(HL) with logβ = 12.04 (logK = 3.50) [38]. The results of the computer analysis of the potentiometric titration data for Ln^{3+}/H_2L systems are listed in Table 7.1.

The equilibrium constants of $Ln(H_2L)_2$, $Ln(HL)_2$ formations were calculated according to the formulas: $\log_{10}K_{Ln(H2L)2} = \log_{10}\beta_{Ln(H2L)2} - 2\log_{10}\beta_{H2L}$, $\log_{10}K_{Ln(HL)2} = \log_{10}\beta_{Ln(HL)2} - 2\log_{10}\beta_{HL}$, respectively. LnL_2 complexes are the result of total ligand deprotonation according to HL ⇆ L + H and 2L + Ln ⇆ LnL_2 equilibria. The potentiometric study for Ln^{3+}/H_2L systems reveals that in solution the complexes of *N,N'*-bis(salicylidene)-4-methyl-1,3-phenylenediamine (H_2L) with lanthanide ions (La^{3+}, Eu^{3+}, Gd^{3+}, Tb^{3+}, Ho^{3+} and Lu^{3+}) appear to have 1:2 lanthanide ion to ligand

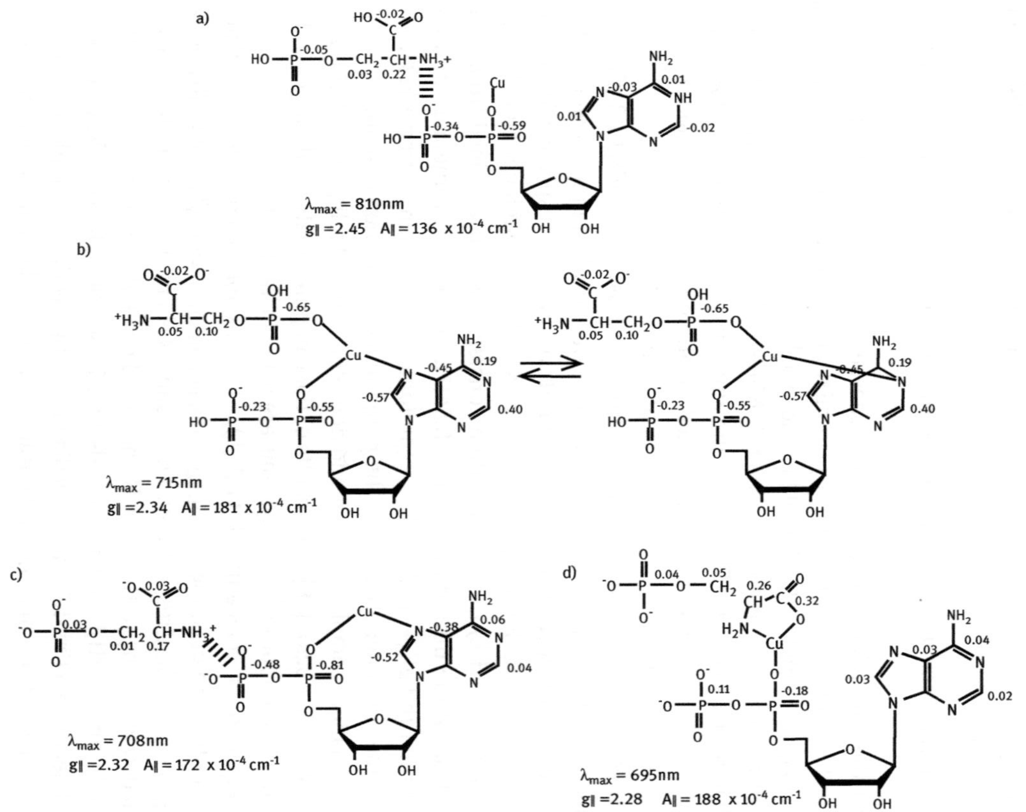

Figure 7.4: Changes in the mode of coordination in the complexes forming in the system Cu/Ser-P/ADP versus pH: (a) Cu(Ser-P)H_5(ADP) pH < 2.0; (b) Cu(Ser-P) H_3(ADP) pH = 4.5; (c) Cu(Ser-P)H(ADP) pH = 7.0; (d) Cu(Ser-P)(ADP) pH = 9.25; (spectral data and differences between NMR chemical shifts for the ligand in the Cu/ADP/Ser-P system in relation to the free ligand [ppm] were added) [32].

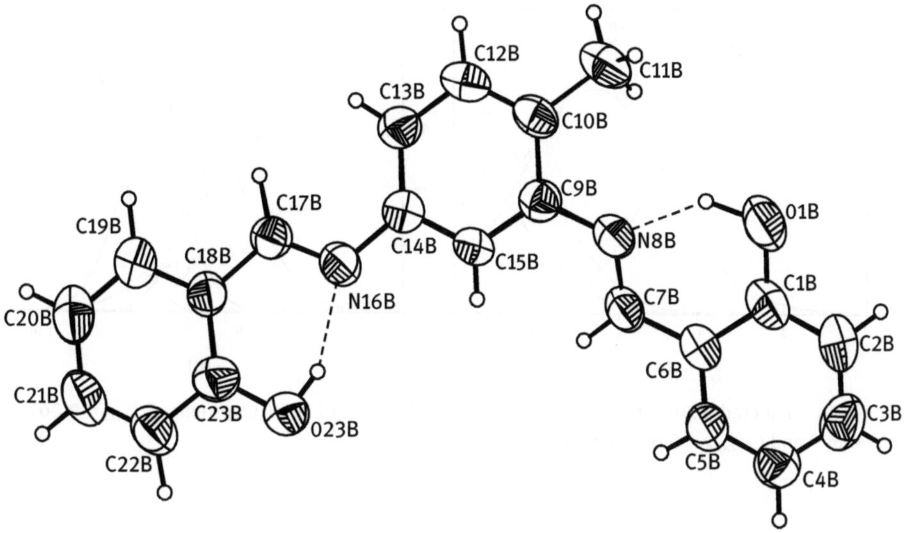

Figure 7.5: Crystal structure of *N,N'*-bis(salicylidene)-4-methyl-1,3-phenylenediamine (H_2L) [37].

Table 7.1: Stability constants ($\log_{10}\beta$) and equilibrium constants ($\log_{10}K_e$) of complexes formed in the Ln^{3+}/H_2L systems. Adapted from Kaczmarek et al. 2009.

System		La	Eu	Gd	Tb	Ho	Lu
$Ln(H_2L)_2$	$\log_{10}\beta$	36.80 (3)	37.39 (6)	37.28 (3)	36.39 (8)	36.32 (8)	36.74 (8)
	$\log_{10}K_e$	7.00	7.59	7.48	6.59	6.52	6.91
$Ln(HL)_2$	$\log_{10}\beta$	27.89 (2)	28.05 (5)	27.91 (3)	27.46 (5)	27.54 (6)	27.97 (6)
	$\log_{10}K_e$	10.81	10.97	10.83	10.38	10.46	10.89
LnL_2	$\log_{10}\beta$	17.72 (2)	17.19 (7)	17.30 (6)	17.30 (4)	17.53 (6)	18.26 (5)
	$\log_{10}K_e$	17.72 (2)	17.19 (7)	17.30 (6)	17.30 (4)	17.53 (6)	18.26 (5)

stoichiometry [39]. Selected species-distribution diagrams are depicted in Figure 7.6. Additionally, potentiometric data has indicated that in solution the lanthanide (Ln) complexes exist as LnL_2, $Ln(HL)_2$ and $Ln(H_2L)_2$ species. Stability constants of lanthanide complexes generally increase monotonically with the increasing atomic number of the lanthanides due to lanthanide contraction (decrease of ionic radii). However, for many lanthanide complexes there is a significant deviation from the expected linear relation around gadolinium. This phenomenon is known as the "gadolinium break" [40–47]. In that case only for the LnL_2 species the depression in the stability constants around the europium ion is observed. The order of stabilities of the LnL_2 species in terms of metal ions is $La^{3+} > Eu^{3+} \approx Gd^{3+} = Tb^{3+} < Ho^{3+} < Lu^{3+}$ with an evident"gadolinium break".

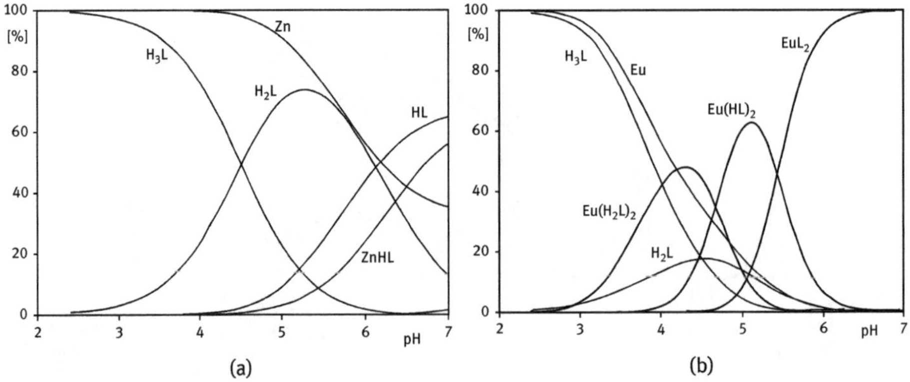

Figure 7.6: Selected distribution diagrams for the (a) Zn/H_2L; (b) Eu/H_2L systems; $C_{[metal]} = 0.001$ $mol \bullet dm^{-3}$ and $C_{[ligand]} = 0.002$ $mol \bullet dm^{-3}$ [38, 39].

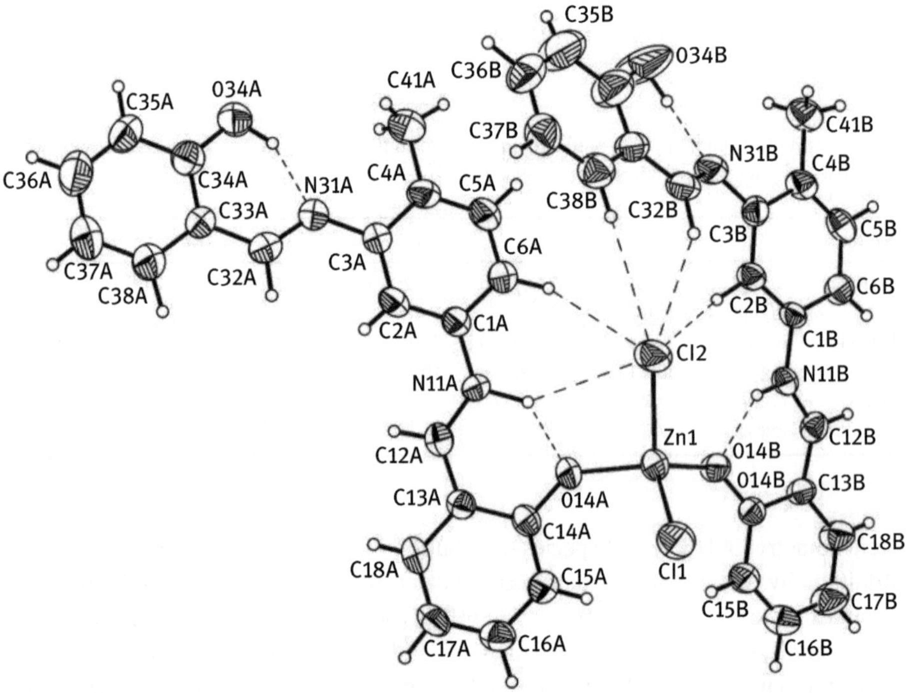

Figure 7.7: Crystal structures of $[Zn(H_2L)_2Cl_2]$ [37, 38, 48].

In both cases 1:1 Zn/ H_2L and 1:2 Ln/H_2L systems these observations appear to be in contrast to solid state behavior of these complexes. In solid state molar ratio for Zn complex is 1:2, for Ln complexes (for example Eu complex) are 2:4, metal to H_2L, respectively, Figure 7.7.

Thus, for Zn/H_2L systems only in solution the presence of the 1:1 complex of monodeprotonated HL ligand was detected. For Ln/H_2L systems complexes with 1:2 metal:ligand stoichiometry were detected also only in solution, and additionally ligands occurred in three forms: fully or partially deprotonated and unionized. Contrary to solution behavior of these complexes, in all solid complexes salicylaldimine ligands remain undeprotonated and exist exclusively in neutral form. These data reveal that the solution and powder solid characteristics are not necessarily reflected in the crystal structure of the complexes [37, 48].

Disclosure statement

No potential conflict of interest was reported by the authors.

References

[1] Durst RA, editors. Ion selective electrodes, NBS spec. pub. 348. Washington: National Bureau of Standards, 1970.

[2] Burger K, editors. Biocoordination chemistry. Coordination equilibria in biologically active systems. New York: Ellis Horwood, 1990.

[3] Martel AE, Smith RM. Critical stability constants. New York: Plenum, 1974. vol. 1, 1975 vol. 2, 1977 vol. 3, 1976 vol. 4, 1982 Supplement vol. 5, 1989 vol. 6.

[4] Sillen LG, Martel AE. Stability constants of metal ion complexes (spec. publ. 17). London: The Chemical Society, 1964.

[5] Bjerrum J. Metal ammine formation in aqueous solution. In: P. Haase and Son. Copenhagen, 1941.

[6] Tsukube H, Furuta H, Odani A, Takeda Y, Kudo Y, Inoue Y, et al. Determination of Stability Constants. In: Atwood JL, Davies JED, MacNicol DD, Vögtel F, Lehn J-M, editors. Comprehensive supramolecular chemistry. vol. 8. Oxford: Pergamon, 1996: 425–482.

[7] Perrin DD, Sayce IG. Complex formation by nickel and zinc with penicillamine and cysteine. J Chem Soc (A). 1968;0:53–57.

[8] Perrin DD, Sharma VS. The stability constants of metal-adenosine triphosphate complexes. Biochem Biophys Acta 1966;127:35–41.

[9] Sillen LG. High-speed computers as supplement to graphical methods. I. Functional behavior of the error square sum. Acta Chem Scand 1962;16:159–172.

[10] Ingri N, Sillen LG. High-speed computers as supplement to graphical methods. II. Some computer programs for studies of complex formation equilibria. Acta Chem Scand 1962;16:173–191.

[11] Sillen LG. High-speed computers as a supplement to graphical methods. III. Twist matrix methods for minimizing the error-square sum in problems with many unknown constants. Acta Chem Scand 1964;18:1085–1098.

[12] Ingri N, Sillen LG. High-speed computers as a supplement to graphical methods. IV. An ALGOL Version of LETAGROP VRID. Arkiv Kemi 1964;23:97–121.

[13] Sayce IG. Computer calculation of equilibrium constants of species present in mixtures of metal ions and complexing agents. Talanta 1968;15:1397–1411.

[14] Albert A, Serjeant EP. The determination of ionization constants. 3rd ed. London: Champan and Hall, 1984.

[15] Leggett DJ, editors. Computational methods for determination of formation constants. New York: Plenum, 2014.
[16] Martell AE, Motekaitis. The determination and use of stability constants. Weinheim: VCH, 1988.
[17] Hartley FR, Burgess C, Alcock R. Solution equilibria. Chichester: Ellis Horwood, 1980.
[18] Sabatini A, Vacca A, Gans P. Miniquad–A general computer programme for the computation of formation constants from potentiometric data. Talanta 1974;21:53–77.
[19] Gans P, Sabatini A, Vacca A. An improved computer program for the computation of formation constants from potentiometric data. Inorg Chim Acta 1976;18:237–239.
[20] Micheloni M, Sabatini A, Vacca A. Nickel(II), copper(II) and zinc(II) complexes of 1,1,1-tris (aminomethyl)propane. A calculation procedure of stepwise formation constants and their standard errors from the Values Obtained for the Cumulative Equilibria. Inorg Chim Acta 1977;25:41–48.
[21] Gans P, Sabatini A, Vacca A. SUPERQUAD: an improved general program for computation of formation constants from potentiometric data. J Chem Soc Dalton Trans 1985;0:1195–1200.
[22] Gans P, Sabatini A, Vacca A. Investigation of equilibria in solution. Determination of equilibrium constants with the HYPERQUAD suite of programs. Talanta 1996;43:1739–1753.
[23] Irving MH, Miles MG, Petit OLD. The stability constants of some metal chelates of triethylenetetraminehexaacetic acid (ttha). Anal Chim Acta 1967;38:475–488.
[24] Gran G. Determination of the equivalent point in potentiometric titrations. Acta Chem Scand 1950;4:559–577.
[25] Lomozik L, Gasowska A. Investigations of binding sites and stability of complexes formed in ternary Cu(II)/adenosine or cytidine/putrescine systems. J Inorg Biochem 1996;62:103–115.
[26] Gasowska A, Jastrzab R, Bregier-Jarzebowska R, Lomozik L. Intermolecular and coordination reactions in the systems of copper(II) with adenosine 5′-monophosphate or cytidine 5′-monophosphate and triamines. Polyhedron 2001;20:2305–2313.
[27] Gasowska A, Lomozik L. Spectroscopic and potentiometric investigation of the solution structure and stability of Ni(II) and Co(II) complexes with adenosine 5′-monophosphate and 1,12-diamino-4,9-diazadodecane (spermine) or 1,11-diamino-4,8-diazaundecane. Polyhedron 2002;21:745–751.
[28] Lomozik L, Jaskólski M, Wojciechowska A. Polish J Chem 1991;65:1797–1807.
[29] Childs CW, Perrin DD. Equilibria in solutions which contain a metal ion and an amino-acid. J Chem Soc (A) 1969;0:1039–1044.
[30] Alderighi L, Gans P, Ienco A, Peters D, Sabatini A, Vacca A. Hyperquad simulation and speciation (HySS): a utility program for the investigation of equilibria involving soluble and partially soluble species. Coord Chem Rev 1999;184:311–318.
[31] Hamilton WC, editor. Statistics in physical science. New York: The Ronald Press Company, 1964.
[32] Jastrzab R. Phosphoserine and specific types of its coordination in copper(II) and adenosine nucleotides systems – Potentiometric and spectroscopic studies. J Inorg Biochem 2009;103:766–773.
[33] Hernández-Molina R, Mederos A, Gili P, Dominguez S, Lioret F, Jano C, et al. Dimer species in dimethyl sulfoxide-water (80:20 w/w) solution of *N,N'*-bis(salicylideneimine)-m-phenylenediamine (H_2sal-m-phen) and similar Schiff base with Cu^{II}, Ni^{II} Co^{II} and Zn^{II}. Crystal structure of $[Co_2(\text{sal-m-phen})_2] \cdot CHCl_3$. J Chem Soc, Dalton Trans 1997;0:4327–4334.
[34] Galić N, Cimerman Z, Tomišić V. Tautomeric and protonation equilibria of Schiff bases of salicylaldehyde with aminopyridines. Analytica Chi Acta 1997;343:135–143.
[35] Podyachev SN, Litvinov IA, Mustafa AR, Shagidullin RR, Habicher WD, Konovalov AI. Synthesis of 1,3-bis(acetylacetonyloxy)- and 1,3-bis(benzoylacetonyloxy)benzene and their complexation with lanthanide ions. Russ Chem Bull 2005;54:623–632.

[36] Montekaitis RJ, Martel AE. Potentiometric determination of the stabilities of cobalt(II) complexes of polyamine Schiff bases and their dioxygen adducts. Inorg Chem 1988;27: 2718–2724.

[37] Kaczmarek MT, Radecka-Paryzek W, Kubicki M. Self-assembly as a route to bimetallic lanthanide complexes with rare coordination pattern of salen-type ligand. Struct Chem 2010;21:779–786.

[38] Kaczmarek MT, Jastrząb R, Hołderna-Kędzia E, Radecka-Paryzek W. Self-assembled synthesis, characterization and antimicrobial activity of zinc(II) salicylaldimine complexes. Inorg Chim Acta 2009;362:3127–3133.

[39] Kaczmarek MT, Jastrząb R, Radecka-Paryzek W. Potentiometric study of lanthanide salicylaldimine Schiff base complexes. J Solution Chem 2013;42:18–26.

[40] Toraishi T, Nagasaki S, Tanaka S. Polynuclear complex formation of trivalent lanthanides by 5-sulfosalicylate in an aqueous system – potentiometric, ^{1}H NMR, and TRLIFS studies. Inorg Chim Acta 2007;360:1575–1583.

[41] Gałęzowska J, Janicki R, Mondry A, Burgada R, Bailly T, Lecouvey M, et al. Coordination ability of *trans*-cyclohexane-1,2-diamine-*N, N, N′, N′*-tetrakis(methylenephosphonic acid) towards lanthanide(III) ions. Dalton Trans 2006;0:4384–4394.

[42] B C D, Tripathy PK, Kanungo BK. Mixed chelates of some trivalent lanthanide ions containing (trans-1,2-cyclohexylenedinitrilo)tetra-acetate and norleucinate. Monatsh Chem 1991;122:341–348.

[43] Pardeshi RK, Palaskar NG, Chondhekar TK. Potentiometric study of lanthanide(III) ion complexes with some Schiff base. J Indian Chem Soc 2002;79:958–959.

[44] Pashchevskaya NV, Bolotin SN, Sokolov ME, Sklyar AA, Panyushkin VT. Potentiometric study of reactions of rare-earth elements with 3-allylpentanedione in a water-dioxane medium. Russ J Gen Chem 2006;76:1011–1014.

[45] Mahalakshmi Sita N. Equilibrium studies of lanthanide(III) complexes of 1-phenyl-3-methyl-4-benzoyl pyrazolone-5 (BMBP) and 1-phenyl-3-methyl-trifluoroacetylpyrazolone-5 (PMTFP). Indian J Chem Sec A 1997;36A:118–120.

[46] Spedding FH, Jones KC. Heat capacities of aqueous rare earth chloride solution at 25°. J Phys Chem 1996;70:2450–2455.

[47] Spedding FH, Csejka DA, DeKock CW. Heat of dilution of aqueous rare earth chloride solution at 25°. J Phys Chem 1996;70:2423–2429.

[48] Kaczmarek MT, Kubicki M, Radecka–Paryzek W. Crystal structure and spectral characterization of rare example of a salen–type zinc complex with neutral monodentate oxygen donor ligands coordination. Monatsh Chem 2006;137:997–1003.

Karolina Wieszczycka

8 Reactive extraction at liquid–liquid systems

Abstract: The chapter summarizes the state of knowledge about a metal transport in two-phase system. The first part of this review focuses on the distribution law and main factors determination in classical solvent extraction (solubility and polarity of the solute, as well as inter- and intramolecules interaction. Next part of the chapter is devoted to the reactive solvent extraction and the molecular modeling requiring knowledge on type of extractants, complexation mechanisms, metals ions speciation and oxidation during complexes forming, and other parameters that enable to understand the extraction process. Also the kinetic data that is needed for proper modeling, simulation and design of processes needed for critical separations are discussed. Extraction at liquid-solid system using solvent impregnated resins is partially identical as in the case of the corresponding solvent extraction, therefore this subject was also presented in all aspects of separation process (equilibrium, mechanism, kinetics).

Keywords: reactive extraction, process modelling, kinetic, resins

8.1 Reactive liquid extraction

8.1.1 Introduction

Liquid extraction is one of the favored separation techniques because it is very simple, rapid, selective and sensitive. By using relatively simple equipment and requiring a few time to perform, extraction procedures offer much to the chemists and engineers. Frequently it appears to be the ideal method of separating trace constituents from large amounts of other substances. Moreover, extraction can be carried out in macro-scale or in micro-scale and that increases its technological area of applying. The alternative technology to liquid extraction is so-called solvent impregnated resins (SIR). This technology synergistically combines adsorption and liquid extraction making classical methods flexible. SIR have been shown to be effective material for the selective recovery of metal ions and organic compounds from aqueous solutions, and even the industry applications of commercial resins

This article has previously been published in the journal *Physical Sciences Reviews*. Please cite as: Wieszczycka, K. Reactive extraction at liquid–liquid systems. *Physical Sciences Reviews* [Online] **2018**, *3*. DOI: 10.1515/psr-2017-0141

https://doi.org/10.1515/9783110482065-008

are limited to the few examples, this technique is very attractive especially from economical point of view.

This article was devoted to the examples of interpretation results obtained during equilibrium and kinetics studies on reactive liquid extraction and on solvent impregnated resin.

8.1.2 Classical solvent extraction

The commonly most important factors during an extraction are the partition of the solute between the phases and the separation between two, or more, different solutes. The main theory describing the classical solvent extraction is the distribution law formulated by Nernst [1] which states that at equilibrium, a given solute will always be distributed between two essentially immiscible liquid in the same proportion.

$$\mu_{org}^{o} + RTln[S_{org}] = \mu_{aq}^{o} + RTln[S_{aq}] \tag{8.1}$$

and after rearrangement

$$\frac{[S_{org}]}{[S_{aq}]} = exp\left\{\frac{\left(\mu_{aq}^{o} - \mu_{org}^{o}\right)}{RT}\right\} \tag{8.2}$$

Thus, in the case of extraction by solvation the distribution of a solute (organic or inorganic compound) between two immiscible solvents in equilibrium is defined by the partition constant (K_D^o) [2].

$$K_D^o = \frac{a_{S,org}}{a_{S,aq}} \tag{8.3}$$

The parameter is a ratio of activity of a given species (S) in the extract to its activity in the other phase with which it is in equilibrium. Its value should not vary with composition but depends on the choice of standard states and on the temperature (and eventually the pressure). However, if the concentration of the extracted compound in both phases is low and composition of aqueous phase is constant, K_D -the constant giving the concentration ratio of the solute between both phases at equilibrium, maybe used in place of the partition constant.

$$K_{D,S} = \frac{[S]_{org}}{[S]_{aq}} \tag{8.4}$$

where $[S]_{org}$ and $[S]_{aq}$ denote the concentration of the distributed solute in the organic and aqueous phase, respectively, at equilibrium. It follows from eq. that $K_{D,S}$ does not depend on the total solute concentration or on the actual phase volumes. The distribution law is only valid if the solute is in both phases in the identical chemical form.

Equation (8.4) can also be derived thermodynamically from the equality of the chemical potentials of the solute in organic and aqueous phases at equilibrium:

$$\mu^o_{org} + RTln[S_{org}] + RTln\gamma_{org} = \mu^o_{ab} + RTln[S_{aq}] + RTln\gamma_{aq} \tag{8.5}$$

where μ^o_{org} and μ^o_{aq} represent the standard chemical potentials of the solute in the organic and aqueous phase, respectively, and γ_{org} and γ_{aq} represent the activity coefficients in the organic and aqueous phases, respectively. From the above relation, the expression for the distribution coefficient at equilibrium can be expressed as follow:

$$K_{D,S} = \frac{[S]_{org}}{[S]_{aq}} = \frac{\gamma_{aq}}{\gamma_{org}} \exp\left(-\frac{[\gamma^o_{org} - \gamma^o_{aq}]}{RT}\right) \tag{8.6}$$

When the solute concentration is low, the activity coefficients γ_{org} and γ_{aq} approach unity and the stoichiometric distribution coefficient $K_{D,S}$ becomes constant because γ^o_{org} and γ^o_{aq} are also constant. Here, the extraction depends on $K_{D,S}$ only.

In many cases the component to be extracted takes part in chemical reactions and the species produced in the reactions are preferred to different extents by the organic solvents. In these cases the distribution of the compound between the two phases is influenced by the chemical composition of the system. Therefore, another factor is used to describe the distribution process – distribution ratio (D):

$$D = \frac{\text{all species concentration in organic phase}}{\text{all species concentration in aqueous phase}} = \frac{[S]_{\text{org}}}{[S]_{\text{aq}}} \tag{8.7}$$

Example, highly hydrophobic organic compounds, which are transferred as neutral species, are dependent not only on the transfer of the neutral species between the aqueous phase, but also on the transfer of the hydrophobic, ionized, organic species as free ions or ion pairs. Therefore, if solute is a weak acid (HA), which dissociates in the aqueous phase with a dissociation constant K_a, eq. (8.8)

$$HL = H^+ + A^- K_a = \frac{[H^+]_{aq}[A^-]_{aq}}{[HA]_{aq}} \tag{8.8}$$

the distribution ratio can be described as:

$$D_{HA} = \frac{[HA]_{\mathrm{T,\ org}}}{[HA]_{\mathrm{aq}} + [A^-]_{\mathrm{aq}}} \tag{8.9}$$

where $[HA]_{\mathrm{T,\ org}}$ is the sum of all species (neutral, ionic and ion pairs) that transfer to the organic phase. But, if we assume that only undissociated form of the compound can be extracted into non-polar organic phase, the distribution ratio depends on the pH and K_a of a weak acid and D_{HA} can be derived as follow:

$$K_{D,HA} = \frac{[HA]_{org}}{[HA]_{aq}} \tag{8.10}$$

and

$$D_{HA} = \frac{[HA]_{org}}{[HA]_{aq} + [A^-]_{aq}} = \frac{[HA]_{org}}{[HA]_{aq}} \times \frac{1}{\left(1 + \frac{[A^-]_{aq}}{[HA]aq}\right)} = \frac{K_{D,HA}}{\left(1 + \frac{K_a}{[H^+]}\right)} \tag{8.11}$$

From the given relationship, it can show that at low pH, when the undissociated form dominates, the compound is extracted with the highest efficiency as $D_{HA} = K_{D,HA}$ and with the lower efficiency until at high pH the compound is completely dissociated ($[HA] = [A^-]$).

In the case of weak base, such as an alkylamine, the relationship among pH and K_a differs from that observed for a weak acid. The difference is from an amine protonation, which occurs at low pH and depends on a basicity of the amine:

$$B + H^+ = BH^+ K_a = \frac{[BH^+]_{aq}}{[B]_{aq}[H^+]_{aq}} \tag{8.12}$$

For this compound the relation between pH and D_B is opposite of a weak acid: D_B increases as pH increases to value $D_B = K_{D,B}$, eq. (8.14)

$$K_{D,B} = \frac{[B]_{org}}{[B]_{aq}} \tag{8.13}$$

and

$$D_B = \frac{[B]_{org}}{[B]_{aq} + [BH^+]_{aq}} = \frac{[B]_{org}}{[B]_{aq}} \frac{1}{\left(1 + \frac{[BH^+]_{aq}}{[B]_{aq}}\right)} = \frac{K_{D,B}}{(1 + K_a[H^+])} \tag{8.14}$$

The relation between pH, Ka and distribution ratio are illustrated by Figure 8.1 and Figure 8.2.

The transport of a solute to organic phase is highly dependent on the solvent and a range of different solvent parameters have been developed in order to estimate solvent effects. When solute is introduced into solvent, it interacts with the solvent particles and generally these interactions are salvation of the solute and much weaker solute-non polar solvent dispersive interactions. Classes of diluents used for solvent extraction purposes are aliphatic molecules e. g. hexane, aromatic ones, e. g. toluene, ketones and alcohols. During a screening, the aliphatic diluents are often considered in order to obtain high distribution ratios, easy handling and easy manufacturing. The solvation of neutral, cationic and anionic pollutes has a

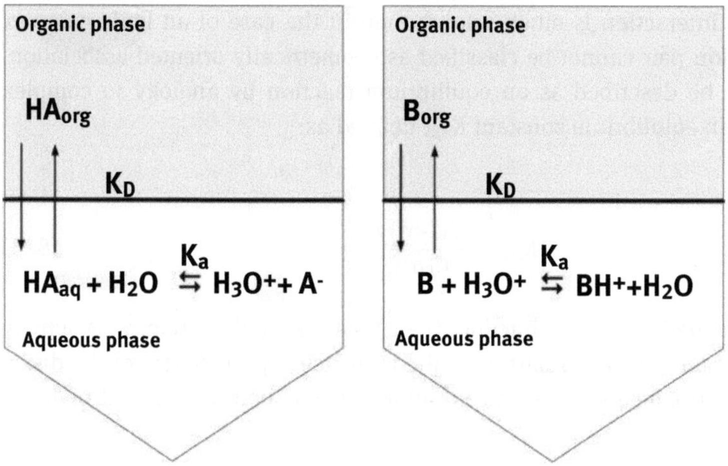

Figure 8.1: Distribution ratio of solute having acidic or basic properties.

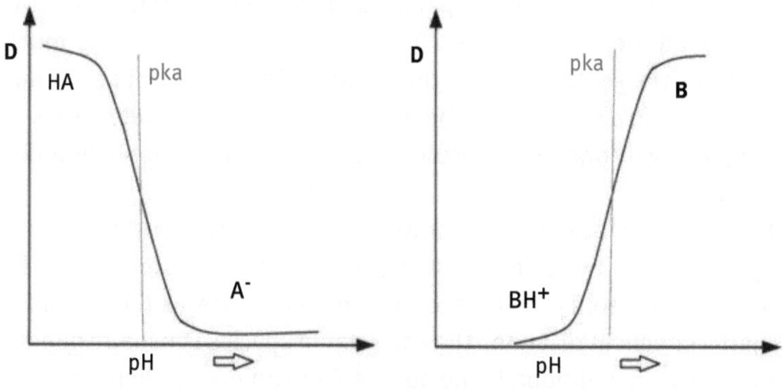

Figure 8.2: Fraction of acid or base in ionized form as function of pH.

significant influence on the chemistry and phase stability of the solution, especially with regard to ion-transfer process. In the case of solvent extraction only dispersive and repulsive forces came into play (ion–ion, ion–dipol and dipol–dipol). In non-polar solvents, with relative permittivity (ε) below 10, ionic dissociation can be completely neglected and solutes behave as if they are nonelectrolytes. Especially aliphatic diluents have a negligible polarity and a similar dipole moment, which makes the cohesive forces in such solvents dependent on the dispersion interactions between the molecules. These interactions are weak, which means that separation of this kind of adducts is relatively easy. Additionally liquids exclusively interacting through dispersion interactions contain a low energy cost for cavity formation and they have no permanent dipole moment, which can increase attractive interactions between complex and diluent. For this system an ion pairing is due to electrostatic forces between ions of opposite charges in a medium of below relative permittivity, $10 < \varepsilon < 40$, and this interaction is much weaker than in the case of an ionic complex formation. The ion pair cannot be classified as geometrically oriented association, although it can be described as an equilibrium reaction by analogy to complex formation with an equilibrium constant K_{ASS} defined as:

$$K_{ASS} = \frac{(1-\alpha)\gamma_N}{C_E \alpha^2 \gamma_\pm^2} \qquad (8.15)$$

where α is dissociated electrolyte fraction, $(1 - \alpha)$ is associated fraction, γ_N is activity coefficient of the undissociated part, $\gamma_\pm$ is the ionic activity coefficient of the dissociated part of electrolyte and C_E is concentration of solute behaves as electrolyte.

8.2 Reactive extraction

Solubility in organic solvents is not a characteristic usually attributed to simple metal salt. As can be expected from their highly ionic nature, most metal salts are strong electrolytes. However in the case of polar organic solvents such as methyl isobutyl ketone, tributyl phosphate or higher amines, the affinity of metals salts is much easier than to an aqueous phase. Much higher dissolution of metals salts in a organic than aqueous phase is due to a solvation of the metal ion of a neutral salt species or by hydrated water replacing by organic molecules, then the result is a more lipophilic adduct complex [3–7].

In the case of metals salts the extraction process is also described by a chemical reactions taking place in the two-phase systems, however, this formula can be applied, when the extracted solute soluble in an organic phase in the chemical form created through a chemical reaction. The extractable species is usually neutral complex, but complexes with additionally the outer-sphere anions (called ion pairs) are also observed in two-phase system [8].

Metal extraction system can be classified based on the nature of extractible species, but inorganic extraction is possible only if the metal-aqua complexes are converted into ion-association complexes, metal chelate or uncharged species (Table 1). Ion-association complexes are uncharged species formed by the association of ions due to purely electrostatic forces of attraction. The extent of such association increases sharply as the dielectric constant of the solvent decreases below 40. This condition not only exists in all of the commonly used organic solvents but also in highly concentrated aqueous solutions of strong electrolytes. Generally, the complexes are formed when an anionic complex of metal ion attracts with cationic species formed by the association of an extractant in an aqueous phase. An example is the extraction of $Fe(SCN)_3$ as $HFe(SCN)_4$ by diethyl ether or octan-1-ol [9], Am(III), Eu(III) and U(VI) from perchloric acid solutions by dialkyl(diaryl) phosphine oxides [10], Zn(II) as $HZnCl_3$ and H_2ZnCl_4 with tributyl phosphate [5] or distribution of ion association complexes of tetrahalogenopalladate(II) ions $PdCl_4^{2-}$ with quaternary ammonium cation [11].

In chelate complexes, organic complexing reagents are employed. These reagents replace hydrated water from metal ions (mainly transition metals) to form neutral, essentially covalent chelate complexes are usually soluble in organic mixture. Extractant forming an uncharged chelate with a metal is a weak acid whose anion participates in the complex formation. In addition, an ideal chelator should have slightly solubility in an aqueous phase. A metal ion (M^{n+}) that is equilibrated with an organic and aqueous phase is first solvated in the aqueous phase by water molecule to form metal-aqua complex. The organic anion (L-) from the dissociated organic reagent (HL) displaces the water molecule from an aqua-metal complex forming neutral metal chelate. The metal chelate distributes itself between the aqueous and organic phase according to Nernst distribution law. The overall solubility of the complex in the organic phase is very high, which is due the decrease in electrostatic interaction solute-water and the hydrophobicity of the chelate molecule. The mechanism is typical for hydroxyoximes [12], 8-hydroxyquinoline and its derivatives [13], and β-diketones [14].

The extraction of mineral salts as uncharged species is generally less complicated than the extraction of mineral acids. Metal salts with monodentate univalent anions like Cl^-, NO_3^-, ClO_4^-, SCN^- are strongly hydrated in the aqueous phase and have quite small, if any, solubility in inert solvents. In order to extract HCl, HNO_3, they must either form adducts with a basic extractant such as trialkyl amine, pyridine derivatives, trialkyl phosphine oxide, or be in solvating solvents such as ethers, ketones, alcohols or esters [23]. Examples of extracted metal salt adducts are $CuCl_2$ (Acorga CLX 50) and $UO_2(NO_3)_2(TBP)_2$.

Most important from application point of view of the extraction process is to use a selective, efficient and easy regenerated extractant. This type of reagent can be selected from commonly commercial extracting agents. However, even though well know mechanism, complex solutions can require the use of optimization procedure. Thus,

Table 8.1: Main classes and subclasses of extracting agents.

Type of complex	Mechanism	Description
Ion Associates	*Simple salts*	Extraction of metals salts using solvents of high coordinating ability, having oxygen donor atoms such as ethyl acetate, methyl isobutyl ketone or diethyl ether. The most frequently used salt systems are the halide, nitrate and thiocyanate ones.
	$M(X)_{n,aq} + mL_o \leftrightarrow M(X)_nL_{m,o}$	Example: extraction of uranyl nitrate by diethyl ether [15]
	Complex metal acids	Solvents of high coordinating ability are preferred. Frequently used systems are the hydrochloric, hydrobromic, hydrofluoric, nitric acid and acidified thiocyanate systems.
	$M^{n+}_{aq} + H^{+}_{aq} + mL^{-}_{o} \leftrightarrow HML_{m,o}$	Example: extraction of zinc(II) nitrate by TBP [5]
	Onium salts $M^{n-}_{aq} + Q^{+}X^{-}_{o} \leftrightarrow MQ_{n,o} + X^{-}_{aq}$	Extraction of various simple or complex metals as anions forming extractable ion pairs with water soluble onium salts (Q-X). Example: extraction of chromium(VI) by ammonium salts [16]
Metal chelates	*Chelate* $M^{n+}_{aq} + nHL_o \leftrightarrow ML_{n,o} + nH^{+}_{aq}$	Extraction of various metal ions using chelate forming agents such as LIX 54, LIX 84-I, LIX 84 [9]
Liquid ion exchangers	*Cation exchangers* $M^{n+}_{aq} + n(HL)_{2,o} \leftrightarrow ML_n(HL)_{2o} + nH^{+}_{aq}$	Extraction of various metal ions like Cu(II), Co(II), Ni(II) using long chain carboxylic acid or acidic phosphorus extractants [17]
	Anion exchangers $M^{n-}_{aq} + Q^{+}X^{-}_{o} \leftrightarrow MQ_{n,o} + X^{-}_{aq}$	Exchange of simple anions or anionic ligands. Frequently extractants: quaternary ammonium salts (Aliquat 336) [18], pyridinium groups extractants (3PC10 and its bromide salts) [19]

(continued)

Table 8.1: (*continued*)

Type of complex	Mechanism	Description
Ternary complexes	*Neutral ligands improving the extraction of ion associates* $M^{n+}_{aq} + mN_o + nL^-_o \leftrightarrow MN_mL_{n,o}$	Neutral ligand such as TBP, MIBK is dissolved in the organic phase to play role of modifier (N)[20]
	Mixed ligand chelate complexes $M^{n+}_{aq} + HL_o + (n-1)A^-_{aq} \leftrightarrow MA_{(n-1)}L_o$	Extraction of metals ions together with chloride ions, solvating reagents and chelate extracting agent [21]
	Extraction with cation exchangers $M^{n+}_{aq} + N + n(HL)_{2,o} \leftrightarrow MN(HL_2)_{2,o} + nH^+_{aq}$	Mechanism is typical for metal amine species and cationic extractant (cation exchangers). Example: extraction of cobalt(II) from ammonia solution with Cyanex 272 [22]
	Extraction with anion exchangers $(m-n)QX_o + MX_{m,aq} \leftrightarrow Q_{(m-n)}MX_{m,o}$	Extraction of negatively charged ionic metal species, in which X can be chloride, bromide, iodide or nitrate ions. Frequently used extractants are using quaternary ammonium groups containing extractant [5]

an experimental work is required to produce the necessary data so that process evaluation and optimization studies can be made [24]. Liquid extraction processes are conventionally designed by carrying out many laborious shakeout tests, producing data for McCabe-Thiele and other diagrams, for evaluation of performance at each set of operating conditions. An easy to use mathematical solvent extraction model that is applicable to practical systems, covers a wide range of conditions, and is accurate, would be very useful. For example, chelating reagents of the hydroxyoxime type have been used commercially for the extraction of copper from acidic solutions. Several attempts have been made at predicting metal extraction with hydroxyoximes from sulfuric acid solutions [13]. The extraction equilibrium of divalent metal ions can be generally expressed by following equation:

$$\mathrm{M_{aq}^{n+}} + \mathrm{nHL_o} \leftrightarrow \mathrm{ML_{n,\,o}} + \mathrm{nH_{aq}^{+}} \tag{8.16}$$

This equation describes the final results of extraction, which is independent of where the compound being extracted is actually formed at the interface of the phases or in the aqueous phase. The omitted constants are K_a and K_D, which are partition coefficient and acid dissociation constant, respectively. If the aggregation of the extractant in the organic phase can be ignored, the concentration of monomeric species HL is nearly equal to its total concentration (C_{HL_o}) in the organic phase. Using K_a and K_D, the concentration of the extractant in the aqueous phase ($C_{HL_{aq}}$) can be described as follow:

$$C_{HL_{aq}} = K_D\left(1 + \frac{K_a}{a_H}\right)C_{HL_o} \tag{8.17}$$

(a_H is an activity of hydrogen ion used interchangeably with pH)

The conducted experiment of the extraction at different pH enables to show the effect of pH on the distribution of the extractant between two immiscible phases (K_D) can differ even for different aqueous phase compositions). If the obtained results show evidently that $C_{HL_{aq}}$ is independent of pH, the acid dissociation of the extractant in the aqueous phase is negligibly small over the studied pH. For these results the eq. (8.17) can be simplified to:

$$C_{HL_{aq}} = K_D C_{HL_o} \tag{8.18}$$

or expressed as follow:

$$logC_{HL_{aq}} = logK_D + logC_{HL_o} \tag{8.19}$$

The value of K_D can be evaluated from intercept of strength lines obtained after plotting $logC_{HL_{aq}}$ versus $logC_{HL_o}$. K_D is a very useful parameter because it delivers information about a solubility of the extractant in the aqueous phase and relations between partition of the extractant and polarity of the diluent.

Now we can study the extraction equilibria of selected metal-extractant system, which can be predicted taking into account effect of an organic diluent and aqueous phase composition on the extraction.

8.3 Extraction equilibria and stoichiometry of extracted metal complex

8.3.1 Chelating extractant

Assuming that divalent metal after dissolution presents in the aqueous phase only as M^{2+} and any complexes with inorganic ions can be negligible, the extraction equilibrium with chelating extractant can be generally expressed by the following equation:

$$M^{2+}_{aq} + (2+x)HL_o \leftrightarrow ML_2(HL)_{x,o} + 2H^+_{aq} \quad (8.20)$$

With the extraction equilibrium:

$$K_{ex} = \frac{[ML_2(HL)_x][H^+]^2}{[M^{2+}][HL]^{(2+x)}} \quad (8.21)$$

or

$$K_{ex} = \frac{[ML_2(HL)_x]\alpha_H{}^2}{[M^{2+}][HL]^{(2+x)}} \quad (8.22)$$

For this equilibrium the distribution ratio (D) is defined as follow:

$$D = \frac{[ML_2(HL)_x]}{C_{M^{2+}}} = \frac{[ML_2(HL)_x]}{[M^{2+}]} \quad (8.23)$$

($[M^{2+}]$ and $C_{M^{2+}}$ are free and total concentration of metal ions, respectively, and at these assumptions $[M^{2+}] = C_{M^{2+}}$)

Thus the eq. (8.23) can be rewritten as

$$\log D \cdot \alpha_H{}^2 = (2+x)\log[HL] + logK_{ex} \quad (8.24)$$

and if $[HL] = C_{HL}$ eq. (8.24) is reduced as follow:

$$\log D + 2\log \alpha_H = (2+x)\log C_{HL} + logK_{ex} \quad (8.25)$$

The value of K_{ex} and x can be evaluated from intercept of strength lines obtained after plotting $logC_{HL}$ versus $\log D + 2\log \alpha_H$ Figure 8.3. Straight line with the slope 2 indicated that x=0 and M^{2+} is extracted as chelate ML_2 and off course if the slope is higher than 2 e. g. 3, 4, 5, then x = 1, 2, 3 and this indicates that M^{2+} is extracted as ML(HL), $ML(HL)_2$ and $ML(HL)_3$, respectively. The equilibrium constants of the reaction (eq. (8.22)) an be evaluated from the interceptions of the strength lines obtained also from plotting $logC_{HL}$ versus $\log D + 2\log a_H$.

The used assumption too much simplifies reaction equation ignoring the possibilities of forming associates by additional the complex extractant molecules or by an aqueous phase ligands (e. g. Cl^-, NO_3^-). Especially soft metal ions such as Zn(II), Cu(II), Cd(II), Co(II) in an aqueous chloride medium forms several chloride complexes according to the overall formula:

$$\left[MCl_i^{2-i}\right] = \beta_i\left[M^{+2}\right]\left[Cl^-\right]^i \tag{8.26}$$

where i = 1, 2, 3 or 4 depending on the number of chloride bonding to the metal ion and $\beta_{i,X}$ is an overall stability constant of ith metal(II) chloro-complexes. Therefore, to obtain a complete chemical model, the formation of various metal-chloride complexes in the aqueous phase should be considered.

The overall stability constant of metal(II) chloro-complexes at a given ionic strength, is related to the equilibrium constant at zero ionic strength, and to activity coefficients ratio as follows:

$$log\beta_{i,Cl} = log\beta_{i,Cl}^{(0)} + log\gamma_{M^{+2}} + log\gamma_{Cl^-} - log\gamma_{MCl_i^{2-i}} \tag{8.27}$$

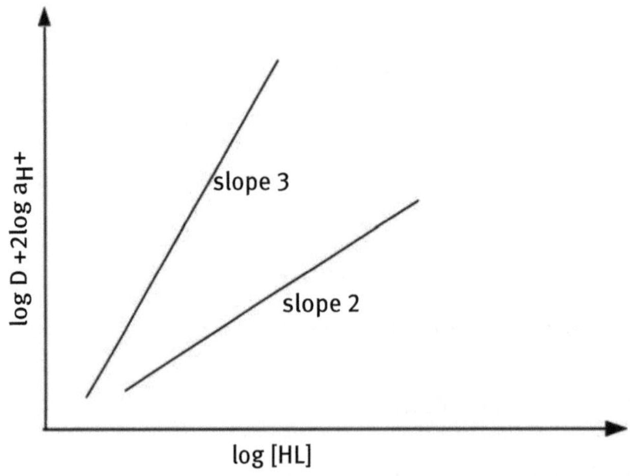

Figure 8.3: Distribution of metal ions presented as relation logC_{HL} and log D+2loga_H.

The complexation studies require the calculation of activity coefficients of all of ions present in the aqueous phase. Several models have been proposed to estimate the activity of coefficients of aqueous multicomponent solutions, but only Pitzer model [25] and Modified Bromley's Methodology [26, 27] can be applied to ionic strength values ranging up to 6 mol/kg.

8.3.2 Modified Bromley's methodology

The equation for the estimation of activity coefficient in electrolyte mixtures according to Modified Bromley's Methodology is as follow:

$$log_{\gamma} = -0.5108|z|^2 I^{0.5} 1 + I^{0.5} + F_i \tag{8.28}$$

For specific cation c and anion a, the term F_i becomes:

$$F_c = \sum_a \left[\frac{(0.06 + 0.6\,B_{ca})|z_c z_a|}{\left(1 + \frac{1.5}{|z_c z_a|} I\right)^2} + B_{ca} \right] Z_{ca}^2 [a] \tag{8.29}$$

$$F_a = \sum_c \left[\frac{(0.06 + 0.6\,B_{ca})|z_c z_a|}{\left(1 + \frac{1.5}{|z_c z_a|} I\right)^2} + B_{ca} \right] Z_{ca}^2 [c] \tag{8.30}$$

$$Z_{ca} = \frac{|z_c| + |z_a|}{2} \tag{8.31}$$

$$I = 0.5 \sum_i m_i z_i^2 \tag{8.32}$$

where z_c and z_a represent cation and anion charge number, respectively. B_{ca} represents the Bromley constant [21].

8.3.3 Pitzer model

The equation proposed by Pitzer is given below:

$$ln_{\gamma_{ca}} = |z_M z_X| F + \frac{v_M}{v} \sum_a m_a \left[2B_{Ma} + ZC_{Ma} + 2\frac{v_X}{v_M} \phi_{Xa} \right] + \frac{v_X}{v} \sum_c m_c \left[2B_{cX} + ZC_{cX} + 2\frac{v_M}{v_X} \phi_{Mc} \right]$$

$$\sum_c \sum_a \frac{m_c m_a}{v} \left[v_M Z_M C_{Ca} + v_M \psi_{Mca} + v_M \psi_{CaX} \right] + \sum_n m_n \left(\frac{v_M \lambda_{nM} + v_X \lambda_{nX}}{v} \right) \tag{8.33}$$

where

$$F_c = f^{\gamma+} \sum_c \sum_a m_c m_a B_{ca} + \sum_{c>c'} m_c m_{c'} \phi'_{cc'} + \sum_{c>c'} m_a m_d \phi'_{ad} \tag{8.34}$$

$$Z = \sum_i m_i Z_i \tag{8.35}$$

$$f^\gamma = -A^\phi \left[\frac{\sqrt{I}}{1 + b\sqrt{I}} + \frac{2}{b} \ln\left(1 + b\sqrt{I}\right) \right] \tag{8.36}$$

$$B = \left(\beta^o \frac{2\beta^1}{\alpha_1^2 I} \right) \left[1 - \left(1 + \alpha_1 \sqrt{I}\right) \right] \exp\left(-\alpha_1 \sqrt{I}\right) \tag{8.37}$$

$$C_{MX} = \frac{C^\phi}{\sqrt{2|z_M z_X|}} \tag{8.38}$$

$$\phi = \phi^S + \phi^E \tag{8.39}$$

$$\phi^E = \left(\frac{z_M z_X}{4I} \right) [J(X_{MN})] - 0.5[J(X_{MM})] - 0.5[J(X_{NN})] \tag{8.40}$$

ϕ^S is the parameter for binary interactions between two ions of the same charge and ψ is the parameter for ternary interactions.

The concentration of free metal ions $[M^{2+}]$ can be expressed using the correlation factor, α, as follow:

$$\left[M^{2+}\right] = \frac{C_M}{\alpha_M} \tag{8.41}$$

The value of α_M depends on the type of metal, formation constants of the metal-inorganic ion complexes and concentration of the inorganic ions in the aqueous phase. For example, taking into account the coordination of M(II) in aqueous chloride solutions, when metal(II) in the aqueous phase exists as $M^{2+}MCl^{+}$, MCl_2, MCl_3^- and as MCl_4^{2-}, α_M can be expressed as follow:

$$\alpha_M = 1 + \beta_i\alpha_{Cl} + \beta_2\alpha_{Cl}^2 + \beta_3\alpha_{Cl}^3 + \beta_4\alpha_{Cl}^4 \tag{8.42}$$

Then, the distribution ratio of M(II) (eq. (8.23)) can be written as:

$$\mathrm{D} \cdot \alpha_\mathrm{M} = \frac{[\mathrm{ML_2}] \cdot \alpha}{\mathrm{C_{M^{2+}}}} = \mathrm{K_{ex}}\left(\frac{\mathrm{C_{HL_o}}}{\alpha_\mathrm{H}}\right)^2 \tag{8.43}$$

and

$$\log D \cdot \alpha = 2\log\left(\frac{\mathrm{C_{HL_o}}}{\alpha_\mathrm{H}}\right) + logK_{ex} \tag{8.44}$$

The value of K_{ex} can be evaluated from intercept of strength lines obtained after plotting $\log D\alpha$ versus $\log\left(\frac{\mathrm{C_{HL_o}}}{\alpha_\mathrm{H}}\right)$. Moreover, using results obtained at a different aqueous phase composition (e. g. different chloride ions concentration), for which the plots are lying on a single straight line with an identical slope (for this consideration the slope should be close to 2), the value of K_{ex} should be nearly the same for each aqueous media.

Similar consideration was done for a modeling of Co(II) extraction with 1-(2-pyridyl)tridecan-1-one oxime from chloride-sulfate solution [28]. Chelating extractant such as 1-(2-pyridyl)tridecan-1-one oxime can extract metal ions from the aqueous solutions by forming a stable chelate. However, Co(II) ions are coordinated simultaneously by deprotonated oxygen of the hydroxyimine group and by an interaction with the pyridine nitrogen. Based on these assumptions, the reactions of cobalt(II) ions complexation by the oxime (HL) can be expressed as follows:

$$Co_{aq}^{2+} + (n+m)HL_{org} = \left(CoL_n(HL)_{m,org}\right)^{(2+)-n} + nH_{aq}^+ \tag{8.45}$$

with

$$K_{ex} = \frac{\left[\left(CoL_n(HL)_m\right)^{(2+)-n}\right][H^+]^n}{[Co^{2+}]\overline{[HL]}^{(n+m)}} \tag{8.46}$$

where free cobalt(II) ions concentration in the aqueous chloride-sulfate solution can be also expressed as follows:

$$[Co^{2+}] = \frac{C_{Co(II)}}{\alpha_{Co}} \tag{8.47}$$

where

$$\alpha_{Co} = 1 + \beta_{SO_4^{2-}} \alpha_{SO_4^{2-}} + \beta_{Cl,1}\alpha_{Cl} + \beta_{Cl,2}\alpha_{Cl}^2 + \beta_{Cl,3}\alpha_{Cl}^3 + \beta_{Cl,4}\alpha_{Cl}^4 \tag{8.48}$$

or by using eq. (8.41) in which inorganic ions activities ($\alpha_X = C_X \cdot \gamma_X$) were replaced by the concentrations.

$$[Co^{2+}] = \frac{C_{Co(II)}}{\left(1 + \beta_{SO_4^{2-}}[SO_4^{2-}] + \sum_{i=1}^{4}\beta_{i,Cl^-}[Cl^-]^i\right)} \tag{8.49}$$

In eqs. (8.40) and (8.41) $\beta_{i,x}$ is an overall stability constant of ith cobalt(II) chloro-complexes or overall stability constant of cobalt(II) sulfate-complex ($\beta_{SO_4^{2-}}$) [29, 30].

Taking into account the cobalt(II) complexation in the aqueous phase, the distribution ratio can be rewritten as follow:

$$D = \frac{\left[(CoL_n(HL)_m)^{(2+)-n}\right]}{[Co^{+2}]_T} = \frac{K_{ex}[HL]^{(n+m)}}{[H^+]^n\left(1 + \beta_{SO_4^{2-}}[SO_4^{2-}] + \sum_{i=1}^{4}\beta_{i,Cl^-}[Cl^-]^i\right)} \tag{8.50}$$

and

$$logD = logK_{ex} + (n+m)log[HL] - nlog[H^+] + log\left(\frac{1}{1 + \beta_{SO_4^{2-}}[SO_4^{2-}] + \sum_{i=1}^{4}\beta_{i,Cl^-}[Cl^-]^i}\right) \tag{8.51}$$

or

$$logD + log\left(1 + \beta_{SO_4^{2-}}[SO_4^{2-}] + \sum_{4}^{i=1}\beta_{i,Cl^-}[Cl^-]^i\right) = logK_{ex} + (n+m)log[HL] + npH \tag{8.52}$$

On the basis of the above considerations, the analysis of the extracted complex stoichiometry was carried out using a graphical method, wherein the experimental values of the distribution of Co(II) are plotted as a function of ($logD + log\left(1 + \beta_{SO_4^{2-}}[SO_4^{2-}] + \sum_{i=1}^{4}\beta_{i,Cl^-}[Cl^-]^i\right)$) versus equilibrium pH. According the equation, the slope of the straight line given in the pH range between 1 and 3 should be close to 1 or its multiple, but it was found to be 0.5. That suggested that cobalt(II) was extracted as a binuclear complex. This theory was confirmed by the logarithmic plot of the cobalt(II) distribution coefficient for the binuclear complex ($D = \frac{[Co]_o^2}{[Co]_{aq}}$) versus equilibrium pH (pH was constant during the experiment). The protonation of the free extractant molecules at studied condition was omitted (pKa = 3.78) [31].

The plot $logD + log\left(1 + \beta_{SO_4^{2-}}[SO_4^{2-}] + \sum_{i=1}^{4}\beta_{i,Cl^-}[Cl^-]^i\right)$ against log[HL] showed that at pH of 3.5 and 5 the slopes were close to 2 (relationship of 2 molecules of the oxime with 1 molecule of Co(II)), but at pH of 0.5 the slope was close to 3 suggesting that three molecule of the oxime per cobalt molecule were involved in the coordination.

It was also indicated that the role of the chloride ions concentration strongly correlates with the cobalt extraction by a coordination of cobalt(II) chloro-complexes, thus the reaction of the extraction was next rewritten to form:

$$Co_{aq}^{2+} + zCl_{aq}^- + (n+m)HL_{org} = \left(CoCl_zL_n(HL)_{m,org}\right)^{(2+)-(n+z)} + nH_{aq}^+ \tag{8.53}$$

with

$$K_{ex} = \frac{\left[\left(CoCl_zL_n(HL)_m\right)^{(2+)-(n+z)}\right][H^+]^n}{[Co^{2+}][Cl^-]^z\overline{[HL]}^{(n+m)}} \tag{8.54}$$

therefore, the distribution ratio was expressed as follow:

$$D = \frac{\left[\left(CoCl_zL_n(HL)_m\right)^{(2+)-n}\right]}{[Co^{+2}]_T} = \frac{K_{ex}[HL]^{(n+m)}[Cl^-]^z}{[H^+]^n\alpha_{Co}} \tag{8.55}$$

and

$$logD + \log\alpha_{Co} = logK_{ex} + zlog[Cl^-] + (n+m)log[HL] + npH \tag{8.56}$$

At different ionic strength the chloride ions concentration was replaced by chloride ions activity.

According to above assumption a plot of $logD + log\alpha_{Co}$ versus $loga_{Cl^-}$ was constructed from experimental data obtained at constant equilibrium pH and at constant cobalt and the extractant concentration. The slope of given straight line indicated the number of chloride molecules participating in the cobalt complexation. In these studies the obtained data have given a slope equal to 1. Thus, on the basis of the conducted graphical analysis, cobalt(II) ions at pH above 3 are transported into the organic phase as a CoL(HL)Cl complex.

8.3.4 Acidic organophosphorus extractant

Acidic organophosphorus extractant, having functional group with strongly electron-donor oxygen atoms have been extensively studied as extractants for numerous metals. The overall distribution of organophosphorus acids between two immiscible liquid phases is a balance of a number of individual equilibria such as the acid dissociation in the aqueous phase, partition of monomeric species between the phases, self – association (mostly dimerization) in the organic phase and adduct formation (extraction together with solvating extractant) [32].

Acidic organophosphorus extractant mostly exist as dimmers, especially in a nonpolar solvent.

$$K_{dim} = \frac{[(HL)_2]}{[HL]^2} \tag{8.57}$$

The dimmer behaves like a chelating reagent having simultaneously two coordinating groups to a metal ion [33–35] and metals are extracted generally as $ML_2(HL)_n$ ($CuL_2(HL)_2$, $CdL_2(HL)_2$ or $CdL_2(HL)_3$).

$$M^{2+}_{aq} + (n+2)(HL)_{org} = ML_2(HL)_{n,org} + 2H^+_{aq} \tag{8.58}$$

In the case of Cyanex 302 (active compound: *bis*(2,4,4-trimethylpentyl)monothiophosphinic acid(79 %)) Alguacil et al. [36] have indicated extraction of zinc(II) in chloride media as $(ZnL_2)_2(HL)_3$ but Benito et al. [37] have suggested that complexation of Zn(II) formed in chloride solutions is much more complicated than has been shown on eq. (8.58) It has been shown that zinc(II) is extracted as

$ZnClL(HL)$,$ZnL_2(HL)_2$,ZnL_2 and in the complex formation other compound of the extracting mixture participates (*bis*(2,4,4-trimethylpentyl)dithiophosphinic acid presented in 2 %).

8.3.5 Procedure of analysis

The zinc(II) extraction by Cyanex 302 may be expressed as follow:

$$yZn^{2+}_{aq} + zCl^-_{aq} + (n+m)HL_{org} = \left(Zn_yCl_zL_n(HL)_{m,org}\right)^{(2+)-(n+z)} + nH^+_{aq} \tag{8.59}$$

with

$$K_{ex} = \frac{\left[(Zn_yCl_zL_n(HL)_m)^{(2+)-(n+z)}\right][H^+]^n}{[Zn^{2+}]\overline{[HL]}^{(n+m)}} \tag{8.60}$$

and the distribution ratio of Zn(II) can be expressed as:

$$D = \frac{\left[\left(Zn_yCl_zL_n(HL)_m\right)^{(2+)-(n+z)}\right]}{[Zn^{+2}]^{(y-1)}} = \frac{K_{ex}[HL]^{(n+m)}}{[H^+]^n \alpha_{Zn}} \tag{8.61}$$

where α_{Zn} is a side reaction coefficient combined with the formation of chloro-complexes with zinc(II) in a concentrated chloride media.

$$\alpha_{Zn} = 1 + \sum_{i=1}^{4} \beta_{i,Cl^-}[Cl^-]^i \tag{8.62}$$

This parameter is very important especially if only a single zinc(II) species is extracted, because the eq. (8.61) can be simplified and after logarithmic procedure it can be written as:

$$logD = logK_{ex} - \log \alpha_{Zn} + (y-1)\log\left[Zn^{+2}\right] + zlog[Cl^-] + (n+m)\log[HL] + npH \tag{8.63}$$

Looking at this equation it can assume that logD should vary linearly with pH, but the obtained curve [38] showed the variation of the slopes, which ranged from 1 to 2 suggesting the formation at least two type of species extracted during the process: $ZnClL(HL)_x$ and $ZnL_2(HL)_x$

Then eq. (8.61) becomes:

$$D = \frac{K_{ex1}[Cl^-][HL]^{(1+m)}}{[H^+]\alpha_{Zn}} + \frac{K_{ex2}[HL]^{(2+m)}}{[H^+]^2\alpha_{Zn}} \tag{8.64}$$

and after rearrangement can be expressed as follow:

$$\frac{D[H^+]^2\alpha_{Zn}}{K_{ex2}[HL]^{(2+m)}} = 1 + \frac{K_{ex1}[Cl^-][HL]^{(1+m)}[H^+]}{K_{ex2}} \tag{8.65}$$

The value of K_{ex1} and K_{ex2} can also be obtained using the graphical analyses according to the experimental results are compared with the theoretical assumptions:

$$logY = \log(1+X) = f(pX) \tag{8.66}$$

where *Y* and *X* are defined as:

$$Y = \frac{D[H^+]^2\alpha_{Zn}}{K_{ex2}[HL]^{(2+m)}} \tag{8.67}$$

$$X = \frac{K_{ex1}[Cl^-][HL]^{(1+m)}[H^+]}{K_{ex2}} \tag{8.68}$$

Finally by potting $logY$ and $logY$ against $\log[HL]$, where [HL] was calculated from eq. (8.61) the values of K_{ex1} and K_{ex2} and coefficients indicating number of undissociated molecules participating in the complex structure.

8.3.6 Extraction with hydrophobic extractant containing pyridinium cation

The quaternary pyridinium ketoximes, similar to quaternary long-chain amines, can form complexes with metal(II) in chloride solutions through ion-pair and ion-exchange mechanisms [15, 16, 38]. The application of the extractant as an ion-exchange extractant clearly correlated with a sufficient concentration of anionic metal-chlorocomplexes, which can be ion-paired with the pyridinium moiety of the extractant. The participation of the hydrogen ion in the complex should also be taken into consideration, as has been observed in the case of the quaternary ammonium ion and the anionic divalent metal complexes [39]. Thus, extraction by the quaternary pyridinium ketoximes (Q^+X^-), may be expressed as follow:

$$\mathrm{MX}_{\mathrm{n,aq}}^{2-\mathrm{n}} + (2-\mathrm{n})\mathrm{Q}^+X_o^- = \left(\mathrm{MX}_{\mathrm{n}}^{2-\mathrm{n}}\right)\mathrm{Q}^+_{(2-\mathrm{n}),\mathrm{o}} + ((2-\mathrm{n}))\mathrm{X}^-_{\mathrm{aq}} \tag{8.69}$$

or ion replacement by a hydrogen cation:

$$\mathrm{MX}_{\mathrm{n,aq}}^{(2-\mathrm{n}+1)} + \mathrm{H}^+_{\mathrm{aq}} + (2-\mathrm{n}+1)\mathrm{Q}^+X_o^- = (\mathrm{MX}_{\mathrm{n}}\,\mathrm{H})^{(2-\mathrm{n}+1)}\mathrm{Q}^+_{(2-\mathrm{n}+1),\mathrm{o}} + (2-\mathrm{n}+1)\mathrm{X}^-_{\mathrm{aq}} \tag{8.70}$$

The alternative mechanism, which is the ion-pair mechanism, does not require the anionic form of the metal species to be present in the aqueous phase and it implies that a neutral metal containing species is transferred to the organic phase. This mechanism can be represented as:

$$\mathrm{M}^{2+}_{\mathrm{aq}} + 2\mathrm{Cl}^-_{\mathrm{aq}}\ \mathrm{nQ}^+X_o^- = (\mathrm{MCl}_2)\left(\mathrm{Q}^+\mathrm{X}^-\right)_{\mathrm{n,o}} \tag{8.71}$$

in the case of the zinc(II) extraction with 3-[1-(hydroxyimino)undecyl]-1-propylpyridinium bromide (3PC10-Br) all of mechanisms were taken into account and in order to find out the composition of the extracted species the experimental data was treated under the assumption of the equilibrium equations [19]. The experimental data obtained at constant ionic strength and water activity (I=4 mol/L; a_w = 0.825) have indicated that the influence of chloride ions concentration depended on the pH of the aqueous feed. The results demonstrated that the relative amounts of the extractable metal ions are directly connected with the concentration of the anionic chlorocomplexes ($ZnCl_3^-$ and both $ZnCl_2$ and $ZnCl_3$ at pH of 0 and 4.5, respectively). Thus, the equilibrium constant of the extraction was generally defined as:

$$K_{ext} = \frac{[ZnCl_p(Q^+)_n][X^-]^{n'}\alpha_{Zn}}{[Zn^{+2}]\beta_i[Cl^-]^p[Q^+X^-]^n} \tag{8.72}$$

where α_{Zn} is a side reaction coefficient combined with the formation of chloro-complexes with zinc(II) in a concentrated chloride media.

$$\alpha_{Zn} = 1 + \sum_{i=1}^{4} \beta_i [Cl^-]^i \tag{8.73}$$

and β_i is the overall stability constant of ith zinc(II) chloro-complexes [40].

$$logD = logK_{ext} + \mathrm{n}\log[Q^+X^-] + \log\beta_\mathrm{i} + (\mathrm{p})\log[\mathrm{Cl}^-] - \mathrm{n'}\log[X^-] - \log\alpha_{Zn} \tag{8.74}$$

Assuming a comparable activity of all inorganic ions (also bromide) [41] and that value of $\log\beta_\mathrm{i}$, Zn(II) and the quaternary pyridinium ketoxime concentrations were constant the eq. (8.66) was simplified to:

$$logD + log\alpha_{Zn} = logK' + (\mathrm{p} - \mathrm{n'})log[\mathrm{Cl}^-] \tag{8.75}$$

where

$$K' = K_{ext}\beta_\mathrm{i}[Q^+X^-]^\mathrm{n} \tag{8.76}$$

Taking into consideration the above equations, it was found that modeling of the extraction process can be achieved by analyzing the influence of the extractant, chloride, zinc(II) and the hydrogen ions concentrations in the aqueous phases. Obtained results should also be compared with the analysis of the chloride ions concentration in the aqueous phase (or bromide, if necessary) and the hydrogen ions concentration in the organic phase.

Based on the above results log-log plots were constructed. The number of chloride participated in the complex was evaluated from intercept of strength lines obtained after plotting $\log D + \log\alpha_{Zn}$ versus $\log[Cl^-]$ at constant Zn(II) and the (Q^+X^-) concentration. Free chloride ions concentration was calculated from equation:

$$C_{Cl^-} = [Cl^-] + \sum_{i=1}^{4} \beta_{i,Cl^-}[Zn^{2+}][Cl^-]^i \tag{8.77}$$

and free Zn(II) ions concentration from equation:

$$C_{Zn^{2+}} = \left[Zn^{2+}\right]\left(1 + \sum_{i=1}^{4} \beta_{i,Cl^-}\left[Cl^-\right]^i\right) \tag{8.78}$$

The obtained correlation was linear with slopes 2 and 3 at pH of 4.5 and 0, respectively.

At constant chloride and zinc(II) concentrations the number of extractant molecules coordinated Zinc-chloride species can be predicted by plotting $logD + log\alpha_{Zn}$ against $log\left[QX^-\right]$ and the obtained values gave straight lines with slopes around 1 and 2 (at pH of 0 and 4.5, respectively).

The values of the slopes mean that zinc(II) is coordinated by one or two molecules of the quaternary pyridinium ketoxime. The slope analysis method revealed that the extraction equilibria of zinc(II) ions could be expressed by four equations:

$$K_{ext} = \frac{[ZnCl_3(Q^+)][Cl^-]}{\left[ZnCl_3^-\right][Q^+X^-]} \tag{8.79}$$

$$K_{ext} = \frac{[ZnCl_3H(Q^+X^-)]}{\left[ZnCl_3^-\right][H^+][Q^+X^-]} \tag{8.80}$$

$$K_{ext} = \frac{\left[ZnCl_4(Q^+)_2\right][Cl^-]^2}{\left[ZnCl_4^{2-}\right][Q^+X^-]^2} \tag{8.81}$$

$$K_{ext} = \frac{[ZnCl_4H(Q^+)][Cl^-]}{\left[ZnCl_4^{2-}\right][H^+][Q^+X^-]} \tag{8.82}$$

Therefore, the confirmation and validation of the extraction models required addition procedure- least square fitting of the theoretical values to the experimental data or by least squares procedure

e. g. using equation:

$$U = \frac{\sum (logD_{exp} - logD_{calc})^2}{N} \tag{8.83}$$

8.4 Kinetic of extraction

Many industrial processes are designed on the basis of their equilibrium behaviors, in which selectivity is combined with too low efficiency. Then, the processes can be significantly enhanced by exploiting the differences in the rate of complexing species. General extraction kinetic is a function of all kinetics of the various reactions that take place in the system and the diffusion rate of the various reactants and product that control the chemistry [24]. They can be described by subsequent partitioning, complexation and diffusion of the initial species from the bulk phase to the reaction interface, and solute-extractant complexes from the interface to the bulk phase [42, 43].

In most extraction process both the organic and aqueous phase are significantly stirred making the concentration of all species throughout a given phase equal. But even the most vigorously stirred systems possess two thin films at the aqueous–organic interface. According to Whitman's double-film theory [44, 45], the phases are separated by an interface and a double film (one of each phase) adheres to this interface. This theory assumes that only mechanism for transport across the laminar sublayers is diffusion and so concentration gradient in that region is linear.

The diffusion layer is the predominating rate-controlling factor in many extraction processes. Therefore, in the field of the extraction kinetics is the control, or at least the minimization, of the diffusive contribution to the overall process. Even when dealing with the most efficiently stirred systems, the chemical species that have to be transported to or away from the interface always have to migrate through the diffusion films. The time required for this process often is longer than that required for the chemical changes. As a consequence, diffusion through these films controls the overall kinetics of extraction. The limits to transport imposed by the diffusion can be minimized by decreasing film thickness or by increasing the mobility of the diffusible species. Both parameters can be regulated by stirring rate of the phases, type of the stirrers used, geometry of the extraction module, and on the viscosity and density of both phases.

The extraction kinetics is mostly studied using a Lewis-type cell [46–48]. It was recognized by Danesi [49] that this technique is characterized by poorly defined hydrodynamics and ultimate film thicknesses of the order of 10 μm. The plateau observed for the rapidity of transfer as a function of the rotation speed of the rotors, which is generally attributed to a kinetic regime, can be misleading. In fact, it can result from a slip of the fluids on the blades of the rotors when their rotation speed is increased above some critical speed, not from a kinetic regime in which the contribution from diffusive transport has been minimized.

Other techniques available for measuring the kinetics of solvent extraction [43]:

1. Stirred tank – technique in which the two phases are so highly stirred that droplets of one phase are dispersed in a continuum of the second phase.

2. Rotating diffusion cell – technique bring the two phases in contact at the surface of a rotating membrane filter having well characterized hydrodynamics.
3. Moving drops – techniques measure the extraction that takes place when a single droplet of one phase travels along a vertical tube filled with the second phase.
4. Short-time phase contact methods – techniques do not stir the two phases, and the contact time between them is very short.

8.4.1 Example of kinetic study

As presented in the previous section, during the metal extraction into an organic phase series reactions occur between metal species and the extracting agent. Some reactions take place in the bulk aqueous phase, some only at interface and other are a combination of both.

Considering the metal extraction using chelating extractant when proton is replaced by metal(II) ion, the reaction is expressed as:

$$\mathrm{M^{2+}_{aq}} + 2\mathrm{HL_o} \leftrightarrow \mathrm{ML_{2,\,o}} + 2\mathrm{H^+_{aq}} \tag{8.84}$$

and for this general description the rate is expressed as:

$$-\frac{\mathrm{d}\left[\mathrm{M^{2+}}\right]}{\mathrm{dt}} = -\frac{1}{2}\frac{\mathrm{d[HL]}}{\mathrm{dt}} = \frac{\mathrm{d[ML_2]}}{\mathrm{dt}} = \frac{1}{2}\frac{\mathrm{d[H^+]}}{\mathrm{dt}} \tag{8.85}$$

However, in the extraction kinetics analysis of the equilibrium relationship of metal (II)-extractant all reactions steps should be taking into account, as expressed below:

$$HL \leftrightarrow HL_o, K_{D,HL} \tag{8.86}$$

$$nHL \leftrightarrow (HL)_n, K_{dim,HL} \tag{8.87}$$

$$HL \leftrightarrow H^+ + L^-, K_a \tag{8.88}$$

$$HL + \mathrm{M^{2+}} \leftrightarrow \mathrm{ML^+} + H^+, K_1 \tag{8.89}$$

$$L^- + \mathrm{M^{2+}} \leftrightarrow \mathrm{ML^+}, K_2 \tag{8.90}$$

$$\mathrm{ML^+} + HL \leftrightarrow \mathrm{ML_2} + H^+, K_3 \tag{8.91}$$

$$\mathrm{ML_2} \leftrightarrow \mathrm{ML_{2,\,o}}, K_{D,ML_2} \tag{8.92}$$

If the reaction proceeds in the bulk aqueous phase then it may be true that K_2 is slow process but K_3 and K_{D,ML_2} are fast. Or if the reaction is interfacial then the relation is similar but the rate is determined by concentration of L_i^- and K_{D,ML_2} at aqueous/interface zone:

$$L_i^- + M^{2+} \leftrightarrow ML^+ \text{ slow } -\frac{d[M^{2+}]}{dt} = k_i S[L_i^-][M^{2+}] \quad (8.93)$$

$$ML^+ + L_i^- \leftrightarrow ML_{2i} \text{ fast} \quad (8.94)$$

$$ML_2 \leftrightarrow ML_{2,\,i} \text{ fast} \quad (8.95)$$

where S represent the specific interfacial area.

From a practical point of view, the apparent values of the rate constants of the liquid extraction reactions are usually evaluated by measuring the rate of extraction of M^{2+} as function of [HL] (at pH constant), of $[H^+]$ (at [HL] constant), and of ML^+ or ML_2 (at pH and [HL] constant). The experimental conditions are usually chosen in such a way that the reaction can be assumed pseudo-first-order.

In the case of an oxime extractant which is a typical chelating reagents Harada and Miyake [50] have assumed that for this reagent the concentration of ML_2 is much larger than that of ML^+ (or even undetectable as in the case of hydroxyoxime extractants [51]). Furthermore, they have assumed that pK_a value is much larger than pH of the aqueous phase, concentration of L^- is much lower than the concentration of HL and HL-consumption rate can be evaluated using instrumental methods (GC-MS, LC-MS, FT-IR). In the agreement with the above description, the rate of ML_2 – formation can be described as:

$$r_{ML_2} = k\left([HL] - \frac{[ML_2][H^+]^2}{K_1K_3[HL][M^{2+}]}\right) \quad (8.96)$$

$$k = \frac{k_1 + \frac{k_2}{\frac{[H^+]}{Ka} + \frac{k_2[M^{2+}]}{Ka}}}{1 + \frac{k_1[H^+]}{K_1k_3[HL]} + \frac{k_2/K_1k_3[H^+]/[HL]}{[H^+]/K_a + k_2[M^{2+}]/k_a}}[M^{2+}] \quad (8.97)$$

$$[ML^+] = kk_3 \quad (8.98)$$

Dobson and Van Der Zeeuw [52] have indicated that oxime molecule is adsorbed at the interface by interacting with disordered water molecules, therefore Harada and Miyake also assumed that only the extractant as monomer interacts with water in the interfacial zone. Thus, taking into account that the polar part of the extractant exists in the interface zone, this space works as aqueous phase. For this system the adsorption equilibrium of HL is expressed by Gibbs adsorption equation with Langmuir's kinetic expression [53]:

$$\Theta_j = \frac{K_J^s[J]_s}{1+\sum_J K_J^s[J]_s} = \frac{K_{J,o}^s[J]_{s,o}}{1+\sum_J K_{J,o}^s[J]_{s,o}} \tag{8.99}$$

where $[J]_s$ and $[J]_{s,o}$ are the concentration of the species J in the aqueous and organic phase, respectively. In the case of the extractant showing a very high interfacial activity the distribution ration can also be expressed as:

$$K_{D,HL} = \frac{[HL]_o}{[HL]} = \frac{K_{HL}^s}{K_{HL,o}^s} \tag{8.100}$$

Next, keeping the last postulates, and additionally assuming that ML_2 is weakly surface active, the stationary state descriptions for ML^+ and L^- are expressed as

$$\frac{2d\theta_{L^-}}{dt} = k_a^*\theta_{HL} - k_a'^*[H^+]_s\theta_{L^-} - k_2^*[M^{2+}]_s\theta_{L^-} + k_2'^*\theta_{ML^+} = 0 \tag{8.101}$$

$$\frac{2d\theta_{ML^+}}{dt} = k_1^*[M^{2+}]_s\theta_{HL} - k_1'^*[H^+]_s\theta_{ML^+} + k_2^*[M^{2+}]_s\theta_{L^-} - k_2'^*\theta_{ML^+} - k_3^*[X_{HR}]\theta_{ML^+} + k_3'^*[H^+]_s\theta_{ML_2} = 0 \tag{8.102}$$

and the rate formation of ML_2 in the interfacial zone is described as:

$$r_{ML_2}^* = \frac{1}{2}k_3^*[X_{HL}]\theta_{ML^+} - \frac{1}{2}k_3'^*[H^+]_s\theta_{ML_2} \tag{8.103}$$

X_{HL} is the concentration of HL species which react with ML^+ leading to ML_2, θ_{ML^+} can be calculated from eq. (8.101) and θ_{ML_2} from below equation:

$$\theta_{ML_2} = \frac{K_{ML_2}^s[ML_2]}{1+K_{HL}^s[HL]} \tag{8.104}$$

The rate of metal extraction is obtained by solving the diffusion equations accompanied by the complex formation. Harada and Miyake [50] assumed that the diffusion coefficients for HL and the metal complexes are equal to each other in the aqueous phase. Finally proposed diffusion equations in linear form were then solved yielding the metal flux *j* as shown in [50]. Finally the experimental results were compared with calculated value confirming that in this case the contribution of the reaction in the interface to the overall reaction is negligibly small. Moreover, the metal complexes are formed in the aqueous stagnant film and the main determining step was formation of ML^+ species (ML_2 is controlled by the reaction shown as eq. (8.94), which proceeds at the interface zone).

8.5 Extraction at liquid-solid system

The application of classical separation techniques such as liquid extraction for processing very dilute solutions is rather impracticable and uneconomical due to high costs and very low yields. Reactive liquid extraction has also drawbacks with phase separation, emulsification and vigorous shaking which leads to the loss of reagent. In addition, reactive extraction is not very feasible for the recovery of solute present in low concentration due to the high excess of diluents and extractant required and not very suitable for recovering selected species from viscous solutions. Another method is to immobilize the selected efficient extractant in porous particles. The solvent impregnated resin (SIR) consists of commercially available macroporous resins (mostly polymer resin) impregnated with an extractant. SIRs can be easy reused by simply rinsing and expensive production of ion exchangers can be eliminated. Furthermore, since the pores are filled with liquid extractant, a higher capacity for solutes can be achieved than with free matrix or ion exchange resins, where only the surface area is available. Possible drawbacks of SIR are re-impregnation of the extractant or blocking of a fixed bed by inorganic ions or excess of the extractant. Important is also a finding a suitable extractant with low water solubility, which is sufficiently retained inside the pores, and selecting an appropriate solid support.

Mechanism of a solute recovery (metals ions or organic compounds) with solvent impregnated resin is partially identical as in the case of the corresponding solvent extraction systems and partially as in the case of ion-exchange resin. Kamio et al. [54] have developed a theoretical model to describe a metal recovery with a microcapsule containing an acidic extractant. The determined kinetic constants for the complex formation reactions are almost the same as that determined for the liquid-liquid extraction system and is supported by Fick's diffusion laws [55, 56]. It was shown, that the extraction mechanism of metal ions into a microcapsule progresses through the three processes (Figure 8.4): (I) diffusion of metal ion across the aqueous liquid film exists near the surface of a microcapsule; (II) intermediated and final extracted complex formation on the surface of a microcapsule, and the last process; (III) diffusion of extracted complex through resin pores. The extracted complex gradually accumulates at the surface of the resin. The surface of the microcapsule then is saturated with the extracted complex.

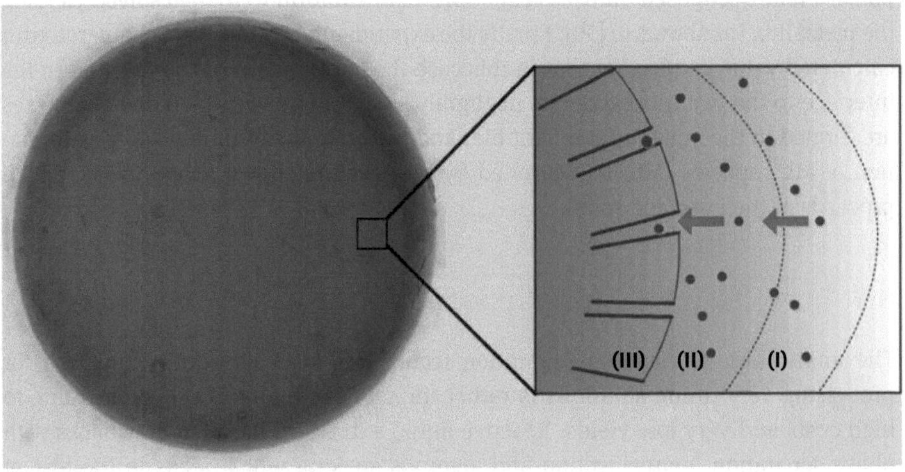

Figure 8.4: Extraction mechanism of metal ions into SIR particle.

The interpretation of the observed equilibrium behavior in the case of the organic compounds removal a model, based on both physical and chemical equilibrium and mass balances, has been proposed by Burghoff et al. [57, 58] and Babić et al. [59]. Using this model the starting point is the characterization of the extraction process at classical solvent extraction conditions (equilibrium between the two immiscible liquid phases). However, the equilibrium should be taken into account not only as a solute transport to an organic phase through chemical reaction but also by physical distribution. In the classical liquid extraction the solute transport is also considered to consist of two mechanisms, but in the case of the metals ions extraction the chemical extraction was mainly considered. The physical equilibrium of a solute between the aqueous phase and the organic extractant phase is described through the physical equilibrium distribution coefficient $K_{D,phys}$

$$K_{D,phys} = \frac{[S]_{org}}{[S]_{aq}} \tag{8.105}$$

But, if the solute reacts with the extractant even to form very weak interaction (e. g. intermolecular hydrogen bonding), the chemical equilibrium of the solute-extractant (SE) bond formation can be described by eq. (8.106)

$$K_{D,chem} = \frac{[SE]_{org}}{[S]_{org}[E]_{org}} \tag{8.106}$$

For the liquid–liquid extraction both chemical distribution and physical distribution are taken into account when deriving the overall distribution coefficient $K_{D,ex}$, which is defined via eq. (8.107), where the total amount of solute in the organic phase consists of the physically dissolved solute and the complexed solute.

$$K_{D,ex} = \frac{[S]_{aq} + [SE]_{org}}{[S]_{aq}} \tag{8.107}$$

In order to calculate the solute equilibrium distribution between the SIR and the aqueous phase it is assumed that the impregnated extractant behaves as a bulk solvent [57]. Since the pore size of the used macroporous particles is <1 mm it is expected that the only difference with liquid–liquid extraction is the organic phase stabilization by the particle matrix. If the polymer matrix has not shown the adsorption properties, it can be considered as an inert matrix. Thus, the solute content in the organic phase is similar to the solute content inside the resin particle and can be calculated from the volume of the organic extractant phase with respect to the volume of the SIR particle. Introduction of these simplifications in eq. (8.107) result in an equation for $K_{D,ex}$, eq. (8.108), in which the distribution coefficient is based on the solute content inside the resin particle.

$$K_{D,ex} = \frac{\left([S]_{ORG}^{SIR} + [SE]_{org}\right)\frac{V_{SIR}}{V_O}}{[S]_{aq}} \tag{8.108}$$

Since the particle can be considered as inert, the constants $K_{D,phys}$ and $K_{D,chem}$ can be directly regarded toward SIR particles and can be expressed as K_{phys} and K_{chem}, respectively. Finally, the concentration of the solute-extractant (SE) inside the SIR particles is calculated from eq. (8.109).

$$[SE]_{org}^{SIR} = \left(\frac{K_{phys}K_{chem}[S]_{aq}[E]_{org}}{K_{phys}K_{chem}[S]_{aq} + 1}\right) \tag{8.109}$$

After Introduction of eq. (8.108) into eq. (8.109) the correlation of the overall distribution coefficient taking into account the chemical and physical extraction equilibrium constants, can be described using eq. (8.110). This overall distribution coefficient $K_{D,SIR}$ describes the distribution of a solute between the extractant phase inside the particle pores and an aqueous phase contacting with the resin particle. It is correlated to the chemical and physical equilibrium constants obtained from classical equilibrium experiments.

$$K_{D,SIR} = K_{phys}\left(1 + \frac{K_{chem}[E]_{org}}{K_{phys}K_{chem}[S]_{aq} + 1}\right) \tag{8.110}$$

The transport of a solute between an immobilized extractant and an aqueous phase can also be expressed by q_s parameter, which describes capacity of the resin:

$$q_s = \frac{\left([S]_{ORG} + [SE]_{org}\right) V_{SIR}}{m_{SIR}} \tag{8.111}$$

However typical physical concentration of solute (physical transport, $K_{D,phys}$) can be described by introducing the physical equilibrium constant K_{phys} as expressed as follow:

$$q_s = K_{phys}[S]_{aq}\frac{V_{org}}{V_{SIR}} \tag{8.112}$$

Therefore, considering both processes, chemical and physical transport, q_s should be described as:

$$q_s = \frac{K_{phys}\left([S]_{aq} + \frac{K_{chem}[E]_{org}}{K_{phys}K_{chem}[S]_{aq}+1}\right)V_{org}}{m_{SIR}} \tag{8.113}$$

Martak and Schlosser [60] have proposed a much higher precise in the modeling of the recovery process of butyric acid from an aqueous solution. In their model four mechanisms have been taken into account: physical extraction of butyric acid by diluent (extraction as monomer and dimmer), competitive extraction of butyric acid as complex with ionic liquids Cyphos IL-104 (Cyphos IL-104 was used as impregnate of polymer matrix), co-extraction of water with butyric acid and nanocompetitive extraction of the acid.

Martak and Schlosser [61, 62] considering the butyric acid extraction by Cyphos IL-104 (IL) divided the model into four parts corresponding to four mechanisms. First was the extraction according to the acid was transported to the organic phase via a coordinating mechanism (association of neutral butyric acid, BAH) with the complexes of structure $(BAH)_p(IL)(H_2O)_2$ (p in the range 1–5). Then, the concentration of BAH in IL phase was calculated using eq. (8.114):

$$[BAH]_{org,IL} = \frac{[IL]_{SIR}\sum_{p=1}^{5} pK_{p,1}\left([BAH]_{aq}\right)^p}{1+\sum_{p=1}^{5} pK_{p,1}\left([BAH]_{aq}\right)^p} \tag{8.114}$$

and the loading parameter of IL was calculated from eq. (8.115):

$$z = \frac{[BAH]_{org,IL}}{[IL]_{SIR}} = \frac{\sum_{p=1}^{5} pK_{p,1}\left([BAH]_{aq}\right)^{p}}{1+\sum_{p=1}^{5} pK_{p,1}\left([BAH]_{aq}\right)^{p}} \tag{8.115}$$

Analyzing the physical extraction of butyric acid by diluent, which in this study was dodecane (dd), the equilibrium concentration of BHA was expressed as follow:

$$[BAH]_{org,dd} = \frac{[BAH]_{phys} V_{phys,BAH}}{V_{BAH}} \tag{8.116}$$

But considering both the reactive (chemical) and physical extraction, the total BAH concentration was calculated from eq. (117):

$$[BAH]_{org,IL} = [IL]_{SIR} \frac{[dd]_o M_{dd} \rho_{BAH}\left(K_{D,BAH}[BAH]_{org} + 2(K_{D,BAH})^2 K_{dim,BAH}[BAH]_{org}^2\right)}{[IL]_{SIR,o}\left[\rho_{BAH} - \left(K_{D,BAH}[BAH]_{org} + 2(K_{D,BAH})^2 K_{dim,BAH}[BAH]_{org}^2\right)\right]\rho_{dd}} + z \tag{8.117}$$

where $[dd]_o$,M_{dd} and ρ_{dd} are initial concentration, molar mass and density of dodecane, respectively, ρ_{BAH} is the density of pure BAH, and $K_{D,BAH}K_{dim,BAH}$ are the distribution ratio and dimmer formation constant, respectively.

The analysis of mass transfer of extraction-stripping proceeding in SIR particles can be done by differential equation describing mass transfer of solute in pores of spherical particle assuming the SIR particle properties. The analysis depends mainly on the procedure of the impregnation [63, 64]. In the case of the impregnated resin prepared by the "dry" method the pores are not completely filled with the organic phase and the transport of water inside of particles can be also noted. Then the model is valid only if the pores of the support are almost filled with organic phase, and the mass transfer in an aqueous phase can be neglected. In the case of the resins prepared by "wet" method the pores are completely filled, and the mass transfer occurs only in an organic phase. For this type of SIR a model which is used to describe the kinetics of a solute extraction combines intraparticle diffusion and chemical reaction. Using this model, the reaction rate in the organic phase can be described according to eq. (8.118).

$$r_{S,org} = k[S]_{org}[E]_{org} - k^{-}[E]_{org} = k\left([S]_{org}[E]_{org} - \frac{1}{K_{chem}}[SE]_{org}\right) \tag{8.118}$$

where k and k^{-} are the reaction rate constants (forward and backward reaction, respectively)

$$K_{chem} = \frac{k}{k^{-}} \tag{8.119}$$

If the molar volumes of the extractant and the complex are comparable and their concentrations are in a similar range, also the flux expressions of the extractant and the complex appear to be dependent (D_S, D_E, D_{SE} are the diffusivities of solute, extractant and created complex, τ is a tortuosity factor).

$$\frac{\partial[S]_{org}}{\partial t} = \frac{D_s}{\tau}\frac{1}{r^2}\frac{\partial}{\partial r}\left(r^2\frac{\partial[S]_{org}}{\partial r}\right) - \gamma_s k\left([S]_{org}[E]_{org} - \frac{1}{K_{chem}}[SE]_{org}\right) \tag{8.120}$$

$$\frac{\partial[E]_{org}}{\partial t} = \frac{D_E}{\tau}\frac{1}{r^2}\frac{\partial}{\partial r}\left(r^2\frac{\partial[E]_{org}}{\partial r}\right) - \gamma_E k\left([S]_{org}[E]_{org} - \frac{1}{K_{chem}}[SE]_{org}\right) \tag{8.121}$$

$$\frac{\partial[SE]_{org}}{\partial t} = \frac{D_{SE}}{\tau}\frac{1}{r^2}\frac{\partial}{\partial r}\left(r^2\frac{\partial[SE]_{org}}{\partial r}\right) - \gamma_{SE} k\left([S]_{org}[E]_{org} - \frac{1}{K_{chem}}[SE]_{org}\right) \tag{8.122}$$

The solute concentration in an aqueous phase can be described with eq. (8.126), where Fick's law of diffusion, which depends on the flux of the solute through the phase interface, is applied at the outer particle surface.

$$\frac{\partial[S]_{org}}{\partial t} = -3\varepsilon_p\frac{D_s}{\tau R}\frac{V_{SIR}}{V_{aq}}\frac{d[S]_{org}}{dr} \tag{8.123}$$

The introduction of the dimensionless radius variable j, eq. (8.123) can be then rewritten as follow:

$$\frac{\partial[S]_{org}}{\partial t} = \frac{D_S}{\tau}\frac{1}{R^2}\frac{1}{\varphi^2}\frac{\partial}{\partial\varphi}\left(\varphi^2\frac{\partial[S]_{org}}{\partial\varphi}\right) - \gamma_s k\left([S]_{org}[E]_{org} - \frac{1}{K_{chem}}[SE]_{org}\right) \tag{8.124}$$

$$\frac{\partial[E]_{org}}{\partial t} = \frac{D_E}{\tau}\frac{1}{R^2}\frac{1}{\varphi^2}\frac{\partial}{\partial\varphi}\left(\varphi^2\frac{\partial[E]_{org}}{\partial\varphi}\right) - \gamma_E k\left([S]_{org}[E]_{org} - \frac{1}{K_{chem}}[SE]_{org}\right) \tag{8.125}$$

$$\frac{\partial[SE]_{org}}{\partial t} = \frac{D_{SE}}{\tau}\frac{1}{R^2}\frac{1}{\varphi^2}\frac{\partial}{\partial\varphi}\left(\varphi^2\frac{\partial[SE]_{org}}{\partial\varphi}\right) - \gamma_{SE} k\left([S]_{org}[E]_{org} - \frac{1}{K_{chem}}[SE]_{org}\right) \tag{8.126}$$

The decrease of the solute in the aqueous phase is the most important aspect of this model and here Fick's law of diffusion, which depends on the flux of the solute through the phase interface, is applied to determine the solute concentration in the aqueous phase. Assumed r = R, the eq. (8.123) can be expressed as:

$$\frac{d[S]_{org}}{dt} = -3\varepsilon_p \frac{D_S}{\tau R^2}\frac{V_{SIR}}{V_{aq}}\frac{d[S]_{org}}{d\varphi}|_{\varphi+1} \tag{8.127}$$

Babić et al. [59] have also applied the modified shrinking core model to predict, which mechanism controls the recovery process. Equations responding to the film diffusion controlled process (X), particle diffusion controlled process (Y) and process controlled by chemical reaction are as follow:

$$X = \frac{3D_f}{a\delta q_E}\int_0^1 [S]_{aq}(t)dt \tag{8.128}$$

$$Y = \frac{X_e}{K}\left[1 - 3\left(1 - \frac{X}{X_e}\right)^{2/3} + 2\left(1 - \frac{X}{X_e}\right)\right] \tag{8.129}$$

$$Z = \frac{1}{K}\left[1 - \left(1 - \frac{X}{X_e}\right)^{1/3}\right] = \frac{K_c b}{a}\int_0^1 [S]_{aq}(t)dt \tag{8.130}$$

According to the modified shrinking core model, the plot X, Y and Z versus $\int_0^1 [S]_{aq}(t)dt$ should give strength line if the sorption process is controlled by that mechanism.

To model the extraction behavior of metals into the SIR particle the appropriate extraction mechanism was considered by Juang and Lin [65], Juang and Chen [66] and by Kamio et al. [54] and the used kinetic model, similar to organic compounds, combined mainly film diffusion or intraparticle diffusion and chemical reaction.

References

[1] Nernst WH. Zur kinetik der in Lösung befindlichen Körper. Z Phys Chem. 1888;2:613–622.

[2] Rice NM, Irving HM, Leonard MA. Nomenclature for liquid-liquid distribution (solvent extraction). IUPAC recommendations. Pure App Chem. 1993;65(11):2373–2396.

[3] Healy TV, McKay HA. The extraction of nitrates by tri-n-butyl phosphate (TBP). II. The nature of the TBP phase. Trans Farad Soc. 1956;52:633–2.

[4] Ferraro JR. The nature of the nitrate in the TBP solvates MfNCTBP, $M(NO])_4 2TBP$ and $MO_2 (NO_3)_2 2TBP$. J Inorg Nucl Chem. 1959;10:319–322.
[5] Hesford E, McKay HA, Scargill D. Tri-n-butyl as an extracting solvent for inorganic nitrates. IV. Thorium nitrate. J Inorg Nucl Chem. 1957;4:321–325.
[6] Majumdar SK, De AK. Liquid-liquid extraction of iron (III) with tributyl phosphate. Talanta. 1960;7:1–6.
[7] Forrest VM, Scargill D, Spickernell DR. The extraction of zinc and cadmium by tri-*n*-butyl phosphate from aqueous chloride solutions. J Inorg Nucl Chem. 1969;31:187–197.
[8] Sayar NA, Filiz M, Sayar AA. Extraction of Zn(II) from aqueous hydrochloric acid solutions into Alamine 336–m-xylene systems. Modeling considerations to predict optimum operational conditions. Hydrometallurgy. 2007;86:27–36.
[9] Bjerrum J. The Iron(III)–thiocyanate system. The stepwise equilibria studied by measurements of the distribution of tris(thiocyanato)iron(III) between octan-2-ol and aqueous thiocyanate solutions. Acta Chem Scan A. 1985;39:327–340.
[10] Chmutova MK, Litvina MN, Nesterova NP, Myasoedov BF, Kabachnik MI. Extraction of Am(III), Eu(III) and U(VI) from perchloric acid and mixtures of acids by dialkyl-(diaryl) [diethylcarbamoylmethyl] phosphine oxides. Solvent Extr Ion Exch. 1992;10(3):439–458.
[11] Motomizu S, Oshima M, Arita A. Liquid – liquid distribution of ion associates of tetraiodopalladate(II) with quaternary ammonium counter ions. Anal Chim Acta. 1990;229:121–125.
[12] Szymanowski J. Hydroxyoximes and copper hydrometallurgy. Boca Raton, USA: CRC Press, 1993.
[13] Albrecht M, Fiege M, Osetska O. 8-Hydroxyquinolines in metallosupramolecular chemistry. Coord Chem Rev. 2008;252(8–9):812–824.
[14] Sekine T, Murai R, Ihara N. Formation and two-phase distribution equilibria of cobalt(II) and zinc(II) complexes with β-diketones in several solvent extraction systems. J Inorg Nucl Chem. 1974;36(11):2569–2574.
[15] Furman NH, Mundy RJ, Harrison GH. The distribution of uranyl nitrate from aqueous solutions to diethyl ether. US. Atomic Energy Commission Report. AECD-2938, Sept. 6, 1960
[16] Lo SL, Shiue SF. Recovery of Cr(VI) by quaternary ammonium compounds. Water Res. 1998;32(1):174–178.
[17] Sole KC, Hiskey JB. Solvent extraction of copper by Cyanex 272, Cyanex302 and Cyanex 301. Hydrometallurgy. 1995;37:129–147.
[18] Loyson P. The solvent extraction of zinc from lithium chloride by Aliquat 336 chloride, bromide and iodide in chloroform: an analytical investigation. Solvent Extr Ion Exch. 2000;18(1):23–39.
[19] Wieszczycka K, Wojciechowska A, Krupa M, Kordala-Markiewicz R. Quaternary pyridinium ketoximes as zinc extractants from chloride solutions. J Chem Eng Data. 2013;58(11):3207–3215.
[20] Kedari CS, Coll T, Fortuny A, Sastre A. Third Phase Formation in the Solvent Extraction System Ir (IV) – Cyanex 923. Solvent Extr Ion Exch. 2005;23(4):545–559.
[21] Hori T, Kawashima M, Freiser H. Mixed ligand chelate extraction of lanthanides: I. 8-Quinolinol systems. Sep Sci Technol. 1980;15(4):861–875.
[22] Parhi PK, Panigrahi S, Sarangi K, Nathsarma KC. Separation of cobalt and nickel from ammoniacal sulphate solution using Cyanex 272. Sep Purif Techn. 2008;59(3):310–317.
[23] Rydberg J, Cox M, Musikas C, Choppin GR. Solvent extraction principles and practice, revised and expanded. Taylor and Francis, 2004:109–205.
[24] Kislik V. Engineering development of solvent extraction processes in Examples of Application of Solvent Extraction Techniques. Elsevier, 2011:157–184.
[25] Pitzer KS. Thermodynamics of electrolytes. 1. Theoretical basis and general equations. J Phys Chem. 1973;77:268–277.

[26] Bromley LA. Thermodynamic properties of strong electrolytes in aqueous solutions. AIChE Journal. 1973;19(2):313–320.

[27] Borge G, Castaño R, Carril MP, Corbillón MS, Madariaga JM. Development of a Modified Bromley's Methodology (MBM) for the estimation of ionic media effects on solution equilibria Part 1. Calculation of the interaction parameters in the molar and molar scales at 25°C. Fluid Phase Equil. 1996;121(1–2):85–98.

[28] Wieszczycka K, Krupa M, Wojciechowska A, Wojciechowska I, Olszanowski A. Equilibrium studies of cobalt(II) extraction with 2-pyridineketoxime from mixed sulphate/chloride solution. J Radioanal Nucl Chem. 2016;307(2):1155–1164.

[29] Sillen LG, Martel AE, Bjerrum J. Stability constants of metal-ion complexes: section I. Inorganic ligands. London: Chemical Society, 1964:241.

[30] Bjerrum J, Halonin AS, Skibsted LH. Studies on cobalt(II) halide complex formation. I. A spectrophotometric study of the chloro cobalt(II) complexes in strong aqueous chloride solutions. Acta Chem Scand. 1975;29:326–332.

[31] Kiliç E, Köseogğlu F, Başgut Ö. Protonation constants of some pyridine derivatives in ethanol-water mixtures. Anal Chim Acta. 1994;294(12):215–220.

[32] Kolaric Z. Critical evaluation of some equilibrium constants involving acidic organophosphorus extractants. Pure Appl Chem. 1982;54(12):2593–6114.

[33] Sole KC, Hiskey JB. Solvent extraction characteristics of thiosubstituted organophosphinic acid extractants. Hydrometallurgy. 1992;*30*(1–3):345–365.

[34] Li LY, Xu SM, Ju ZJ, Zhang Z, Liao FH, Li GB. Dialkyl phosphinic acids: synthesis and applications as extractant for nickel and cobalt separation. Trans Nonferrous Met Soc China. 2010;20(1): s205–10.

[35] Sole KC, Hiskey JB. Solvent extraction of copper by Cyanex 272, Cyanex 302 and Cyanex 301. Hydrometallurgy. 1995;37:129–147.

[36] Alguacil FJ, Cobo A, Caravaca C. Study of the extraction of zinc(II) in aqueous chloride media by Cyanex 302. Hydrometallurgy. 1992;163–174.

[37] Benito R, Menoyo B, Elizalde MP. Extraction equilibria of Zn(II) from chloride medium by Cyanex 302 in toluene. Hydrometallurgy. 1996;51–63.

[38] Juang RS, Kao HC, Wu WH. Analysis of liquid membrane extraction of binary Zn(II) and Cd(II) from chloride media with Aliquat 336 based on thermodynamic equilibrium models. J Membrane Sci. 2004;228:169–177.

[39] Daud H, Cattrall RW. The extraction of Cd(II) and Zn(II) from acidified lithium chloride solutions by Aliquat 336 dissolved in chloroform. J Inorg Nucl Chem. 1981;43:599–601.

[40] Cote G, Jakubiak A. Modelling of extraction equilibrium for zinc(II) extraction by a bibenzimidazole type reagent (ACORGA ZNX-50) from chloride solutions. Hydrometallurgy. 1996;43: 265–276.

[41] Jee JG, Lee SH. The pressure effect on the activity coefficient of sodium chloride and bromide ions. Bull Korean Chem Soc. 1986;7:163–166.

[42] Danesi PR, Chiarizia R, Coleman CF. The kinetics of metal solvent extraction. In: Campbell B, editor. Critical reviews in analytical chemistry. Boca, Florida: CRC Press, 1980.

[43] Stevens GW, Perera JM. Kinetics of solvent extraction processes. Min Process Extract Metal Rev. 1997;17(1–4):205–226.

[44] Whitman WG. The two-film theory of gas absorption. Chem Metal Eng. 1923;29(4):146–150.

[45] Whitman WG, Davies DS. Comparative absorption rates for various gases. Ind Eng Chem. 1924;16:1233–1237.

[46] Niemczewska J, Cierpiszewski R, Szymanowski J. Mass transfer of zinc(II) extraction from hydrochloric acid solution in the Lewis cell. Desalination. 2004;162(1):169–177.

[47] Yang X, Wang X, Wei C, Zheng S, Sun Q, Wang D. Extraction kinetics of tantalum by MIBK from pulp using Lewis cell. Hydrometallurgy. 2013;1(34):131–132.

[48] Weigl M, Geist A, Gompper K, Kim JI. Kinetics of Lanthanide/Actinide Coextraction with N,N'-Dimethyl-N,N'-Dibutyltetradecylmalonic Diamide (DMDBTDMA). Solvent Extr Ion Exch. 2001;19:215–229.

[49] Danesi PR. Chapter 5, solvent extraction kinetics. In: Rydberg J, Cox M, Musikas C, Choppin GR, editors. Solvent extraction principles and practice. New York: Marcel Dekker, 2004:203–251.

[50] Harada M, Miyake Y. Formulation of metal extraction rates in solvent extraction with chelating agents. J Chem Eng Japan. 1986;16:187–207.

[51] Harada M, Miyake Y, Kayahara Y. Kinetic mechanism of metal extraction with hydroxyoximes. J Chem Eng Japan. 1989;22(2):168–177.

[52] Dobson S, Van Der Zeeuw AJ. Hydrocarbon solvent diluents in hydroxyoxime solvent extraction processes. London: Chem Ind, 175–181.

[53] Staszak M. A linear diffusion model of adsorption kinetics at fluid/fluid. Interfaces J Surfactants Deterg. 2016;19:297–314.

[54] Kamio E, Matsumoto M, Kondo K. Theoretical development of metal extraction mechanism into an extractant-impregnated microcapsule. Ind Eng Chem Res. 2007;46:255–265.

[55] Juang RS, Lin HC. Metal sorption with extractant-impregnated macroporous resins. 1. Particle diffusion kinetics. J Chem Technol Biotechnol. 1995;62(2):132–140.

[56] Juang RS, Lee SH. Metal sorption with extractant-impregnated macroporous resins. 2. Chemical reaction and particle diffusion kinetics. J Chem Technol Biotechnol. 1995;62(2):141–147.

[57] Burghoff G, Zondervan E, De Haan AB. Phenol extraction with Cyanex 923: Kinetics of the solvent impregnated resin application. React Funct Polym. 2009;69:264–271.

[58] Burghoff G, Goetheer EL, De Haan AB. Solvent impregnated resins for the removal of low concentration phenol from water. React Funct Polym. 2008;68:1314–1324.

[59] Babi´C K, Van Der Ham L, De Haan AB. Recovery of benzaldehyde from aqueous streams using extractant impregnated resins. React Funct Polym. 2006;66:1494–1505.

[60] Martak J, Schlosser Š. New mechanism and model of butyric acid extraction by phosphonium ionic liquid. J Chem Eng Data. 2016;61:2979–2996.

[61] Martak J, Schlosser Š. Pertraction of organic acids through liquid membranes containing ionic liquids. Desalination. 2006;199:518–520.

[62] Martak J, Schlosser Š. Liquid-liquid equilibria of butyric acid for solvents containing a phosphonium ionic liquid. Chem Papers. 2008;62(1):42–50.

[63] Jerabek K, Hankova L, Strikovsky AG, Warshawsky A. Solvent Impregnated Resins: relation between impregnation process and polymer support morphology I. Di(2-ethylhexyl)-dithiophosphoric acid. React Funct Polym. 1996;28:201–207.

[64] Kabay N, Cortina JL, Trochimczuk A, Streat M. Solvent-impregnated resins (SIRs) – Methods of preparation and their applications. React Funct Polym. 2010;70:484–496.

[65] Juang RS, Lin HC. Metal sorption with extractant- impregnated macroporous resins. 1. Particle Diffusion Kinetics. J Chem Technol Biotechnol. 1995;62:132–140.

[66] Juang RS, Chen ML. Comparative equilibrium studies on the sorption of metal ions with macroporous resins containing a liquid ion-exchanger. Sep Sci Technol. 1997;32:5:1017–1035.

Katarzyna Staszak

9 Membrane processes

Abstract: The membrane processes have played important role in the industrial separation process. These technologies can be found in all industrial areas such as food, beverages, metallurgy, pulp and paper, textile, pharmaceutical, automotive, biotechnology and chemical industry, as well as in water treatment for domestic and industrial application. Although these processes are known since twentieth century, there are still many studies that focus on the testing of new membranes' materials and determining of conditions for optimal selectivity, i. e. the optimum transmembrane pressure (TMP) or permeate flux to minimize fouling. Moreover the researchers proposed some calculation methods to predict the membrane processes properties. In this article, the laboratory scale experiments of membrane separation techniques, as well their validation by calculation methods are presented. Because membrane is the "heart" of the process, experimental and computational methods for its characterization are also described.

Keywords: membrane process, modeling, fouling, retention

9.1 Basic principle of membrane process

The performance or efficiency of a given membrane is determined by two parameters – its selectivity and the flow through the membrane. The latter, often denoted as the flux (J) or permeation rate, is defined as the volume flowing ($\dot{V}$) through the membrane per unit area (A) and time (t):

$$J = \frac{\dot{V}}{A \times t} \tag{9.1}$$

The selectivity of a membrane is generally expressed by one of two parameters – the retention (R) or the separation factor (α). For dilute aqueous mixtures, consisting of a solvent (mostly water) and a soluble, it is more convenient to express the selectivity in terms of the retention towards the solute. The solute is partly or completely

This article has previously been published in the journal *Physical Sciences Reviews*. Please cite as: Staszak, K. Membrane processes. *Physical Sciences Reviews* [Online] **2017**, *2* (12). DOI: 10.1515/psr-2017-0142

https://doi.org/10.1515/9783110482065-009

retained while solvent molecules pass freely through the membrane. The retention is given by equation:

$$R = \frac{c_f - c_p}{c_f} = 1 - \frac{c_p}{c_f} \tag{9.2}$$

where c_f is the concentration in the feed, c_p – concentration in the permeate.

The separation factor generally is used for gas or organic liquids mixtures and it is expressed as:

$$\alpha_{A/B} = \frac{\frac{y_A}{y_B}}{\frac{x_A}{x_B}} \tag{9.3}$$

where y is the concentration in the permeate, x – concentration in the feed.

9.1.1 Retention

Mostly in the mathematical description and modeling of membrane processes permeate fluxes are taken into account. Because in the process of membrane filtration the flux usually changes over time due to fouling, so its modeling is described in detail in Section 9.1.2. Much less work can be found on retention modeling. It is not possible to talk about a generalized retention model because of the fact that, depending on the membrane process considered, there are different mechanisms of separation. The membrane separation methods can be divided into classes according to their separation characteristics: (i) separation by sieving action, in porous membranes; (ii) separation due to a difference in affinity and diffusivity, in dense membranes; (iii) separation due to a difference in charge of molecules, in charged membranes; (iv) carrier-facilitated transport; (v) the process of time-controlled release by diffusion [1]. Therefore, only the selected modeling methods of membrane filtration are presented below. An interesting example is nanofiltration. It is a separation technique where selectivity is governed by size selectivity, electrical surface charge and diffusion mechanisms. Thus some interactions that affect rejection as steric hindrance effects (sieving effect), Donnan exclusion and electrostatic repulsion (charge effect) and hydrophobic–adsorptive interaction should be considered in this process [2].

Moreover, very important aspect in the modeling is the characterization of physicochemical properties of compounds and characterization of the membrane materials. These properties are helpful in understanding the transport and retention of compounds during the membrane process.

In the nanofiltration process the charged and uncharged compounds can be rejected, thus the models for their retention are divided also for those for charged and uncharged substances. In the literature, several models for maximal retention of uncharged compounds are proposed, i. e. the steric hindrance pore (SHP) model, the model of Zeman and Wales, the log-normal model and an adapted version of the log-normal model [3]. The maximal retention is defined as the calculated retention corresponds to the retention at an infinite pressure. Because pressure dependence of retention is not included in the models, in real systems the retention is lower, due to the contribution of diffusion to the transport process. These models can be used not only in nanofiltration but also in all filtration process where sieving effect of retention is considered, such as microfiltration or ultrafiltration.

Transport of uncharged compounds in nanofiltration is a combination of diffusion (first term in eq. (9.4)) and convection (second term in eq. (9.4)). Thus, equation for flux of dissolved component (J_s) is presented as [4, 5]:

$$J_s = -P\Delta x \frac{dc}{dx} \pm (1 - \sigma) J_v c \tag{9.4}$$

J_v is the water flux and is equal to:

$$J_v = L_p\,(TMP - \sigma\Delta\pi) \tag{9.5}$$

Integration of eq. (9.4), with following boundary conditions [6]: $c_s = c_0$ for $x = 0$ and $c_s = c_p$ for $x = \Delta x$ and using eq. (9.5) the retention (R) is directly related to volumetric flux and can be calculated as:

$$R = \frac{\sigma(1 - F)}{1 - \sigma F} \tag{9.6}$$

where

$$F = \exp\left(-\frac{1-\sigma}{P} J_v\right) \tag{9.7}$$

At higher values of J_v, the exponential term in eq. (9.7) tends to zero and R will be equal to σ. Thus σ is a reflection coefficient of a given component and means the maximal possible retention for that component.

In eqs (9.4)–(9.7) P is a permeability, which described the transport of a molecule by diffusion; $\Delta\pi$ is osmotic pressure difference across the membrane; Δx is the membrane thickness; subscripts 0, p and s in c are concentrations in the feed solution, in the permeate and in the membrane, respectively.

SHP model is the model where the reflection coefficient is calculated from the pore size of the membrane and the diameter of the molecule. In this model there is big simplification that all membrane's pores have the same size [7]. According to the SHP model membrane is represented as a bundle of cylindrical pores, with the same diameter. During the transport of solution through membrane the molecules with bigger or the same size as the pore diameter are completely retained. Moreover, there is a partially retention of molecules with smaller diameter than pore size due to a certain amount of sterical hindrance and interactions with the pore wall. Thus reflection coefficient can be related to the effect of the pore wall (wall correction parameter H_F) and sterical hindrance during transport through the pore (S_F) according to the equation:

$$\sigma = 1 - H_F S_F \tag{9.8}$$

where

$$H_F = 1 + \frac{16}{9}\eta^2 \tag{9.9}$$

$$S_F = (1-\eta)^2\left[2-(1-\eta)^2\right] \tag{9.10}$$

$$\eta = \frac{d_m}{d_p} \tag{9.11}$$

The parameters d_m and d_p are the diameter of a molecule and the diameter of a membrane's pore, respectively.

The model of Zeman and Wales [8, 9], similar to the SHP model, assumes that the pores have a uniform cylindrical diameter. Furthermore, parabolic velocity dependence in the pore is considered. With the assumption that transport of molecules through membrane is the transport of sphere through capillary, the reflection coefficient is represented by simply expression:

$$\sigma = 1 - \left(\eta(\eta-2)^2\right) \tag{9.12}$$

Because during the filtration the steric hindrance gives rise to a hydrodynamic lag in the membrane pores, Zeman and Wales, based on the experimental results, proposed modification of eq. (9.12) to:

$$\sigma = 1 - \left(\eta(\eta-2)^2\right)e^{-\alpha\eta^2} \tag{9.13}$$

where α is a dimensionless constant.

Log-normal model is the first from proposed models which assumes that pore size is not constant and the log-normal distribution for the pore sized is proposed in Ref. [10, 11]. In this model authors assumed that the only criterion of the transport of molecules through membranes' pores is their sizes. Thus only molecule with smaller diameter than the diameter of pore can permeate through pores in membrane filtration process. Some effects are negligible in this model: steric hindrance in the pores, hydrodynamic lag and contribution of diffusion to transport. With these assumptions and simplifications, the reflection coefficient is represented by expression:

$$\sigma(r^*) = \int_0^{r^*} \frac{1}{S_p\sqrt{2\pi}}\frac{1}{r}\exp\left(\frac{-[\ln(r) - \ln(\bar{r})]^2}{2S_p^2}\right)dr \tag{9.14}$$

In eq. (9.14) S_p is the standard deviation of the distribution of the pore size, $\bar{r}$ is a mean pore size.

Adapted version of the log-normal model is the model in which to the log-normal model the hydrodynamic lag is taken into account. Thus, it could be assumed that if the molecule has larger diameter than pore size than the 100 % retention is obtained. The partition retention is obtained in the case when the pore diameter is larger than the molecular diameter and could be explained by difference in velocity of molecule (v_m) and water in the pore (v_p). According to the Zeman and Wales model the ratio of these velocities can be calculated from the equation:

$$\frac{v_m}{v_p} = \exp(-\alpha\eta^2) \tag{9.15}$$

The combination of both models: log-normal and Zeman and Wales, allows calculating the reflection coefficient as:

$$\begin{aligned}\sigma(r^*) &= \int_0^{r^*} \frac{1}{S_p\sqrt{2\pi}}\frac{1}{r}\exp\left(\frac{-[\ln(r) - \ln(\bar{r})]^2}{2S_p^2}\right)dr \\ &+ \int_{r^*}^{\infty} \left(1 - \exp\left(-\alpha\left(\frac{r^*}{r}\right)^2\right)\right)\frac{1}{S_p\sqrt{2\pi}}\frac{1}{r}\exp\left(\frac{-(\ln(r) - \ln(\bar{r}))^2}{2S_p^2}\right)dr \\ &= 1 - \int_{r^*}^{\infty} \exp\left(-\alpha\left(\frac{r^*}{r}\right)^2\right)\frac{1}{S_p\sqrt{2\pi}}\frac{1}{r}\exp\left(\frac{-(\ln(r) - \ln(\bar{r}))^2}{2S_p^2}\right)dr\end{aligned} \tag{9.16}$$

Independently of the using model to their experimentally validations the size of the molecules is necessary. For the calculation of the molecular diameter the HyperChem or Gaussian programs can be used successfully.

Van der Bruggen et al. [3] compared the experimental results with models described above by fitting the relevant equations to the experimental data using a least-squares method. They used three nanofiltration membranes NF70 (crosslinked aromatic polyamide), NTR 7450 (sulfonated polyethersulfone), UTC-20 (polyamide) to recovery the small organic molecules. Obtained results show that all models have deviation from experimental data. The main reason of that are too many simplifications. The both log-normal models gave better results than SHP model and the model of Zeman and Wales. Better fitting is observed for high value of TMP. Acceptable results of log-normal model, with deviation from experimental results up to 20 %, were presented by the same authors for separation of low molecular organic compounds from aqueous solution using nanofiltration NF70, NTR 7450 and UTC-20 membranes [12] They suggested that the log-normal model appeared to be most useful to predict reflection coefficients in practical applications. Moreover utility of these models to predict the retention specific active pharmaceutical ingredients (API's) from toluene, methylene chloride, and methanol using solvent resistant nanofiltration process are presented in the literature [13]. SHP model, the Zeman-Wales model, the Verniory model [14] and the log-normal model were used to calculate the pore diameters of membranes. The order of magnitude of estimated pore diameter is the same for each model, about 1 nm and it is compatible with experimental results.

Wherein, in Verniory model in cylindrical membrane pores the frictional drag force is included and the reflection coefficient can be calculated as:

$$\sigma = 1 - g(\eta) S_F \tag{9.17}$$

where

$$g(\eta) = \frac{1 - \frac{2}{3}\eta^2 - 0.2\eta^5}{1 - 0.76\eta^5} \tag{9.18}$$

$$S_F = (1 - \eta)^2 \left(1 - (1 - \eta)^2\right) \tag{9.19}$$

Martin-Orue et al. [15] used Zeman and Wales model in their study of nanofiltration process. It would not be surprising, but this was a study of the nanofiltration process of various charged amino acids and peptides. As was mentioned above, the Zeman and Wales model is the model for uncharged molecules. The theoretical retentions, estimated form the model, were different from the experimental results. Based on this results authors suggested that charge effect, both repulsion of coions and attraction of the counterions, was more important in description transport through membrane than size effects.

As was mentioned above (see eq. (9.6)) retention in membrane filtration is directly related to volumetric flux. Therefore, the membrane transport models could be used to describe the efficiency of separation in membrane process. The most popular is solution diffusion model (SD), solution-diffusion-imperfection model (SDI), preferential-sorption-capillary flow model (PSCF), Donnan exclusion model (DE) and extended Nernst–Planck model (ENP). In SD model [16] the main assumption is that solute and solvent dissolve and then diffuse through the homogeneous non-porous membrane due to chemical potential gradient. Therefore, the separation process occurs due to the difference in the solubility and diffusivities of compounds in the feed solution. The gradient (driving force of the process) results from concentration (due to concentration polarization) and pressure difference across the membrane. The SDI model is extension SD model. Model takes into account the pore flow. In the PSCF model, there is assumption that separation mechanism is based on surface phenomena and fluid transport through pores in the membrane. The membrane has such chemical properties that the sorption of solvent and repulsion of solutes are preferred. The Donnan exclusion model is proposed when a charged membrane is used in separation process of charged compounds. In this case concentration of opposite to membrane charge ions from the feed solution is higher and concentration of ions with the same charge as membrane is lower in comparison to the concentration in the bulk solution. Thus the counterions diffuse from the membrane phase to the bulk solution, while coion in the opposite direction.

The transports of ionic species through membrane, as well as their retentions are well described by the ENP equation [17–19]. This phenomenological model is also used for description of molecules transport through the charged membranes [20]. The model assumes each of the mechanisms described above. Both solubility and diffusivity of solute and solvent in the nonporous and homogeneous surface layers of the membrane due to the chemical potential gradient which is the result of concentration and pressure difference across the membrane, as well as transport of ions through pores with convective, diffusive and electrostatic migration forces are taken into account. Electro neutrality between ions among each other and between ions in the solution and in the membrane are guaranteed by adding the Donnan condition. ENP equation is given by:

$$J_i = -D_{i,p}\frac{dc_i}{dx} - z_i c_i D_{i,p}\frac{F}{RT}\frac{d}{dx} + K_{i,c}c_i J_v \tag{9.20}$$

where J_i is the solute flux, $D_{i,p}$ – the hindered diffusivity, c_i – the solute concentration, z_i – the ion valence, F – the Faraday constant, Ψ – the electric potential, $K_{i,c}$ – the hindrance factor for convection.

The first term in eq. (9.20) represents the flux component due to the diffusion, the second term accounts for flux due to the Donnan potential, and the last term describes flux due to convection. The solution of this equation was discussed by Bowen and Mukhtar [21]. The hindered diffusion coefficient and the convective hindrance factor could be estimated from numerical calculations using the ratio of ionic radius over pore radius, and the bulk diffusion coefficient.

The solution of eq. (9.20), with following boundary conditions using $c_i = c_{m_i}$ for $x = 0$ and $c_i = c_{p_i}$ for $x = \Delta x$, allows calculating the retention according to the relationship [18]:

$$R = 1 - \frac{1 - \sigma}{1 - \sigma \cdot \exp\left[-\frac{(1-\sigma)J_v}{P_s}\right]} \tag{9.21}$$

where P_s is solute membrane permeability, σ – salt reflection coefficient, subscripts m and p in c – the concentration in membrane and permeate, respectively.

The ENP model was used to fit the experimental data of NaCl and $CuSO_4$ transport in nanofiltration process with polyamide membrane [17]. From the ENP model, with good agreement to the experimental results in different conditions ($R^2 \approx 0.99$), authors estimated diffusive and convective flow. Similar results for $Ca(NO_3)_2$, Cd $(NO_3)_2$, $Cu(NO_3)_2$ and $ZnCl_2$ rejection in process with nanofiltration Nanomax50 membrane (polyamide arylene on polysulfone support) were obtained by Chaabane et al. [18]. Authors showed good agreement between theoretical and experimental data.

In the literature, several authors proposed the combination of ENP equation with other equations whose describe the transport. By a combination of ENP model and film theory equations the thickness of the boundary layer, solute concentration of membrane surface and concentration profiles in the polarization layer could be estimated as proposed in Ref. [22]. By this approach three principal parameters are determined: reflection coefficient (σ), the solute membrane permeability (P_s) and the layer polarization thickness (δ). The authors' assumptions were verified on the experimental results of removal of phosphorus ions using nanofiltration membrane NF90 (polyamide thin-film composite with a microporous polysulfone supporting layer). The results showed that the convective transport dominated at higher values of TMP and pH, while diffusive transport dominated at higher temperature. The results from the model are in agreement with the experimental observations. Another approach was proposed by Hua et al. [23] Authors used ENP model and film theory to calculate the retention of xylo-oligosaccharides syrup in nanofiltration process using aromatic polyamide membrane ((HDS-12-2540). The model proposed very well fitted the experimental results. Two model parameters: reflection coefficient (σ) and the solute membrane permeability (P_s) were estimated by curving fitting using genetic algorithm method. Moreover, using steric-hindrance pore model (SHP) two parameters which describe the membrane structural were estimated – ratio of solute radius to pore radius (see eq. (9.8)) and ratio of effective membrane porosity to membrane thickness.

To clarify experimental results of membrane separation, authors support their work by quantum calculations, based on density functional theory (DFT). Zhao et al. [24] studied the role of calcium ions in the process of removal perfluorooctane sulfonate (PFOS) by nanofiltration using NF270 membrane (semiaromatic piperazine-based polyamide thin-film composite with a microporous polysulfone supporting layer.). The results indicated that with increasing calcium chloride concentrations in the feed solution the rejection of PFOS also increased. To explain these results authors checked the possible interaction between PFOS and calcium ions. Calculations were carried out using gradient-corrected DFT with the Becke three-parameter nonlocal exchange functional and the Lee–Yang–Parr correlation functional (i. e. B3LYP). The geometries of PFOS interacting with calcium ion were optimized using the 6-31G(d) basis set. Low-spin and restricted closed-shell formulae were applied during the structural optimization. The formation of $CF_3(CF_2)_7SO_3Ca^+$ structure was confirmed by DFT calculation. This structure with higher molecular polarity is favorable to interact with the charged membrane surface. Moreover calcium ions by neutralization PFOS anions (with formation $CF_3(CF_2)_7SO_3CaO_3S(CF_2)_7CF_3$) and the negative charged NF270 membrane promoted sorption of PFOS on membrane surface. Surface sorption and formation described above structure (Ca^{2+} ion linked with two PFOS molecules) caused the aggregation of PFOS at the surfaces and increase the size of the molecules (from 10.88 to 26.17 Å). Thus increase of retention could be explained by size exclusion. Moreover, Ca^{2+} could bridge the negatively charged membrane surface and PFOS. It could cause enhancing the adsorption/deposition of PFOS on the membrane, which hinders the passage of water and PFOS molecules. The results obtained from DFT calculation were confirmed by SEM and AFM images. These images exhibited that with the increasing of calcium concentration, the membrane surface had more precipitation and higher surface roughness, and PFOS accumulation on the membrane increased, all off which correspond to flux decline and retention change.

DFT methods were used also to investigate the adsorption configurations of natural organic matter (NOM) compound on the nanofiltration, polyamide membrane surface [2]. The four compounds: p-coumaric acid (hydrophobic phenolic molecule), L-leucine (hydrophilic amino acid), acetic acid (nonionic hydrophilic carboxylic molecule), L-tryptophan (hydrophobic amino acid) and two types of membranes: NE90 (fully aromatic polyamide based on trimesoyl chloride and 1,3-benzenediamine) and NE70 (semi aromatic polyamide based on trimesoyl chloride and piperazine with polyvinyl chloride (PVA) coating) were considered. To calculation highest occupied molecular orbital (HOMO) and lowest unoccupied molecular orbital (LUMO) with frontier orbital gap were carried out with the B3LYP/6-31G* level theory. This method is helpful in estimated adsorption energy between organic molecules and the membrane materials, which is relative to retention of compounds in the process, as well as membrane fouling. The prediction of the interfacial phenomenon between compounds from feed solution and membranes could be relative to membrane fouling and retention. The compounds with high-energy gap have got higher tendency to adsorb on the membrane surface. Moreover, this tendency is depended on the kind of membrane's material (energy of bound). Kaewsuk and Seo [25] compared the results obtained by DFT calculation with experimental results and shoved good agreement between them. The experimental results showed that in the process of nanofiltration with the membrane NE90 the lower permeate flux and higher retention (especially carboxylic compound) were obtained in comparison to process with NE70 membrane. It could be explained by adsorption of organic compounds on the membrane surface. The calculation results confirmed these assumptions and showed that carboxylic compound has high-energy gap and tend to adsorb on the membrane surface than the other compounds (phenolic and acetic acid) and it bound higher energy with NE90 than NE70. Consistency of experimental and computational results indicates on possibility to predict the efficiency of membrane separation process by quantum calculations. Thus, without the experiments it can be assumed that the membrane will be or not suitable for retention of compounds considered. This is important at the planning stage of the synthesis and modification of the membrane surface as well as its selection for the process.

The novelty approach of rejection modeling of charged and uncharged organic compounds by nanofiltration and reverse osmosis membranes was presented by Khaouane et al. [26]. In their work authors used bootstrap aggregated neural networks (BANN) or bootstrap aggregated multiple linear regressions (BAMLR) to predict the value of retention. These methods are better than classical artificial neural networks (ANN) when only a limited amount of data is available. In this case, ANN model gives the result with significant errors. The aggregated neural network is a technique that improves the generalization ability of a model through training a number of neural networks and them combining them. Both BANN and BAMLR model are presented schematically in Figure 9.1. BANN model consists of several individual neural network models (INN) and BAMLR model – several MLR models, which are both developed to model the same relationship. Authors used results from literature and created database containing 436 rejections of 42 charged and uncharged organic compounds. From all results 350 data points (80 %) were used for the training phase, 43 points (10 %) for the validation step and 43 data points (10 %) for the testing phase of the model. The inputs model variables were divided into three groups which describe the compounds properties, membrane properties and process condition. To the first group belongs: molecular weight, compound

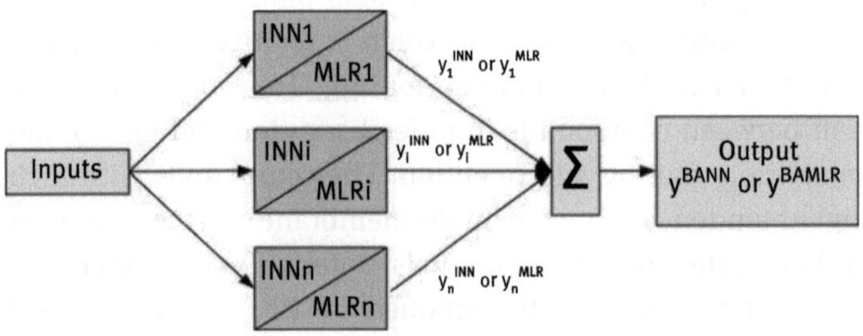

Figure 9.1: Schema of the bootstrap aggregated neutral networks (BANN) or bootstrap aggregated multiple linear regressions (BAMLR).

hydrophobicity (logD), dipole moment, molecular length and equivalent molecular width. The membrane properties are described by the following inputs variable: sodium chloride salt rejection "SR (NaCl)" or magnesium sulfate salt rejection "SR ($MgSO_4$)", the molar weight cut off, surface membrane charge (represents as zeta potential) and membrane hydrophobicity (represents as contact angle). The last group is: pH, TMP and temperature. The output variable in the model was rejection. The authors pointed out that the most commonly used parameter, the molar weight cut off, was insufficient alone to determine retention of charged and uncharged organic compounds by nanofiltration or reverse osmosis. In the case of RO process, with dense membrane, the most appropriate parameter is salt retention [27] while in NF process the molecular weight cut off and ionic retention of salts (mainly $MgSO_4$ "SR($MgSO_4$)" or NaCl "SR(NaCl)).

From the simulation, it can be concluded that BANN model is the best one to predict the rejection of charged and uncharged organic compounds in membrane filtration process, in comparison to the single neural network (SNN) and the BAMLR. The following results of the root mean squared errors were obtained 5.33 %, 6.45 % and 18.78 %, respectively.

Without using neural networks, the simple MLR was also proposed to modeling of RO/NF membrane rejections of pharmaceutical compounds and organic compounds [28]. In the statistical analysis 64 cases representing rejections of 14 compounds were considered. At first principal component analysis (PCA) was used to data reduction – replacement larger number of original variables (physical-chemical properties) by small number of derived variables to simplify subsequent analysis of the data. Then MLR was used to determine the strength of the relationship between a set of explanatory variables known as independent variables, and a single response or dependent variable using stepwise method for linear regression. Based on the principal component analysis results authors suggested that the most important parameters for prediction of membrane rejection were dipole moment, molar volume, hydrophobicity/hydrophilicity, molecular length and equivalent width. Molecular weight was found to be a poor variable to rejection simulation. MLR could predict well the rejection of uncharged molecules (R^2>95 %) while this method is not suitable to modeling the retention of charged molecules (R^2<60 %), due to the important influence of charge repulsion between the membrane and those charged compounds.

9.1.2 Membrane fouling

Generally, during the membrane process the decrease of the permeate flux is observed. This is mainly caused by the substances deposition on the membrane surface or into membrane pores. Depends of the kinds of containments two processes can occur – fouling or scaling. Fouling of membranes is due to the suspended or emulsified materials, such as colloidal (clay, flocs, surfactants), biological (bacteria, fungi),

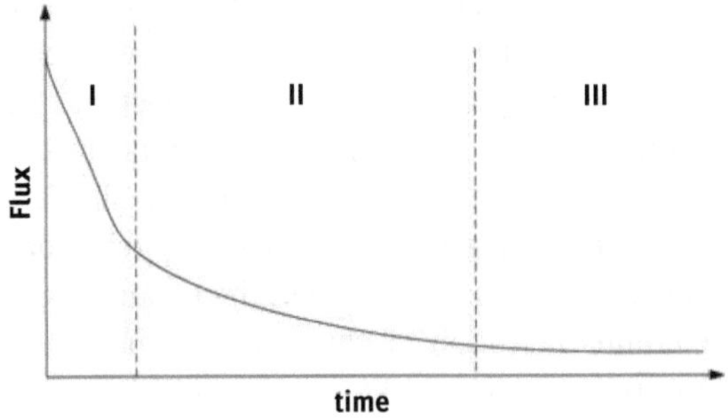

Figure 9.2: Schema of flux reduction: (I) initial rapid, (II) long-term gradual flux decline and (III) steady-state flux.

organic (oils, polyelectrolytes, humics) compounds. The mineral precipitates, mainly calcium, cause the scaling effect. When the concentration of salts in the process is above solubility equilibrium the deposition of particles on a membrane in the form of solid is occurred [29].

Typical relationship between permeate flux and time of flirtation process, with three stage, is presented in Figure 9.2 [30]. Stage (I) is the rapid reduction of flux in relation to flux obtained for pure water filtration. The next stage is slowly decreasing the flux in the filtration process. This stage exists always in membrane fouling system regardless the operation conditions. In stage (III) flux is in a steady state.

Membrane fouling is negative phenomenon in membrane technology. It increases the costs of the process by increasing energy consumption, system down time, necessary membrane area and construction, labor, time and material costs for backwashing and cleaning processes.

9.1.2.1 Fouling – experimental methods

Fouling results in higher membrane resistance and affects permeate quality. The simplest experimental method of fouling description is to measure of permeate flux decreasing during the process. The flux reduction can be related to fouling using mass transfer and fluid mechanics concepts. Of course, such approach allows only to determine whether the phenomenon of fouling occurs or not, without any explanation of it. To characterize fouling on membrane is very important to identify where it is occurring and how much is being deposit. Several experimental methods are proposed. Lots of them are similar to these used in characterization of membrane structure (described in detail in Section 9.2.2). For example scanning electron microscopy (SEM), environmental scanning electron microscopy (ESEM) transmission electron microscopy (TEM), confocal microscory, radiolabelling, X-ray photoelectron spectroscopy (XPS) and infrared spectroscopy (IR) are very helpful to see the place of foulants deposit [31–34] However internal deposition with the pores and low foulant levels remain difficult to detect. The fouled membranes can be also analyzed for any changes in pore size distribution using a polydisperse dextran solution sieving test [35].

9.1.2.2 Fouling modeling

To explain the flux decline many different models have been proposed in the literature. The most often used ones are: A. the standard blocking model, B. intermediate blocking model, C. complete blocking model and D. cake filtration model (Figure 9.3) [36, 37]. These models are successfully evaluated to explain fundamental mechanisms involved in flux decline during filtration process as microfiltration, ultrafiltration[30], nanofiltration and reverse osmosis [38].

Standard blocking model assumes that particles accumulate inside the membrane on the pore walls and the resulting smaller size of pores the membrane's permeability is reduced. In the model of intermediate blocking the reason of the decreasing of flux is interpreted by the accumulation portion of particles in the pores, while the rest accumulate on the top of the deposited particles. In the complete blocking model there is assumption that particles are larger than the pore size and can seal of the membrane. Cake filtration model is based on the hypothesis that the particle accumulation on the membrane surface, in a permeable cake, causes the thickness increasing.

These mechanisms are helpful in interpretation of experimental results. In the standard blocking model the time dependency of volume flow (J_V) is expressed as [37]:

$$J_V(t) = \frac{J_V(0)}{(1 + B \cdot t)^2} \tag{9.22}$$

where B is equal to:

$$B = K_B \cdot u_0 \tag{9.23}$$

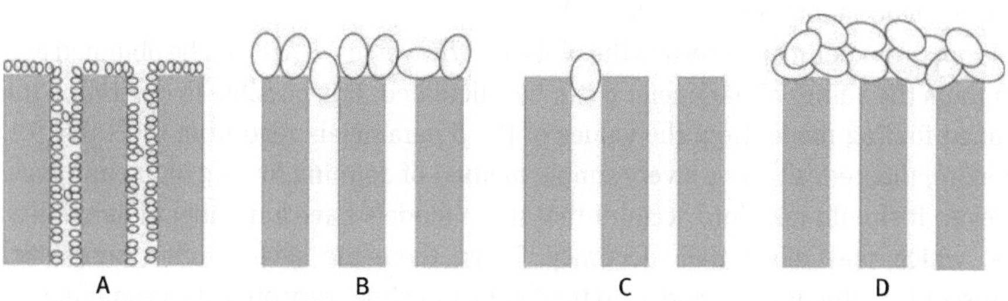

Figure 9.3: Schema of fouling models: A. Standard blocking model, B. Intermediate blocking model, C. Complete blocking model and D. Cake filtration model.

K_B is the reduction in the cross section area of the pores per unit of total permeates volume, caused by the adsorption on the pores walls. Parameter u_0 is the mean initial velocity of the filtrate and is calculated according to the relationship:

$$u_0 = \frac{J_V(0)}{A_0} \tag{9.24}$$

where A_0 is a porous surface of the membrane.

Total permeate volume is changed with time and is expressed as:

$$V(t) = \frac{J_V(0) \cdot t}{1 + B \cdot t} \tag{9.25}$$

Bowen et al. [37] suggested that characteristic equation of the blocking processes is given by:

$$\frac{d^2t}{dV^2} = \frac{2B}{\sqrt{J_V(0)}} \left(\frac{dt}{dV}\right)^{3/2} \tag{9.26}$$

The same considerations were made for other fouling models and general pattern of the characteristic equation was proposed:

$$\frac{d^2t}{dV^2} = \alpha \left(\frac{dt}{dV}\right)^{\beta} \tag{9.27}$$

Values of parameters α and β are depended of the fouling model according to the Table 9.1, as was proposed in [37].

In Table 9.1, K_A is the membrane surface blocked per unit of total volume permeated through the membrane; K_C is area of the cake per unit of permeate volume; R_r is the hydraulic resistance of the cake divided by the resistance of the initial or clean membrane ($R_r = R_c/R_0$).

From the experimental results the plots d^2t/dV^2 versus dt/dV can be obtained and from them the values of parameter β can be calculated. It is possible to determine the assumed fouling model from the values of the β parameters read from the Table 9.1. Therefore, theoretically, it is a very simple method of defining fouling on membranes. However, it should take into account that these models have a number of simplifications, which may affect their accuracy. Thus, there are several other simplistic, macroscopic mathematical models to predict flux decline, very often as a combination of above models presented [39–41]. The main problem of these approaches is that there is no certainty that estimated parameters for one set of membrane operating conditions

Table 9.1: Parameters in equation (27).

Fouling model	α	β
Standard	$\left(2K_B/A_0^{1/2}\right) \cdot u_0^{1/2}$	1.5
Intermediate	K_A/A_0	1
Complete	$K_A \cdot u_0$	2
Cake	$\left(R_r K_C/A_0^2\right) \cdot u_0^{-1}$	0

can be applied to another set. Moreover, commonly used membrane blocking models (i. e .standard, intermediate, complete or cake blocking) do not adequately describe fouling phenomena for membranes. Instead, of one model the phenomena can occur in successive or simultaneous coexistence [42]. This causes that experimental validation is necessary for new process conditions (i. e. kind of membrane, composition of the feed solution or operating conditions). For example there are few propositions of description of fouling for the filtration process of the same compounds – bovine serum albumin (BSA). Ho and Zydney [43] developed a mathematical model (combined pore blockage and cake filtration model) for the microfiltration of protein. The model showed excellent agreement with the experimental data obtained during the filtration of BSA solutions operated at constant pressure through polycarbonate membrane. Simply model based on deposition mechanism and validation of microfiltration process of BSA using polyethersulfone membrane at different pH and pressures and using cellulose acetate was described in [44] and [45], respectively. Bolton et al. [46, 47] proposed combined caking and complete blockage models of membrane fouling in the microfiltration process of BSA with PVDF membrane. Duclos-Orsello et al. [48] showed that fouling model depends of the kind of membrane material for the same feed solution containing BSA. The model predictions were validated by experimental results from polystyrene microsphere solution through PCTE membranes (representing complete external fouling), prefiltered BSA solution through Durapore membranes (representing internal fouling), and standard BSA solution through Durapore membranes.

Because macroscopic models have their limitations, Wessling [49] proposed two-dimensional stochastic modeling of membrane fouling. The algorithm assumes the deposition of a particle with defined length on a microfiltration membrane having a cylindrical pore. Particle motion is described by diffusion limited aggregation (DLA) model. In DLA, there is a stationary sees particle on a lattice and second mobile particle at a random location on the gird. This second particle walks on the lattice until it meets the seed particle and becomes immobile. A new mobile particle is added at a new random position. This new particle has got new walks until meeting the immobile cluster. Schematic drawing representing the basic simulation algorithm is presented in Figure 9.4.

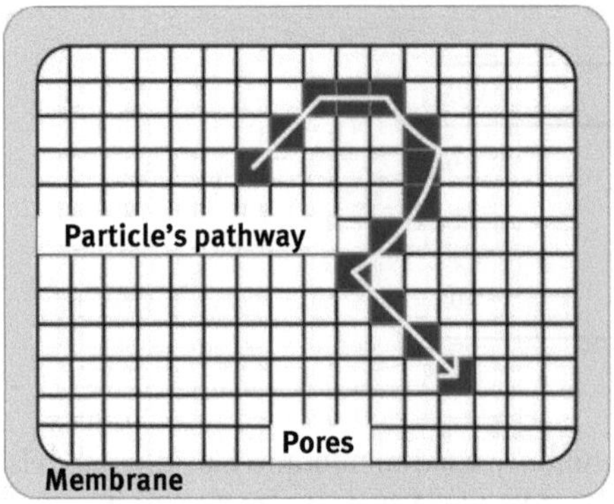

Figure 9.4: Schema of simulation algorithm.

The effect of pore diameter, flow conditions, membrane thickness on the permeate flux, resistance of membrane was simulated by this method. Moreover model is helpful to description of retention characteristics. Counting the numbers of particles passing the pore as a function of the number of particles deposited allows characterizing the change in retention behavior. As a result of simulation, presented in work [49], it could be concluded that the aggregate density above the flat adsorption surface was bigger in comparison to the aggregate density above the pore opening. Two distinct regimes are visible during the flux decline as a function of number of particles deposited. Initially, in the filtration process, the flux decline is determined by internal fouling and membranes with the same initial flux but different pore diameter show different flux decline: the membrane with the smaller pore has a more rapid flux decline.

As an alternative to the above-described simple mathematical and stochastical models, often based on the experimental observation, a completely different approach based on quantum methods is proposed. This method allows describing the interaction between compounds of foulants and membrane's material. For example in the process of ultrafiltration of proteins the non-covalent intermolecular interactions are responsible for adsorption and the packing of macromolecules, like proteins, on the membrane surface [50]. The distribution of the electrostatic charges on the protein surface controls the interactions between protein molecules, thus it affects the adsorption on the membrane. This is where the Quantum Mechanics (QM) approach, based on DFT, is useful – methods for determination of the partial charges on the atoms of any molecular system. De Luca et al. [50] proposed to use quantum mechanics approaches to calculate the effective diameter of bovine serum albumin (BSA) and the electrostatic surface charges (by Lowdin's approach and the ElectroStatic Potential method (ESP)). Lodwin's approach is based on the weighing of each atomic orbital belonging to the overall molecular orbitals. ESP method is based on the fitting of molecular quantum mechanics electrostatic potentials obtained from the electron density function and the nuclear geometry. Authors of citied work used in their calculations the Hybrid energy functional, X3LYP in conjunction with a Gaussian-triple Double-ζ orbital basis set added with polarization function (6-311 g*). Results obtained from the quantum calculation were used as the starting point for multi-scale model based on unsteady-state mass balance equation. Solution of the mass balance as well as force balances for each the layers constituting the deposit allows to estimate the mass of the particles constituting the deposit in the membrane and deposit specific resistance. Presented model was experimentally validating (ultrafiltration of BSA under different TMP using polyethersulfone membrane). A rather good agreement was obtained only after about 20 min of process (end time – 1 h). At the beginning of the process percentage errors exceeded of model 20 %.

By using models described above, permeate flux can be predicted for various process variables, such as such as TMP, composition of the feed solution, kind of membrane (material and pore size). However, there are various complex phenomena effecting flux in a membrane filtration process and until now, none of the developed models were fully and satisfactorily described the membrane filtration process. Thus, to predict membrane fouling and permeate flux decay as a function of process operating parameters the empirical models, based on ANN, are proposed briefly in literature [51–56]. ANN, the so-called "black-box" model, is simple and effective predictive instrument for solving non-linear problems. Generally in citied literature the permeate flux is ANN output, while the operating variables are ANN input (Figure 9.5). During the training process of the networks the four steps are followed: i. analysis and elaboration of the experimental data; ii. building of the neural model; iii. training of the network and post-training analysis; iv. post-simulation analysis. The results obtained from ANN modeling showed the excellent agreement between experimental data and predicted values of permeate flux vs. time.

Liu and Kim [54] compared the performance of purely mathematic and mechanical model (see Figure 9.3) with the so-called "black-box" ANNs model. The models were evaluated based on bench-scale experiments with synthetic water (mixture of Georgia kaolinite, alumina, Aldrich humic acid (AHA), $NaHCO_3$, $CaCl_2$) in this study using PVDF hollow fiber membrane. Authors used all blocking models – standard, intermediate, complete blocking model and cake filtration model. Unfortunately, models could not fit the experimental data well in whole experimental period, but they fitted much better in separate specific experimental periods. The combined cake-complete and the cake-intermediate models demonstrated relative high consistency with experimental TMP data. The excellent agreement between experimental data and prediction has been obtained with ANN model what confirmed the results presented in literature [51–53].

Interesting solution of fouling modeling was proposed by Chew et al. [57] A novel approach combining first principle equation of Darcy's law on cake filtration and ANN predictive models were utilized to represent the dead-end ultrafiltration process. The first principle model allows establishing the relationship between permeate flux, resistance and TMP. ANN model was used as an alternative to predict the

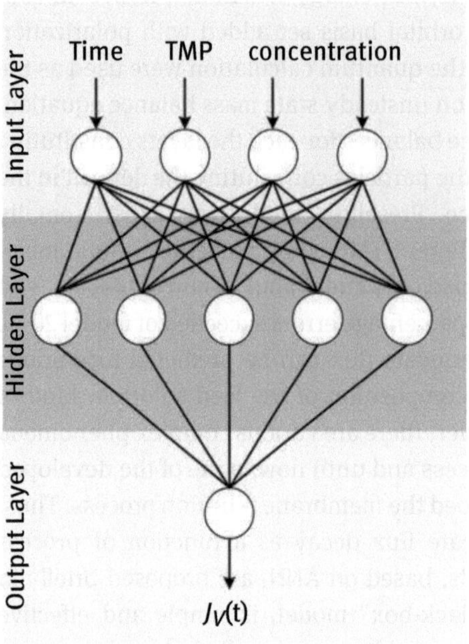

Figure 9.5: Architecture of neural networks.

specific cake resistance which requires complicated laboratory analysis procedures using the conventional method. This hybrid model allows to rapid estimation of the specific cake resistance with common on-line data such as feed water turbidity, filtration time and TMP.

Presented above consideration suggests the potential application of model to characterization of efficiency of membrane separation processes.

9.2 Membrane characterization

9.2.1 Introduction

Type of the membrane and its properties plays very important role in membrane technology. Membrane material, with specific morphologies and transport properties, determines in which process it can be used. Membrane selection depends on a variety of factors, including the composition of the feed solution, operating parameters (temperature, pH, pressure etc.), application type, and separation goals.

Membranes are made from inorganic (ceramic, glass, metal) or organic materials (polymer). Usually, the synthetic, polymeric membranes are used in the membrane separation processes. Their classification by the membrane morphology, geometry, preparation method is presented in Figure 9.6 [58].

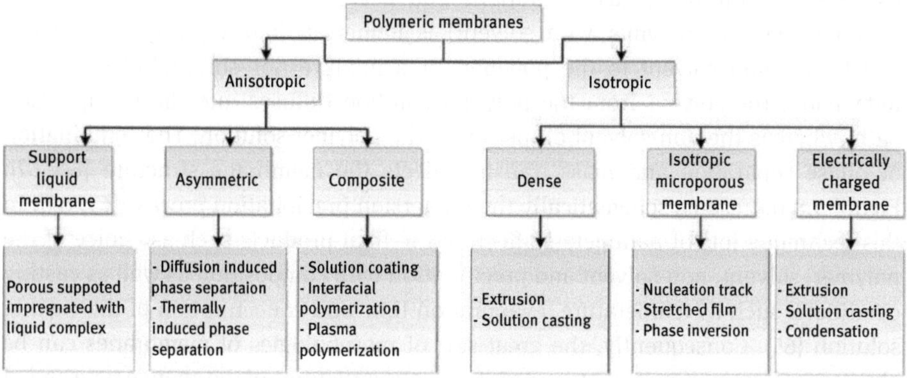

Figure 9.6: Polymeric membrane classification.

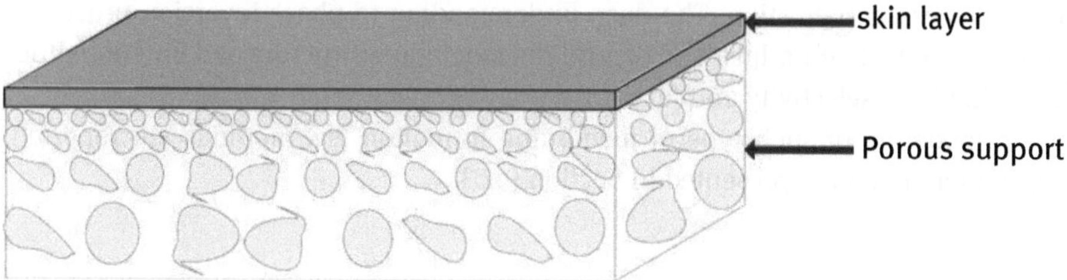

Figure 9.7: Layer structure of membrane.

In terms of morphology polymeric membranes are classified into anisotropic and isotropic membranes. Anisotropic membranes are layer structures. It is means that the porosity and pore size are changing in cross-section (Figure 9.7) and thin surface layer is supported on a thick microporous substrate. The skin layer is responsible for separation while porous support provides the mechanical strength. Nonporous dense membranes, isotropic microporous membranes, and electrically charged membranes are the examples of isotropic membranes. Dense membranes are made by one kind of polymer without porous, isotropic microporous membranes have randomly distributed interconnected pores. Electrically charged membranes (anion-exchange or cation-exchange membranes) have got dense or microporous structures, with fixed positive or negatively charged ions.

Depending of the kind of membranes (their morphology, structure) different method of their fabrication (see Figure 9.6), experimental characterization and modeling approach is proposed, which is described in detail in the following sections.

Most commercial membranes are formed via casting process called immersion precipitation [59]. This type of membranes is used in such processes as reverse osmosis, ultrafiltration, gas separations and pervaporation. In this technique, a polymer solution (polymer with solvent) is immersed into a precipitation bath, which is a non solvent to the polymer, or a mixture non-solvent/solvent. After immersion, the solvent from the polymer solution diffuses into the precipitation bath, whereas the non-solvent diffuses into the polymer solution. The combination of phase separation and mass transfer affects the membrane structure [60–62]. Figure 9.8 represents schematically the immersion precipitation process [63, 64]. In this technique lots of parameters affect on the final products such as choice of the polymer, solvent, non-solvent and precipitation bath composition as well as casting conditions such as temperature, evaporation time and concentration of the casting solution [65]. Consequently, the great sort of morphologies of membranes can be obtained – from non-porous structures to porous structures of the sponge type and of the finger/macrovoid type. Because membranes with various morphologies show different mechanical and transport properties, the ability to predict final membrane structures through analytical or computational methods is very important and helpful in process optimization. The deep understanding to phase inversion process is valuable to effectively adjust and control the membrane structure and functionalities for antifouling, selectivity and flux.

In modeling of such type of membrane formation few methods are reported. These approaches are presented in Section 9.2.3.

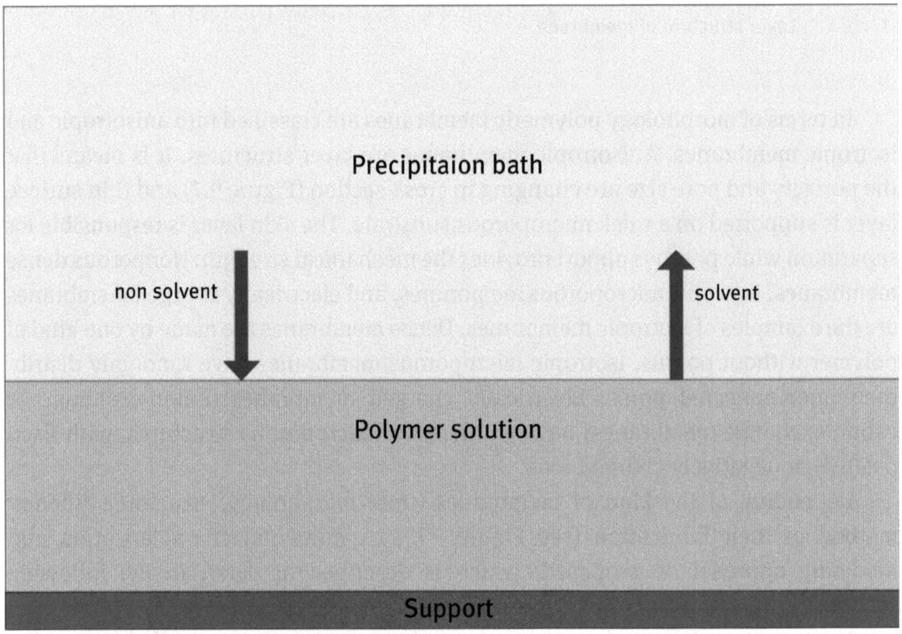

Figure 9.8: Schema of the immersion precipitation process.

9.2.2 Experimental characterization of membranes

The characterization of membranes is important for membrane selection, membrane process diagnosis, and new membrane material design. Their description can be classified into three categories: physical morphology, chemical composition and membrane fouling characterization. The characterization of membrane depends on its morphology [66]. Because, mainly, the porous membranes are used, the typical way of characterizing the membrane is to determine pore size and pore size distribution of the membrane. In the case of dense membranes the parameters of their characterization are volume size and free volume distribution [67]. Moreover it is very important to characterize of membrane surface morphology for both porous and dense membrane. The knowledge about the chemical composition of the membrane surface is necessary to know the chemical changes of the surface before and after surface modification or to know what kind of foulants is adhered to the membrane surface. Some experimental methods of membrane characterization proposed in literature are listed below.

Pore size distribution measurements (porous membranes) [67–75]:

- bubble gas transport method (bubble point method) – measurement of the pressure necessary to blow gas (mainly air) through porous membrane filled with water;
- mercury porosimetry – variation of above method, instead of water the mercury is used to fill membrane;
- adsorption-desorption method – measurements of adsorbed gas (mainly nitrogen) versus relative pressure (pressure/saturation vapor pressure of the adsorbent);
- permporometry (gas liquid equilibrium method) – measurements of the gas flux through the membrane, method based on the controlled blocking of pores by condensation of vapor;
- thermoporometry – performance of the freezing or melting thermograms using differential scanning calorimeter (DSC);
- water permeability – measurements of hydraulic permeability ($L_p = \frac{J_V}{TMP}$, where J_v is the permeate flux, TMP – transmembrane pressure) and calculate the pore radius using capillary pore diffusion model and Hagen-Poiseuille equation;
- electron microscopy (scanning electron measurements (SEM), transmission electron microscopy (TEM), scanning transmission electron microscopy (STEM), field effect scanning electron microscopy (FESEM), atomic force microscopy(AFM)) – methods for the viewing of cross sections of membranes, which is helpful for analyzing of pore size distributions;
- nuclear magnetic resonance (NMR) – methods based on comparison between the NMR characteristics of a material trapped within a porous network depending on whether it is in the liquid or solid state.

Characterization of free volume size and free volume distribution (dense membranes) [67, 76–79]:

- density measurements and the Bondi group contribution theory – methods based on difference between the sample's weight in air and in a nonsolvent with known density; from the density data the fractional free volume (FFV) can be calculated from equation: $FFV = \frac{V-1.3V_w}{V}$, where $V = \frac{1}{\rho}$ is the polymer's specific volume, ρ is the polymer's density and V_w is the estimated van der Waals volume calculated by the group contribution method of Bondi [76];
- scattering methods (small-angle X-ray scattering (SAXS), small-angle neutron scattering (SANS)) – methods for determination of volume or open cavities in membrane materials;
- positron annihilation spectroscopy (PAS) – measurements of positron lifetime and doppler broadening [77];
- wide-angle X-ray diffraction (WAXD) – the corresponding d-spacing values, which provide a measure of intersegmental distances between polymer backbones, is calculated from the diffraction peak maximum through the Bragg equation: $d = \frac{\alpha}{2sin\theta}$, where α is the wavelength of the radiation and 2θ is the angle of maximum intensity in the amorphous halo exhibited by the polymer [78];
- ellipsometry – measure the small changes in index of refraction [79];
- sorption and transport of low-molecular-weight molecules – method based on comparison of mass transfer of low-molecular-weight molecules with known properties through the membrane.

Membrane surface morphology measurements (dense and porous membranes) [58, 80]:

- microscopic method (scanning electron microscopy (SEM), transmission electron microscopy (TEM), atomic force microscopy (AFM) or X-ray photoelectron microscopy (XPS)) – technique allowing to get a tri-dimensional image of membrane surface;
- spectroscopic method (Infrared (IR) and Fourier transform infrared (FTIR) spectroscopy, Raman spectroscopy, electron spin resonance (ESR), nuclear magnetic resonance (NMR), wide angle X-ray scattering (WAXS), photoelectron spectroscopy (XPS), small angle neutron scattering (SANS), positronium annihilation lifetime spectroscopy (PALS), ultrasonic spectroscopy);
- differential scanning calorimetry (DSC);
- thermogravimetry (TGA);
- contact angle measurements – characterization of membrane hydrophilic/hydrophobic properties, its wettability;
- zeta potential measurements – technique to determine surface charge of membrane.

Polymer membrane, during the process, can swell. This phenomenon depends on the kind of polymer materials. Among experimental methods of characterization of membrane swelling there are very simple ones – like comparison of weight or thickness between dry and wet membrane [81]. Moreover the adaptations of techniques used in morphology characterization are proposed: thermoporometry [82], microscopy techniques [83] (SEM, TEM, fluorescence microscopy), spectroscopy methods (spectroscopic ellipsometry [84], Raman and high-resolution magic angle spinning NMR spectroscopy [85]), X-ray powder diffraction.

9.2.3 Calculation method of membrane characterization

To predict the structural changes of the effective layer for membrane prepared under different conditions the two-parameter model of ENP equation (see Section 9.1.1, eq. (9.20)) is proposed [17]. This method is generally used for modeling the mass transport in terms of diffusive and connective flow, but indirectly could be helpful in characterization of membranes morphology, independent of the way of their preparation. In citied work authors prepared thin film composite polyamide membrane with microporous polysulfone support under different trimesoyl chloride (TMC) content and different reaction time. Depending of the TMC concentration as well as reaction time different membranes were obtained, i. e. high concentration of TMC – membrane with higher thickness and diffusion resistance, long reaction time – membrane with narrower pore size. The proposed model very well described the experimental results. The same, ENP model was proposed by Diaper et al. [86] to develop new polyacrylic acid based membranes. The experimental results (rejection and permeate flux) of separation single electrolyte solutions by nanofiltration were interpreted based on ENP model. This approach let to predict the effective membrane charge density, porosity and thickness of membrane. Haasan et al. [87] proposed the combination of ENP equation with other equations to characterize the asymmetric, nanofiltration membrane. At first, based on pore flow, SD and ENP equations the modeling of experimental data (rejection) was done. Then the membranes parameters such as reflection coefficient, solute permeability and steric hindrance effects were estimated using Spielger–Kedem equations. Moreover, SHP model and Teorell–Meyer Sievers (TMS) model were proposed to estimate effective pore radius, effective charge density and ratio of effective membrane thickness to membrane porosity. From the modeling results, it was found that the polymer concentration can influence the membrane performances by varying of structural details. These conclusions were supported by the observation using scanning electron microscopy (SEM).

Because, as mentioned in Section 9.2.1, mainly membrane are prepared via phase inversion thus studies of thermodynamics and kinetics of this process, as well as computational method of membrane characterization are described briefly in literature. To calculation of thermodynamics phase diagrams of the system of polymer-solvent-nonsolvent (see Figure 9.8) few approaches are proposed: Flory–Huggins theory [88], binodal of ternary system with the consideration of the concentration dependence on the interaction parameters [89] and constant specific volume formulations [90]. Also, to calculation of kinetics of phase inversion some methods are presented in literature: simply mass transfer model known as Cohen's model [91] with assumption of equilibrium boundary condition at the interface between polymer solution and coagulation bath; extended Cohen's model by interface diffusion and frictional coefficients for the components [92] and by spinodal decomposition [93]; dissipative particle dynamics (DPD) simulation method [94].

Moreover, there is possibility to simulate precipitation process using multi-phase and multi-component Lattice–Boltzmann (LB) model to simulate precipitation process [95]. Generally, LB method is a mesoscopic model using to describe the macroscopic behavior of fluid flows. This method allows simulating a time-dependent structural formation during the membrane casting process of immersion precipitation. Two-dimensional (2D) simulation very well characterizes asymmetric membrane formation such as membrane compaction and formation of a selective skin layer. Moreover, the simulation of viscosity and forcing interactions effect on the final membrane structure agree with the experimental reports. Zhou and Powell [96] extended the simulation to 3D. They used ternary Cahn–Hilliard formulation incorporating a Flory–Huggins homogeneous free energy function to simulate the liquid–liquid demixing stage of the immersion precipitation process, which determines much of the final morphology of membranes. To simulation of actual membrane fabrication conditions in 2D and 3D authors used two-layer polymer–solvent/nonsolvent as initial conditions. Moreover, this system is coupled with the Navier–Stokes equations to model hydrodynamics in 2D system. The model results were similar to experimental data. He et al. [97] proposed simulation with Monte Carlo method applying polymer bond fluctuation lattice model with the exchange algorithm using probability density. This method allows simulating the open and wide diffusion layer in coagulation bath.

To characterize the polymer materials DFTs also can be applied. DFT is able to account for microscopic details such as the molecular excluded-volume effects, associating interactions, van der Waals attraction, Coulomb forces, and inter- and intra- molecular correlations that are important for understanding interactions of polymers with other substances [98]. The numerical results show that the DFT predictions are in good agreement with experiments and molecular simulations for the polymer structure and surface properties. Although authors used DFT to explore the antifouling properties of polymer brushes and polymer nanocomposites, this method could also be used in polymer membrane characterization.

Presented above consideration suggests the potential application of model described to process design for optimizing membrane morphology and performance.

Funding: This research was supported with 03/32/DS-PB/0701 grant.

References

[1] Van den Berg GB, Smolders CA. Diffusional phenomena in membrane separation processes. J Membr Sci 1992;73:103–118.
[2] Kaewsuk J, Seo GT. Computational study of NF membrane removal in rejection of specific NOM compounds. Desalination Water Treat 2013;51:31–33, 6218–23.
[3] Van der Bruggen B, Schaep J, Wilms D, Vandecasteele C. A comparison of models to describe the maximal retention of organic molecules in nanofiltration. Sep Sci Technol 2000;35: 169–182.
[4] Spiegler KS. Thermodynamics of hyperfiltration (reverse osmosis): criteria for efficient membranes. Desalination 1966;1:311–326.
[5] Nakao SI, Kimura S. Models of membrane transport phenomena and their applications for ultrafiltration data. J Chem Eng Jpn 1982;15:200–205.
[6] Mehiguene K, Garba Y, Taha S, Gondrexon N, Dorange G. Influence of operating conditions on the retention of copper and cadmium in aqueous solutions by nanofiltration: experimental results and modeling. Sep Purif Technol 1999;15:181–187.
[7] Anderson JL, Quinn JA. Restricted transport in small pores: a model for steric exclusion and hindered particle motion. Biophys J 1974;14:130–150.
[8] Zeman L, Wales M. Steric rejection of polymeric solutes by membranes with uniform pore size distribution. Sep Sci Technol 1981;16:275–290.
[9] Zeman L, Wales M. Polymer solute rejection by ultrafiltration membranes, Synthetic Membranes: volume II. Washington, DC: ACS, 1981, Chapter 23:411–434.
[10] Zydney AL, Aimar P, Meireles M, Pimbley JM, Belfort G. Use of the log-normal probability density function to analyze membrane pore size distributions: functional forms and discrepancies. J Membr Sci 1994;91:293–298.
[11] Bowen WR, Welfoot JS. Modelling of membrane nanofiltration – pore size distribution effects. Chem Eng Sci 2002;57:1393–1407.
[12] Van der Bruggen B, Vandecasteele C. Modelling of the retention of uncharged molecules with nanofiltration. Water Res 2002;36:1360–1368.
[13] Geens J, De Witte B, Van der Bruggen B. Removal of API's (Active Pharmaceutical Ingredients) from organic solvents by nanofiltration. Sep Sci Technol 2007;42:2435–2449.
[14] Verniory A, Dubois R, Decoodt P, Gassee JP, Lambert PP. Measurement of permeability of biological membranes – application to glomerular wall. J Gen Physiol 1973;62:489–507.
[15] Martin-Orue C, Bouhallab S, Garem A. Nanofiltration of amino acid and peptide solutions: mechanisms of separation. J Membr Sci 1998;142:225–233.
[16] Wijmans JG, Baker RW. The solution-diffusion model: a review. J Memb Sci 1995;107:1–21.
[17] Ahmad AL, Ooi BS. Characterization of composite nanofiltration membrane using two-parameters model of extended Nernst–Planck equation. Sep Purif Technol 2006;50: 300–309.
[18] Chaabane T, Taha S, Ahmed T, Maachi R, Dorange G. Coupled model of film theory and the Nernst–Planck equation in nanofiltration. Desalination 2007;206:424–432.
[19] Kumaran M, Bajpai S. Application of extended Nernst Planck model in nanofiltration process – A critical review. Int J Eng Res Rev 2015;3:40–49.
[20] Tsuru T, Nakao SI, Kimura S. Calculation of Ion rejection by extender Nernst-Planck equation with charge reverse osmosis membranes for single and mixed electrolyte solutions. J Chem Eng Jpn 1991;24:511–517.

[21] Bowen WR, Mukhtar H. Characterization and prediction of separation performance of nanofiltration membranes. J Membr Sci 1996;112:263–274.
[22] Cathie Lee WP, Mah SK, Leo CP, Wu TY, Chai SP. Phosphorus removal by NF90 membrane: optimization using central composite design. J Taiwan Inst Chem Eng 2014;45:1260–1269.
[23] Hua X, Zhao H, Yang R, Zhang W, Zhao W. Coupled model of extended Nernst–Planck equation and film theory in nanofiltration for xylo-oligosaccharide syrup. J Food Eng 2010;100:302–309.
[24] Zhao C, Zhang J, He G, Wang T, Hou D, Luan Z. Perfluorooctane sulfonate removal by nanofiltration membrane the role of calcium ions. Chem Eng J 2013;233:224–232.
[25] Kaewsuk J, Seo GT. Verification of NOM removal in MIEX-NF system for advanced water treatment. Sep Purifi Technol 2011;80:11–19.
[26] Khaouane L, Ammi Y, Hanini S. Modeling the retention of organic compounds by nanofiltration and reverse osmosis membranes using bootstrap aggregated neural networks. Arab J Sci Eng 2017;42:1443–1453.
[27] Kimura K, Toshima S, Amy G, Watanabe Y. Rejection of neutral endocrine disrupting compounds (EDCs) and pharmaceutical active compounds (PhACs) by RO membranes. J Membr Sci 2004;245:71–78.
[28] Yangali-Quintanilla V, Sadmani A, McConville M, Kennedy M, Amy G. A QSAR model for predicting rejection of emerging contaminants (pharmaceuticals, endocrine disruptors) by nanofiltration membranes. Water Res 2010;44:373–384.
[29] Warsinger DM, Swaminathan J, Guillen-Burrieza E, Arafat HA, Lienhard VJ. Scaling and fouling in membranes distillation for desalination applications: a review. Desalination 2015;356: 294–313.
[30] Song L. Flux decline in crossflow microfiltration and ultrafiltration: mechanisms and modeling of membrane fouling. J Membr Sci 1998;139:183–200.
[31] Chan R, Chen V. Characterization of protein fouling on membranes: opportunities and challenges. J Membr Sci 2004;242:169–188.
[32] Maruyama T, Katoh S, Nakajima M, Nabetani H, Abbott TP, Shono A, et al. FT-IR analysis of BSA fouled on ultrafiltration and microfiltration membranes. J Membr Sci 2001;192: 201–207.
[33] Rabiller-Baudry M, Le Maux M, Chaufer B, Begoin L. Characterization of cleaned and fouled membrane by ATR–FTIR and EDX analysis coupled with SEM: application to UF of skimmed milk with a PES membrane. Desalination 2002;146:123–128.
[34] Reichert U, Linden T, Belfort G, Kula MR, Thommes J. Visualizing protein adsorption to ion-exchange membranes by confocal microscopy. J Membr Sci 2002;199:161–166.
[35] Bakhshayeshi M, Kanani DM, Mehta A, Van Reis R, Kuriyel R, Jackson N, et al. Dextran sieving test for characterization of virus filtration membranes. J Membr Sci 2011;379:239–248.
[36] Sampath M, Shukla A, Rathore AS. Modeling of filtration processes – microfiltration and depth filtration for harvest of a therapeutic protein expressed in pichia pastoris at constant pressure. Bioeng 2014;1:260–277.
[37] Bowen WR, Clavo JI, Hernández A. Steps of membrane blocking in flux decline during protein microfiltration. J Membr Sci 1995;101:153–165.
[38] Mohammadi T, Kazemimoghadam M, Saadabadi M. Modeling of membrane fouling and flux decline in reverse osmosis during separation of oil in water emulsions. Desalination 2003;157:369–375.
[39] Kim J, Di Giano FA. Fouling models for low-pressure membrane systems. Sep Purif Technol 2009;68:293–304.
[40] Miller DJ, Kasemset S, Paul DR, Freeman BD. Comparison of membrane fouling at constant flux and constant transmembrane pressure conditions. J Membr Sci 2014;454:505–515.

[41] Huang H, Young T, Jacangelo JG. Unified membrane fouling index for low pressure membrane filtration of natural waters: principles and methodology. Environ Sci Technol 2008;42:714–720.
[42] Jacob K, Pradanos P, Calvo JI, Hernández A, Jonsson G. Fouling kinetics and associated dynamics of structural modifications. Colloids Surf A 1998;138:173–183.
[43] Ho CC, Zydney AL. A combined pore blockage and cake filtration model for protein fouling during microfiltration. J Coll Interface Sci 2000;232:389–399.
[44] Velasco C, Ouammou M, Calvoa JI, Hernández A. Protein fouling in microfiltration: deposition mechanism as a function of pressure for different pH. J Coll Interface Sci 2003;266:148–152.
[45] Sun X, Kanani DM, Ghosh R. Characterization and theoretical analysis of protein fouling of cellulose acetate membrane during constant flux dead-end microfiltration. J Membr Sci 2008;320:372–380.
[46] Bolton GR, La Casse D, Kuriyel R. Combined models of membrane fouling: development and application to microfiltration and ultrafiltration of biological fluids. J Membr Sci 2006;277: 75–84.
[47] Bolton GR, Boesch AW, Lazzara MJ. The effects of flow rate on membrane capacity: development and application of adsorptive membrane fouling models. J Membr Sci 2006;279:625–634.
[48] Duclos-Orsello C, Li W, Ho CC. A three mechanism model to describe fouling of microfiltration membranes. J Membr Sci 2006;280:856–866.
[49] Wessling M. Two-dimensional stochastic modeling of membrane fouling. Sep Purif Technol 2001;24:375–387.
[50] De Luca G, Bisignano F, Paone F, Curcio S. Multi-scale modeling of protein fouling in ultrafiltration process. J Membr Sci 2014;452:400–414.
[51] Curcio S, Scilingo G, Calabrò V, Iorio G. Ultrafiltration of BSA in pulsating conditions: an artificial neural networks approach. J Membr Sci 2005;246:235–247.
[52] Shetty GR, Chellam S. Predicting membrane fouling during municipal drinking water nanofiltration using artificial neural networks. J Membr Sci 2003;217:69–86.
[53] Liu QF, Kim SH, Lee S. Prediction of microfiltration membrane fouling using artificial neural network models. Sep Purif Technol 2009;70:96–102.
[54] Liu QF, Kim SH. Evaluation of membrane fouling models based on bench-scale experiments: a comparison between constant flow rate blocking laws and artificial neural network (ANNs) model. J Membr Sci 2008;310:393–401.
[55] Curcio S, Calabrò V, Iorio G. Reduction and control of flux decline in cross-flow membrane processes modeled by artificial neural networks. J Membr Sci 2006;286:125–132.
[56] Aydiner C, Demir I, Yildiz E. Modeling of flux decline in crossflow microfiltration using neural networks: the case of phosphate removal. J Membr Sci 2005;248:53–62.
[57] Chew CM, Aroua MK, Hussain MA. A practical hybrid modeling approach for the prediction of potential fouling parameters in ultrafiltration membrane water treatment plant. J Ind Eng Chem 2017;45:145–155.
[58] Ren J, Wang R. Preparation of polymeric membranes. In: Wang LK, Chen JP, Hung Y-T, Shammas, editors. Volume 13 of the series handbook of environmental engineering – membrane and desalination technology. New York: Springer, 2010:47–100.
[59] Baker RW. Membrane technology and applications. New York: McGraw-Hill, 2000.
[60] Reuvers AJ, Van den Berg JW, Smolders CA. Formation of membranes by means of immersion precipitation: part I. A model to describe mass transfer during immersion precipitation. J Membr Sci 1987;34:45–65.
[61] Akthakul A, Scott CE, Mayes AM, Wagner AJ. Lattice Boltzmann simulation of asymmetric membrane formation by immersion precipitation. J Membr Sci 2005;249:213–226.
[62] Yong SK, Hyo JK, Un JK. Asymmetric membrane formation via immersion precipitation method I. Kinetic effect. J Membr Sci 1991;60:219–232.

[63] Pereira CC, Nobrega R, Borges CP. Spinning process variables and polymer solution effects in the die-swell phenomenon during hollow fiber membranes formation. Brazilian J Chem Eng 2000;17:4–7.

[64] Di Luccio M, Nobrega R, Borges CP. Microporous anisotropic phase inversion membranes from bisphenol-A polycarbonate: study of a ternary system. Polymer 2000;41:4309–4315.

[65] van de Witte P, Dijkstra PJ, Van den Berg JW, Feijen J. Phase separation processes in polymer solutions in relation to membrane formation, J. Membr Sci 1996;117:1–31.

[66] Khulbe KC, Feng CY, Matsuura T. Membrane Characterization in Desalination and Water Resources Membrane Processes, vol. 1, ed. DMK Al-Gobaisi. Paris, France: EOLSS, 2010: 131–172.

[67] Tung KL, Chang KS, Wu TT, Lin NJ, Lee KR, Lai JY. Recent advances in the characterization of membrane morphology. Curr Opin Chem Eng 2014;4:121–127.

[68] Tylkowski B, Tsibranska I. Overview of main techniques used for membrane characterization. J Chem Technol Metall 2015;50:3–12.

[69] Hernández A, Calvo JI, Prádanos P, Tejerina F. Pore size distributions in microporous membranes. A critical analysis of the bubble point extended method. J Membr Sci 1996;112:1–12.

[70] Kaneko K. Determination of pore size and pore size distribution: 1. Adsorbents and catalysts. J Membr Sci 1994;96:59–89.

[71] Calvo JI, Hernadez A, Pradanos P, Martinez L, Bowen WR. Pore size distributions in microporous membranes II. Bulk characterization of track-etched filters by air porometry and mercury porosimetry. J Colloid Interface Sci 1995;176:467–478.

[72] Cao GZ, Meijerink J, Brinkman HW, Burggraaf AJ. Permporometry study on the size distribution of active pores in porous ceramic membranes. J Membr Sci 1993;83:221–235.

[73] Zhao C, Zhou X, Yue Y. Determination of pore size and pore size distribution on the surface of hollow-fiber filtration membranes: a review of methods. Desalination 2000;129:107–123.

[74] Glaves CL, Smith DM. Membrane pore structure analysis via NMR spin-lattice relaxation measurements. J Membr Sci 1989;46:167–184.

[75] Jeon JD, Kim SJ, Kwak SY. ^{1}H nuclear magnetic resonance (NMR) cryoporometry as a tool to determine the pore size distribution of ultrafiltration membranes. J Membr Sci 2008;309: 233–238.

[76] Vanegas ME, Quijada R, Nunes SP, Yave W. Syndiotactic polypropylene copolymer membranes and their performance for oxygen separation. J Membr Sci 2010;348:34–40.

[77] Zhao CT, Rosário Ribeiro M, De Pinho MN, Subrahmanyam VS, Gil CL, De Lima AP. Structural characteristics and gas permeation properties of polynorbornenes with retained bicyclic structure. Polymer 2001;42:2455–2462.

[78] Recio R, Palacio L, Prádanos P, Hernández A, Lozano AE, Marcos A, et al. Gas separation of 6FDA–6FpDA membranes. Effect of the solvent on polymer surfaces and permselectivity. J Membr Sci 2007;293:22–28.

[79] Baker EA, Rittigstein P, Torkelson JM, Roth CB. Streamlined ellipsometry procedure for characterizing physical aging rates of thin polymer films. J Polym Sci B 2009;47:2509–2519.

[80] Xu Z, Huang X, Wan L. Surface Engineering of Polymer Membranes. Hangzhou, China ang Berlin, Germany: Zhejiang University Press and Springer; 2009.

[81] Ho WS, Sirkar KK, editors. Membrane handbook. New York: Van Nostrand Reinhold, 1992.

[82] Quinson JF, Mameri N, Guihard L, Bariou B. The study of the swelling of an ultrafiltration membrane under the influence of solvents by thermoporometry and measurement of permeability. J Membr Sci 1991;58:191–200.

[83] Colby AH, Colson YL, Grinstaff MW. Microscopy and tunable resistive pulse sensing characterization of the swelling of pHresponsive, polymeric expansile nanoparticles. Nanoscale 2013;5:3496–3504.

[84] Ogieglo W, Werf H, Tempelman K, Wormeester H, Wessling M, Nijmeijer A, et al. n-Hexane induced swelling of thin PDMS films under non-equilibrium nanofiltration permeation conditions, resolved by spectroscopic ellipsometry. J Membr Sci 2013;437:313–323.
[85] Mele A, Castiglione F, Malpezzi L, Ganazzoli F, Raffaini G, Trotta F, et al. HR. MAS, NMR, powder XRD and Raman spectroscopy study of inclusion phenomena in CD nanosponges. J Incl Phenom Macrocycl Chem 2011;69:403–409.
[86] Diaper C, Correia V, Judd S. Characterisation of zirconium/poly(acrylic acid) low pressure dynamically formed membranes by use of the extended Nernst-Planck equation. J Membr Sci 1998;138:135–140.
[87] Hassan AR, Ali N, Abdull N, Ismail AF. A theoretical approach on membrane characterization: the deduction of fine structural details of asymmetric nanofiltration membranes. Desalination 2007;206:107–126.
[88] Tompa H. Polymer solutions. London: Butterworths Scientific Publications; 1956.
[89] Altena FW, Smolders CA. Calculation of liquid-liquid phase separation in a ternary system of a polymer in a mixture of a solvent and a nonsolvent. Macromolecules 1982;15:1491–1497.
[90] Tsay CS, McHugh AJ. Mass transfer modeling of asymmetric membrane formation by phase inversion. J Polym Sci Polym Phys Ed 1990;28:1327–1365.
[91] Cohen C, Tanny GB, Prager S. Diffusion-controlled formation of porous structures in ternary polymer system. J Polym Sci Polym Phys Ed 1979;17:477–489.
[92] Reuvers AJ, Van den Berg JW, Smolders CA. Formation of membranes by means of immersion precipitation. Part I. A model to describe mass transfer during immersion precipitation. J Membr Sci 1987;34:45–65.
[93] Kim YD, Kim JY, Lee HK, Kim SC. A new modeling of asymmetric membrane formation in rapid mass transfer system. J Membr Sci 2001;190:69–77.
[94] Wang XL, Qian JJ, Chen LJ, Lu ZY, Li ZS. Dissipative particle dynamics simulation on the polymer membrane formation by immersion precipitation. J Membr Sci 2008;311:251–258.
[95] Akthakul A, Scott CE, Mayes AM, Wagner AJ. Lattice–Boltzmann simulation of asymmetric membrane formation by immersion precipitation. J Membr Sci 2005;249:213–226.
[96] Zhou B, Powell AC. Phase field simulations of early stage structure formation during immersion precipitation of polymeric membranes in 2D and 3D. J Membr Sci 2006;268:150–164.
[97] He X, Chen C, Jiang Z, Su Y. Computer simulation of formation of polymeric ultrafiltration membrane via immersion precipitation. J Membr Sci 2011;371:108–116.
[98] Xu X, Cao D, Wu J. Density functional theory for predicting polymeric forces against surface fouling. Soft Matter 2010;6:4631–4646.

Irene Tsibranska, Serafim Vlaev and Bartosz Tylkowski

10 The problem of fouling in submerged membrane bioreactors – Model validation and experimental evidence

Abstract: Integrating biological treatment with membrane separation has found a broad area of applications and industrial attention. Submerged membrane bioreactors (SMBRs), based on membrane modules immersed in the bioreactor, or side stream ones connected in recycle have been employed in different biotechnological processes for separation of thermally unstable products. Fouling is one of the most important challenges in the integrated SMBRs. A number of works are devoted to fouling analysis and its treatment, especially exploring the opportunity for enhanced fouling control in SMBRs. The main goal of the review is to provide a comprehensive yet concise overview of modeling the fouling in SMBRs in view of the problematics of model validation, either by real system measurements at different scales or by analysis of the obtained theoretical results. The review is focused on the current state of research applying computational fluid dynamics (CFD) modeling techniques.

Keywords: membrane fouling, computational fluid dynamics, submerged membrane bioreactors

10.1 Introduction

Integrating biological treatment with membrane separation has found a wide area of applications. Submerged membrane bioreactors (SMBR) containing membrane modules immersed in the bioreactor, or side stream ones connected in recycle have been used in different biotechnological processes for separation of thermally unstable products, including two-phase systems [1–3]. The membrane bioreactor technology (MBR) has been efficient in biological treatment of waste flows [4] with high content and variety of organic contaminants, high strength industrial wastewater [5, 6], as well as an effective barrier against many active pharmaceutical ingredients and resistant to degradation organic pollutants [7, 8]. Innovative MBR designs attempt the use of submerged membrane modules to achieve better membrane performance, less energy consumption and effective and green/sustainable fouling control. SMBRs

This article has previously been published in the journal *Physical Sciences Reviews*. Please cite as: Tsibranska, I., Vlaev, S., Tylkowski, B. The problem of fouling in submerged membrane bioreactors – Model validation and experimental evidence. *Physical Sciences Reviews* [Online] **2018**, *3* (1). DOI: 10.1515/psr-2017-0143

https://doi.org/10.1515/9783110482065-010

compete successfully with conventional continuous reactors with suspended biomass and agitation [9], the research interest in them [10, 10–13] includes creating conditions for high product quality at reduced energy consumption and beneficial economic estimates such lower operating costs of the integrated operation.

Membrane fouling remains the main limitation in integrated systems combining membrane separation and bioreactor. A number of works are devoted to fouling analysis and its treatment, especially exploring the opportunity for enhanced fouling control in SMBR. Fouling is related to MBR hydrodynamics and the advantage of fouling control in SMBR to be ensured, thorough examination of the bioreactor operating conditions is essential. Along with the theoretical and experimental approaches, computational fluid dynamics (CFD) is the important third approach to systems hydrodynamic analysis and resolution. CFD modeling effectively helps the design and optimization of SMBR, based on fluid flow and separation dynamics of submerged membranes, mass transfer rate and fouling conditions [1, 14]. Combined knowledge on the reactor performance is obtained through models of various complexities, accounting for the interrelationship between bio-kinetics, filtration and hydrodynamics of the integrated system with application of CFD [15]. Hydrodynamic models are mainly focused on fouling, as decrease and/or control of fouling can be achieved through process hydrodynamics and especially by creating favorable hydrodynamic conditions near the membrane surface. By inducing shear stress, promoting turbulence of fluid and enhancing the convective flow of particles away from the membrane, an essential impact on fouling reduction is obtained. Special interest is given to *membrane fouling* (including multidimensional [16]) in SMBRs with different geometries of the membrane surface (flat sheet, tubular, hollow fiber [11, 17–20]) and shear enhancement effect of the geometry [20], gas bubbles [21–23], stirring [24], fiber motion [25], etc. CFD is largely applied for such investigations [26] with emphasis on the understanding of shear-stress distribution on the membrane wall. The latter is essentially affected by: stirring (the flow pattern created by the impeller); fluid viscosity (especially in cases of complex rheology); gas bubbling in aerated SMBRs (size of the bubbles, gas flow rate).

Mechanical stirring contributes to the uniform distribution of the reaction system and affects membrane fouling by the shear force. CFD simulations on the influence of mechanical stirring on membrane fouling show that the magnitude and direction of velocity vectors on the membrane surface change with stirring rate variation [24], thus creating different shear rate intensity and distribution on the membrane surface. CFD was applied to demonstrate the effect of reactor configurations (the relative position between membrane and impeller, effect of impeller design) on the fluid flow pattern in submerged membrane reactors [27]. Supposing negligible effect of permeation on the hydrodynamic conditions at the membrane surface, the impeller-induced cross flow was examined for tubular membrane module submerged in a conventional stirred tank reactor with radial Rushton turbine [28]. In the CFD simulations, the rate-of-deformation tensor was targeted, as determined by the local gradients of the component velocities near the membrane interface. Shear-stress field and inhomogeneity on the membrane surface are discussed together with bulk fluid shear, the latter being at least an order of magnitude lower that the one at the wall [28]. Thus, it is expected that the operational regimes that ensure fouling control would not be harmful for shear-sensitive microorganisms. Mixing conditions in an unbaffled stirred tank coupled to an external microfiltration module were investigated by CFD [19], the bioreactor behaving as a submerged membrane device. In case of two-phase systems the combined influence of stirring and gas sparging on mixing and membrane fouling is studied by CFD in view of uniformity of shear and velocity distributions in the volume of the reactor and along the membrane surface [29]. SMBR with stirring and aeration divided into two zones with internal circulation was operated in a continuous mode and used for cultivation of granular sludge [30]. In this case the membrane module is situated in the aeration zone; both aeration and internal hydrodynamic circulation creating conditions to mitigate membrane fouling.

Using combined impeller and membrane designs allows application in highly *viscous and complex fluids* where both interface and transport problems occur [24]. In case of complex fluids (Newtonian fluid containing solutes, colloids and particulates), such as fermentation broths and natural organic matter, fouling in membrane bioreactors is also affected by the penetration of macromolecules into the membrane structure and foulant cake structure [31]. CFD modeling for SMBR, incorporating the non-Newtonian fluid behavior is reported in the field of wastewater treatment applications [32], such as activated sludge systems [33–35]. Shear stress and area weighted average shear stress relationships were made giving error less than 8 % compared with the CFD results [33]. Nevertheless complex fluid rheology remains a challenge to measurement and CFD modeling with regard to relationships between wall shear rate and filtration flux, for both simple and complex fluid systems [15, 36]. CFD modeling of membrane bioreactors including non-Newtonian rheology models (ex. power-law model [28, 33, 35], Herschel–Bulkley model [29]) is often met with the necessity to incorporate a specific rheological model which is difficult to choose because of the heterogeneity of the system (ex. activated sludge [15]). An interpolation method may be needed [37] or concentration dependence of viscosity has to be included (ex. total solid concentration of the sludge) [16, 38], as well as the concentration limit up to which the liquid phase may be treated as Newtonian [32]. With regard to non-Newtonian behavior in full-scale MBR, some CFD studies report no effect on mixing, but bubble formation or air distribution within the membrane module have to be included to correctly comment on the dependency of non-Newtonian behavior on mixing [14, 15]; the latter could affect oxygen transfer efficiency and local shear effects. The presence of solids has also an impact on oxygen transfer coefficients and leads to changes in the fouling propensity of the mixed liquor in SMBR [39]. One of the most efficient strategies to limit fouling is the use of a *gas/liquid two-phase flow* [40] and especially in case of SMBRs [22]. In aerated submerged MBR, enhanced shear stress on the deposited cake layer is achieved through upward blowing of air bubbles.

By higher aeration rates the created shearing stress at the membrane surface leads to increased back transport of foulants thus resulting in reduced concentration polarization and rate of cake layer formation [41]. Relationships are reported between the filtration flux and the two-phase flow velocity and the bubbling air flow rate per unit membrane area [42]. CFD results on the distribution of shear stress on membrane surface during bubble rising in a flat sheet MBR, proved to be in good agreement with electrochemical experimental results [23]. More important effect on membrane shear stress has bubble diameter than air flow rate, found by CFD modeling of submerged membrane airlift reactor with regard to cake layer fouling control [38]. In any case the balance has to be found between the effects of increased shear stresses on fouling reduction and on mixed liquor integrity preservation [22]. Optimum operation conditions for submerged in an aerated bioreactor membrane modules are searched [3] in direction of: positioning of the membrane modules, gas/liquid ratio, bubble size, transmembrane pressure, obtaining sufficient wall shear stress to create friction on the membrane surface. Approached by CFD modeling, the relation between shear rate and permeate flux is confirmed qualitatively, but effective relationships between them for simple and complex fluid systems are still lacking. Membrane fouling models are usually limited to simple hydraulic profiles. Modeling approaches accounting for shear profile and dynamic linking of flux and transmembrane pressure are relatively new [16].

Whatever the model structure is in terms of level of abstraction, model assumptions and number of parameters involved, the question of how accurately the model calculations correspond to the experimental evidence is of primary importance. Whether and how the model assumptions are reasonable with respect to real systems is an object of model validation, which for CFD is in the focus of data-model comparison rather than parameter estimation or calibration [32].

Over the years there have been a number of extensive surveys published that have covered/dealt with/similar aspects of MBR research. The most recent ones demonstrating the CFD ability to solve problems in connection with SMBR [14, 15, 17, 43–46] are noteworthy. This article aims to summarize works on SMBRs with CFD simulation of fouling with focus on model validation experience.

This article aims to provide a comprehensive yet concise overview of modeling the fouling in SMBRs in view of the problematics of model validation, either by real system measurements at different scales or by analysis of the obtained theoretical results. The review is focused on the current state of research applying CFD modeling techniques.

10.2 Membrane fouling modeling and validation

Although MBR is of increasing importance, its primary drawback remains the fouling phenomena, leading to higher membrane resistance and lower permeate flux. The fouling mechanisms in a SMBR were investigated [31, 47, 48] with special accent on understanding [4, 31, 45, 48–50], monitoring [51] and control of membrane fouling [26, 38, 52]. Controlled hydrodynamic conditions in the vicinity of the membrane surface are crucial for the flux stability, as well as in minimizing the membrane area requirement [43]. Enhancement of membrane shear rates is considered as one of the most efficient factors for fouling control. Shear forces generated by pumps, bubbling or impellers are used to remove the fouling layer, yet care has to be taken for shear sensitive ingredients that may undergo deactivation [53, 54]. From this point of view, the shear stress field and the homogeneity of the shear environment near the membrane surface have to be revealed. So far systematically obtained information on flowing matter shear stress in integrated stirred SMBRs is lacking. Aspects of flow internal circulation, bioreactor residence time distribution (RTD), as well as shear-related differences in soluble exopolysaccharides production and the membrane fouling potential have been studied [1, 54]. The effects of dynamic pressure [55] and fluid velocity at the membrane surface [27] upon the permeate flux have been highlighted by CFD. The great number of research in the last years has contributed to the progress in the field of understanding and controlling the fouling in membrane bioreactors, further investigation being needed in the field of pilot and full-scale applications, as well as reactor performance optimization according to the CFD modeling [56].

10.2.1 Fouling modeling

Fouling mechanisms in MBR are related to the interactions between the liquid phase and the membrane, namely the size of the membrane pores in relation to the size of particles, suspended in the reactor liquor. Soluble and micro-colloidal substances may cause pore narrowing if adsorbed on the pore walls; pore blocking appears when particles of the same size as the pores interact with the latter; cake layer is formed when larger particles are deposited on the membrane surface and this fouling mechanism is recognized as the most important in SMBR. In the case of complex liquors containing all soluble, colloidal and suspended particles a synergistic effect on the fouling process is observed and investigated, including redistribution of fouling resistance between the cake layer and the membrane pores and predicting flux decline [48].

CFD studies of SMBR are mainly focused on bubble flow (effect of scouring) and/or impact of the membrane module on the mixing behavior of the reactor [15]. Hydrodynamic conditions in SMBR are highly dependent on the module design and geometry, the application of CFD models to individual SMBR configurations with accent on shear-stress calculations and fouling being discussed [14, 15] in a number of publications (hollow fiber [57–59], flat sheet [17, 20, 23, 60], or tubular [1, 27, 28] membrane modules). They include continuity and momentum balances, extended with a turbulence model for the reactor vessel hydrodynamics, membrane separation and fouling being accounted for by incorporating sub-models. Both the standard k-ε

(SKE) [16, 60, 61] and RNG k-ε [38, 62] turbulence models have been employed. Submodels cover different aspects of fouling, e. g. different drag–force relationships, namely Shiller–Nauman [38], Morci–Alexander [29] or Grace drag model [63], the filtration Darcy's law flux–pressure relationship [16], the resistance-in-series model [16, 64], the porous media model [35] and diffusion-convection [64] models. Mostly two-phase models [20, 25] were employed, but single-phase flow was also studied [62, 65, 66]. The two-phase gas–liquid Euler–Euler approach [38], but also the volume-of-fluid (VOF) model [67] has been used. The presence of solid phase has been considered in a new fouling model [68]. Concerning the reactor hydrodynamics, the major approaches to model multiphase flow in membrane bioreactors with CFD are discussed in Ref. [15] together with an overview of the reported investigations in the field. CFD modeling of MBRs is reviewed with respect to the reactor configuration. In the case of SMBR with membrane modules immersed inside the reactor vessel in presence of aeration Eulerian multiphase models or mixture models are recommended [15]. Euler–Euler approach was widely used in SMBR hydrodynamic models used for shear-stress analysis on the membrane surface, effect of air flow rate on fouling control, particle-size distribution in relation to particle deposition propensity [26, 38, 60, 69]. The model for the reactor vessel usually assumes that the permeate velocity can be neglected (less than 10 %) with regard to the cross flow velocity of the membrane surface, so non-permeability of the latter is supposed [26]. The arguments for including the permeability of the membrane in the model are that its effect at micro-scale has to be checked, as well as the transmembrane pressure could be used as validation variable for these models [15].

The submodel considers the fouling mechanism (permeate flux evolution according to the cake layer formation, pore blocking, or combined pore constriction, pore blockage and cake formation mechanisms [48], membrane resistance calculation [60]) and the relation of shear rate to flux and resistance data. Mean membrane wall shear stress is calculated by a surface integral over the area of the membrane surface [26]. Experimental data are used to find the critical flux value and, operating under it, to describe the relationship between flux and transmembrane pressure development [26].

The *relation between flux and shear stress* created by stirring are usually qualitatively defined [27, 28], or limits of influence are defined, where membrane fouling could be effectively controlled by shear stress [69]. The effect of shear on flux decline is approached by membrane resistance versus shear modeling [60]. The authors assumed that each resistance (cake, adsorption, pore blocking, intrinsic resistance) is a linear function of gas and liquid shear stresses on membrane surface and proposed empirical multiple linear regression equations connecting experimental resistance data with by CFD evaluated shear stresses. In case of complex fluids the experimentally observed sludge rheology is described as a function of shear rate and mixed liquor solid concentration [38]. The impact of viscosity can be either in direction of lower shear rate on the membrane and retarding the flow, or the opposite if the increase of sludge viscosity compensates the decrease of shear rate. CFD model of submerged hollow fiber membrane module was reported, accounting for the non-Newtonian rheology of the mixed liquor by incorporating the Ostwald-de Waele rheology model [35]. The liquid velocities profile obtained by CFD calculations was confirmed by the experimentally observed using particle image velocimetry (PIV).

With the evolution of computational technique lately 3-D hydrodynamic models are used [20, 25, 35, 38, 66], but early 2-D simulations are still practiced [64, 65]. In cases, species equations or source terms accounting for momentum or species transport [64, 70] are added.

The *shear rate distribution* on the membrane surface *in connection to applied aeration rate* is attained by CFD, the results being important with regard of fouling control [26]. Assuming bubble flow with constant size and regular shape of the bubbles and no interactions between them [26], the model confirms the experimentally observed positive effect air flow rate on the mean shear stress on the membrane surface. The model helps the understanding of this effect, indicating that aeration increases the liquid velocity near the membrane, the latter leading to higher shear stress. The main factors affecting the shear stress distribution in presence of aeration are found to be (in order of significance) mixed liquor suspended solids fraction, bubble diameter, and aeration intensity [38].

In several recent publications, the *effect of particle size distribution on fouling* is investigated by CFD simulations [14, 26, 38]. The limits of dependency on shear stress (back transport mechanisms respectively) are set within 1 and 100 μm particle diameters [14, 38]. The significance order of factors affecting the particle deposition propensity was found to be: mixed liquor suspended solids fraction > aeration intensity > bubble diameter [38]. CFD calculations combined with Population Balance Equation have a great potential in describing particle size distribution dynamics due to aggregation and breakage phenomena, remaining an open field for further investigations [15].

Modeling SMBR fouling performance at the vessel dimension is still rare and reduced to several recent studies [16, 20, 35, 38, 50, 58, 60]. Mostly models of fouling at the scale of *membrane modules* of various type have been validated [64, 70–87]. Design optimizations by SMBR modeling relevant to fouling are still rare. Recently, optimization studies related to membrane tube design were reported by [35, 62, 66]. A side-stream SMBR was studied [75]. In intermediate position is the rotating-disc membrane module [88].

10.2.2 Larger-scale applications

One important issue from the review on fouling modeling is the need of more modeling studies focused on larger scale applications [15, 56], which supposes sufficient experimental evidence including long-term operation [52, 89]. The latter is important in view of type of foulants and fouling mechanisms which intervene during long-term operation (long-term fouling evolution, irreversible resistance mechanism, change in apparent viscosity), as well as validation of the model (short-term and long-term validation data) in order to prove that it covers the process dynamics over a range of operating conditions [90, 91].

Recent developments in the field of full scale membranes bioreactors are critically reviewed in [89], including SMBR fouling control. The following approaches are discussed: membrane scouring through bubble aeration, introduction of a granular medium into submerged MBRs to mitigate membrane fouling; use of baffles, inserted into the membrane compartment of submerged MBRs to optimize hydraulic conditions; introduction of periodic high-frequency vibrations (for hollow fibers SMBR) etc. Nevertheless, fouling remains the key challenge in view of the market penetration of the MBR technology and SMBR in particular. Strategies to limit fouling like manipulating bioreactor conditions, adjusting hydrodynamics and flux and optimizing module design [31] can benefit from CFD applications to larger-scale membrane bioreactors.

Pilot plant hydrodynamics in a bioreactor with submerged hollow fiber module was satisfactorily predicted by CFD simulations in view of liquid velocity and gas holdup [92], but lower average velocities were calculated for both liquid and gas phase in the full-scale plant. The results were essentially improved by a small design modification (enlarged size). The CFD approach used Eulerian multiphase model, the hollow fibers zone being modeled as porous. Effect of suspended solids on liquid viscosity was accounted for, as well as the change of the flux along the fibers (due to the pressure drop).

RTDs in two different full-scale MBRs (with hollow-fiber and flat sheet membrane modules) were successfully predicted by CFD [29]. Because of the high sludge viscosity, the CFD model includes an experimentally calibrated rheological model. The minor role of the latter on the reactor hydrodynamics was observed due to the high turbulent viscosity.

CFD modeling was used to analyze the membrane surface shear in a pilot plant and full-scale SMBR [35]. Sub-models for the rheology of the mixed liquor (Ostwald-de Waele) and for the membrane module (porous media) were incorporated within the Eulerian model to simulate the three-dimensional flow field of the filtration chamber (hollow fiber membranes) of the pilot scale MBR. The CFD simulated shear rate for the pilot scale reactor was in the range of the experimentally measured in the mixed liquor suspended solids. In the full-scale simulations (under the same model setup) lower values for the area-weighted average shear stress were observed, but more even shear-stress distribution on the membrane surface. The full-scale simulations with larger aeration intensity achieved the same level of shear stress as in the pilot scale reactor.

Three-dimensional two-phase CFD simulations (Eulerian model, and RNG k-ε model), incorporating rheology models for the mixed liquor and porous media models for the membrane zone, accurately predict the hydrodynamics and aeration patterns in a pilot scale membrane bioreactor fitted with vertically orientated commercially available hollow fiber modules [93]. The authors point out the limits of the model toward small-scale turbulence, where increasing the grid resolution or using different turbulence models is not in help of accuracy. Nevertheless CFD results on area-averaged membrane surface shear gave important information for fouling control along the sections of the hollow fiber membrane: vertical orientation gives less resistance to the fluid flow and larger shear on the membrane surface); baffles around the membrane modules promote higher turbulence and increased shear in the upper section of the membrane module; increasing bubble size gives higher shear for constant bubble density; smaller bubbles increase shear for fixed aeration rate.

Novel strategies to achieve efficient fouling control in lab-, pilot- and full-scale MBRs are reviewed in Ref. [56]. The authors point out the potential of CFD modeling in aeration optimization, such as *intermittent or cyclic aeration* and *mechanically-assisted aeration scouring*, in particular with regard to fluidization and scouring behavior of particles in membrane bioreactors [56].

10.2.3 Validation experience

Whether we can have confidence in the simulation results is a question of primary importance. Model verification and validation is required. In principle, referring to the various types of chemical engineering models, e. g. experimental/empirical, structural/cell models and theoretical CFD models based on the laws of fluid dynamics, the latter are most reliable and require least validation effort in order to imply confidence. In fact, according to the authors' experience, a well-designed computational grid for a particular geometry, appropriate selection of the turbulence approximation, and carefully selected sub-model correlations ensure solutions that are enough accurate and acceptable for chemical and biochemical engineering problems that in general exhibit 10 to 30 %% experimental errors and more than 50 % parameter deviation by extrapolation. At least so it is seen from the review of the validation examples. Yet as every model has its level of abstraction, the model assumptions have to be verified for being correctly implemented by the model and the results have to provide a valid representation of the real system. Accordingly, following the above brief overview of the basic models practiced in membrane separation research, this section attempts an overview of validation examples and the validation techniques employed.

Validation involves qualitative correspondence between model and reality and quantitative comparison between actual system and the model. The philosophy of verification and validation for modeling and simulation in computational engineering applications as well as useful guidelines regarding the design of validations experiments in a way as to allow precise and conclusive comparison of calculations and experimental data are given in Ref. [94]. The predictive capability of the model, i. e. the extrapolation beyond the understanding gained from the validation data is emphasized. A comprehensive approach to verification and validation of CFD simulations is commented [95] in view of its applicability to a broad range of CFD codes, including Navier–Stokes, RANS, Euler, boundary-element methods, and others. The validation database should represent reproducible evidence that the model has achieved a certain level of accuracy, i. e. the validation experiments are focused on the assessment of modeling capability [94]. With regard to CFD applications to fouling investigations in SMBR, the need of more process model validation is recognized [15]. On the other hand, the limits of what is measureable are constantly enlarged, though data collection for validation remains both time and resource consuming [15, 32]. The following validation techniques denoted as qualitative (A) and quantitative (B) ones have been experienced.

(A) Qualitative validation is carried out by comparison of membrane fouling patterns. This has been carried out by micrographs shot by SEM and even by ordinary photography [71]. Zones' images of experimental fouling patterns obtained by PIV were compared with CFD-generated patterns [74]. In general, the images show zones in bright color of reduced fouling and zones in dark representing intensive fouling [71]. Related to the fundamental parameter of fouling condition, these zones are interpreted as zones of high and low shear, respectively.

Qualitative validation has helped to show the potential of CFD simulation to interpret near-wall flow patterns with reference to fouling resistances in SMBR in parallel to hydrodynamic data obtained otherwise separately by pressure and velocity measurements [71].

(B) Quantitative validation is the procedure to collect, by experiments or by reference, experimental data for the real system and to juxtapose them with CFD model predictions in terms of the allowable error margin.

(B1) The oldest method for quantitative validation compares predicted and experimental membrane filtration parameters, i. e. pressure drop [16] and permeate flux profiles [62, 64, 66, 72, 96–98]. The method remains an overall assessment technique good for validation based on global membrane filtration parameters [71]. In a recent paper [68] flux versus time predictions compared with experimental data at various conditions showed relative error as low as 4 %. Permeate flux obtained in experiments of membrane filtration of sodium chloride solutions at various cross-flow velocity and TMP was compared [72] with computed permeate flux data and showed good agreement between experiment and calculations. The error spread range was 7–11 % related to different cross flow velocities and transmembrane pressures [72]. Experimental vs predicted global pressure drop and global resistance validation were carried out in the case of hollow fiber filtration system in Ref. [64]. Fairly good agreement has been obtained of experimental flux versus predicted CFD fluxes in different micro- and ultrafiltration membrane channels [70]. Based on the adequacy of measured and predicted pressure drop/flux data, authors derive empirical equations for membrane coefficients, e. g. friction loss of fibers [35] and incorporate them into the CFD model-seeking CFD validation.

(B2) Because the early design of MBR was based on RTD data, RTD has been an early parameter for MBR evaluation and model validation. Since RTD has been described originally by empirical structural models, e. g. compartmental models, it has been natural to compare experimental RTD functions with the solutions of compartmental models and further with the more advanced CFD models. This has been done recently [29]. Unlike CFD modeling, it has been shown that compartmental/structural modeling, such as the tank-in-series model, is unable to predict the behavior of the complex flows observed in MBR [99].

(B3) Quantitative validation is done also by comparing reference data, either experimental or computational ones to validate models [65, 68, 96] and vice versa [74]. In an attempt to simulate the effect of uniform sinusoidal perturbations on shear, an "extreme" case of validation has been attempted in [65] by comparing electro-osmotic flow perturbation velocity profiles against profiles reported in a reference paper of a different nature, *viz.* reduced order model introduced as being in close similarity to the original one.

(B4) Quantitative validation based on concentration allows improved assessment of adequacy, since concentration is measured easily locally. This version is practiced to validate advanced models including species transport. Suspension concentration validation was carried out in [96] and by comparing dissolved oxygen (DO) concentration [38], both showing reproducible results (errors of less than 5 % are reported). Dimensionless mass transfer data obtained both numerically and by experiment have been compared to validate models for spacer-filled channels [100].

(B5) Validation by comparing velocity and velocity gradients is most advanced and also an intensive instrument, as it allows measurement and comparison of local flow parameters. PIV combined with CFD modeling was successfully used to investigate the effects of membrane module configuration and aeration patterns on the local velocities and shear forces in MBR. It has been used in the validation practice for determination of liquid velocity in bubbly flow [35, 93]. PIV [20, 38] allowed 3-D viewing of motion of tracer particles. Measured liquid flow velocities from bench scale reactor were compared with results of CFD simulations for a model reactor of identical geometry (membrane module, tank and aerator) and operating parameters. The results were obtained with porous media model for the hollow fiber filtration zone. The CFD model based on two-phase Eulerian approach, RNG k-ε model coupled with sludge rheology models and porous media model was found to underestimate the randomness of turbulence, but capable of quantifying the grid averaged flow variables in hollow fiber SMBR [93]. Both wall shear stress and liquid flows along the membrane surface were investigated by filtration experiments coupled with PIV technique and CFD simulation [101]. PIV measurements of liquid velocities were used for hydrodynamic validation of a flat sheet SMBR CFD model [38]. Good agreement between CFD simulations and PIV measurements were achieved. By reference, PIV has been practiced in SMBR fouling model validation with error margins as low as 6 % [87, 93], 5 % [38], and also [74]. Recently even higher precision was reported [20] showing relative error as low as 1.4 % found between simulated and experimental data, "much smaller than the literature reported values of 8–30%".

(B6) Considering the emerging fouling control methods, including mechanically assisted aeration scouring, in-situ chemical cleaning, enzymatic and bacterial degradation of foulants, electrically assisted fouling mitigation, and nanomaterial based membranes [56], advanced measurement techniques have been used in membrane bioreactors. Three different scanning electron microscopy (SEM) techniques were applied to investigate the detailed fouling morphology on the membrane surface in flat sheet membrane SMBR (environmental SEM (ESEM), cryo-SEM, and freeze-drying SEM) together with in situ observation using optical coherence tomography (OCT), which allowed to acquire 3D data in real-time and non-destructivity [102]. The use of novel non destructive methods for in situ observation has increased, giving access to the fouling spatial distribution and 3D mapping of its development over time [103]. Along with SEM, EDX (energy dispersive X-ray), FTIR (Fourier transform infrared) or CLSM (confocal laser scanning microscopy) techniques, used for investigation of fouling deposition on membrane surface and within the membrane pores, the new effective and non-invasive methods are used for *in situ* monitoring of membrane fouling and local filtration characteristics [51].

(B7) Validation based on local measurement of shear has been used. The electrochemical-diffusion method once practiced for mass transfer measurements [104] has been applied as measurement of velocity gradients in mixing vessels [105] and in stirred SMBR [28, 44, 100, 106].

Bubble-induced shear profiles in submerged hollow fiber membrane modules were measured using electrochemical method [107] and a high speed imaging system. The shear profiles resulting from different bubble geometries and flow paths under different operating conditions are commented in view of better understanding of the hydrodynamic conditions and fouling control via gas sparging. In case of membrane bioreactors with immersed flat sheet membrane module the measured enhancement of the shear stress in a two-phase flow relative to the single-phase was in accordance with CFD simulations of the single bubble rise [108]. The authors use electro-diffusion method to determine the shear stress in a flat sheet membrane system in conditions of bubbly flow and non-Newtonian liquid together with high speed imaging to measure the gas hold-up.

The electro-diffusion method was used to validate CFD simulations of stirred SMBR in case of non-Newtonian fluids comparing membrane surface shear rate [28]. The experimental configuration includes radial Rushton turbine and submerged mono-tubular membrane module. Experimental wall shear rate data at the solid-liquid interface were obtained using a spherical probe with sensor adjacent to the membrane tip. Referring to membrane studies, a similar technique of shear measurement at the surface of the flat bottom of a stirred cell for frontal filtration with embedded sensor electrodes was used in [106]. Computed shear rates compared

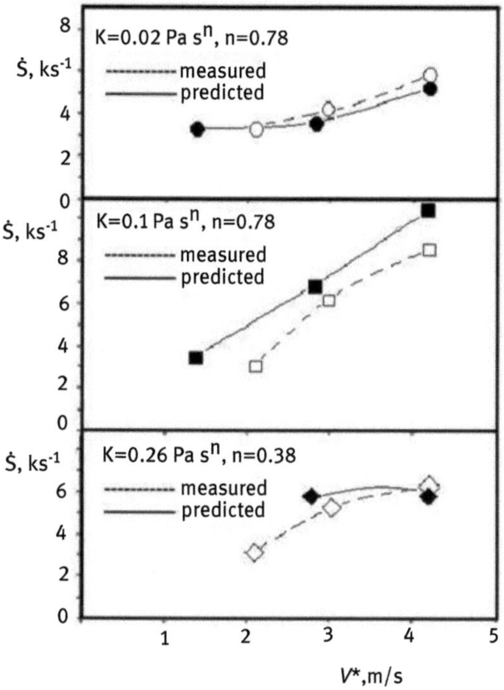

Figure 10.1: Predicted (black symbols) versus measured (void symbols) wall shear rates at the membrane section facing the impeller [28].

well with the measured values, Figure 10.1; Referring to the three model fluids, solutions of starch (flow index n = 0.78, consistency coefficient K = 0.02 Pa s^n), CMC (n = 0.78, K = 0.1 Pa s^n), and xanthan gum (n = 0.34, K = 0.26 Pa s^n), CFD produced values in the range $\dot{S}$~3–10 ks^{-1} that compared well with $\dot{S}$~2–9 ks^{-1} measured by the probe. The largest deviation of theory to experimental data of 17 % was registered for the low Re number case of viscous CMC.

Visualization of the velocity contour and vector plots illustrated in Figure 10.2 shows zones of adverse gradients and distorted flow laminae of the membrane surface that imply that partial separation may occur. Yet a portion of disagreement between measured and predicted values might be due also to the flow field changes imparted by the probe displacement.

In another approach, model validation was carried out through analysis of the obtained theoretical results by running the SKE in parallel with the "k-ω"-SST model. The low Re range at K = 0.1 Pa.s^m, n = 0.78 was targeted. The membrane area-average shear stress obtained at N = 400 rpm was 97 N/m^2 (model SKE) vs. 91 N/m^2 (model SST), a deviation of 6 %, and the deviation grew up to 14 % at higher rotational speed (N = 600 rpm). Hence, no significant difference between the two models' solutions was registered.

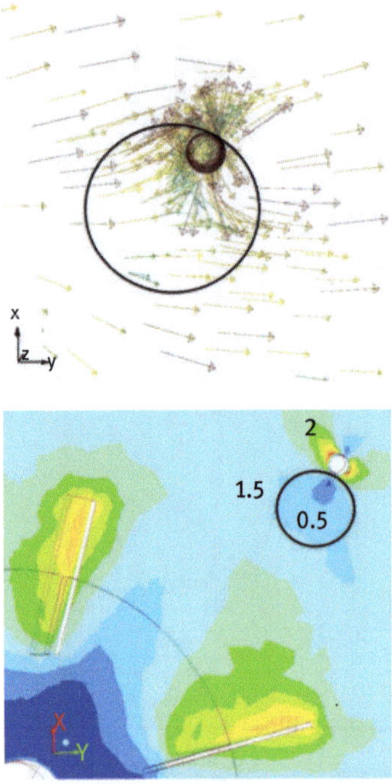

Figure 10.2: Velocity (m/s) magnitude and vector plots at membrane cross-section plane in case z/H = 0.52 at N = 400 rpm, n = 0.78, K = 0.02 Pa.s^n [28].

The effect of gas presence on shear distribution along the membrane interface in SMBR (Figure 10.3) – stirred tank reactor (STR) Biostat equipped with membrane immersed in a non-Newtonian model solution, was studied by CFD [109]. Figure 10.4 illustrates the shear distribution in gas presence in comparison with the one at non-aerated conditions. Gas presence causes the shear field to be more uniform. The zones of moderate shear in areas where shear is low in gas absence (marked in red) are enlarged. The effect is more clearly assessed by XY plots along the Z-coordinate adhering to membrane external front and proves the positive effect of the gas in achieving more uniform distribution of velocity, shear and viscosity close to the membrane surface.

To summarize, further validation effort is required to combine advanced 3-D visualization of SMBR flow field parameters with CFD 3-D simulation predicted values. In parallel to the ever improving measurement techniques, the CFD models have also to be representative enough. Because membrane fouling is related to compression, permeability, adsorption, cake formation in series of resistances, validation depends on many empirical constants that require separate experiments to be determined. Since conditions of these experiments and the original ones may vary, this brings doubt if the CFD model or the coupled sub-model is the cause of good

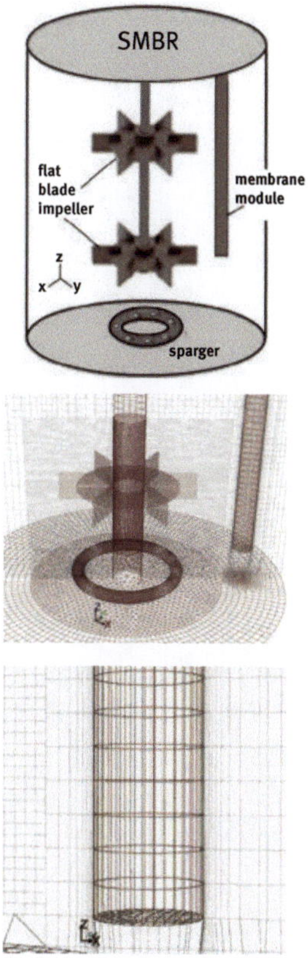

Figure 10.3: Experimental reactor and simulation grid [44].

prediction. Consequently, simple models with less constraint to describe component processes of physical nature seem more reliable and should be practiced. In another aspect, quantitative validation based on parameters of state that are measurable locally, such as velocity, shear and concentration is of a value.

10.3 Conclusions

Fouling remains a significant problem in the integrated SMBR, although understanding of the phenomenon and the possibility of its control are greatly improved.

SMBRs being designed mainly based on biokinetic and fouling considerations, the present state of understanding of membrane fouling developed by models ***approaches*** the level of conditions and scales to be used confidently by industry.

CFD-modeling shows high potential for fouling analysis and controls provided required model adequacy is ensured.

Recommendations for fouling analysis by modeling *point at* the need to ensure better correspondence of model to reality, such as for example is the correspondence

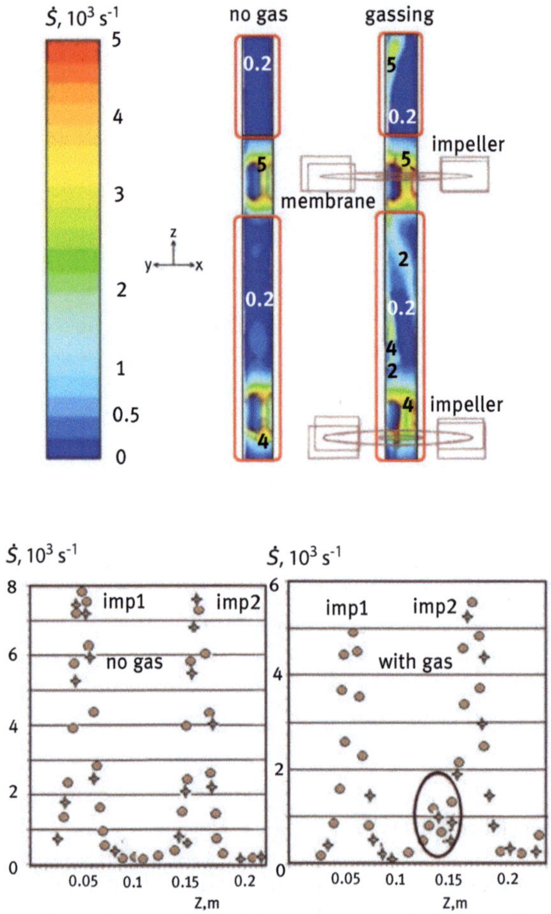

Figure 10.4: Membrane contour plots showing the effect of gassing on the distribution of shear rate *S* (in ks-¹) [44].

of k-ε turbulence models (SKE, realizable or RNG) and other viscous versions to specific flow conditions. The experience in model versions, such as SST, registered in the literature is insufficient.

A major drawback of most models is the misuse of parameters, such as permeability constants, specific cake resistances, lift coefficients, drag coefficients, sludge stickiness parameters, permeability, compression coefficients, friction loss coefficients that have been determined previously at conditions that differ from the conditions of simulation. On the other hand, curve fitting by model validation with such a lot of parameters is always ineffective to bring confidence in the validity of the specific model results. Further effort is needed for replacing some empirical relationships with multiple experimental constants by appropriate simple source terms in the balance equations, as undertaken with the porous-media sub-model [35, 110].

Validations should be quantitative and based on comparisons of measurable local variables, such as velocities (e. g. PIV), velocity gradients (electrochemical shear) rather than global parameters, such as global TMP or membrane flux. Qualitative validation, such as comparison of photo images with CFD flow patterns helps interpretation but is insufficient to bring confidence in SMBR CFD analysis.

Funding: This work was financially supported by the National Science Fund at the Bulgarian Ministry of Education and Science, Contract No DN 07/11/15.12.2016.

References

[1] Brannock MWD, De Wever H, Wang Y, Leslie G. CFD Simulations of MBRs: inside submerged versus outside submerged membranes. Desalination. 2009;236:244–51.

[2] He Y, Bagley DM, Leung KT, Liss SN, Liao B-Q. Recent advances in membrane technologies for biorefining and bioenergy production. Biotechnol Adv. 2012;30:817–58.

[3] Wibisono Y, Cornelissen ER, Kemperman AJB, Van Der Meer WGJ, Nijmeijer K. Two-phase flow in membrane processes: a technology with a future. J Memb Sci. 2014;453:566–602.

[4] Judd S. The status of membrane bioreactor technology. Trends Biotechnol. 2008;26(2):109–16.

[5] Mutamim NSA, Noor ZZ, Hassan MAA, Yuniarto A, Olsson G. Membrane bioreactor: applications and limitations in treating high strength industrial wastewater. Chem Eng J. 2013;225:109–19.

[6] Mutamim NSA, Noor ZZ, Hassan MAA, Olsson G. Application of membrane bioreactor technology in treating high strength industrial wastewater: a performance review. Desalination. 2012;305:1–11.

[7] Tambosi J, De Sena R, Favier M, Gebhardt W, José H, Schröder H, et al. Removal of pharmaceutical compounds in membrane bioreactors (MBR) applying submerged membranes. Desalination. 2010;261:148–56.

[8] Nguyen LN, Hai FI, Kang J, Price WE, Nghiem LD. Removal of emerging trace organic contaminants by MBR-based hybrid treatment processes. Int Biodeterior Biodegradation. 2013;85:474–82.

[9] Bakonyi P, Nemestóthy N, Simon V, Bélafi-Bakó K. Fermentative hydrogen production in anaerobic membrane bioreactors: a review. Bioresour Technol. 2014;156:357–63.

[10] Mohammadmahdi M, Stickel J, Wickramasinghe SR. Investigation of a submerged membrane reactor for continuous biomass hydrolysis. Food Bioprod Process. 2015;96:189–97.

[11] Jiang H, Qu Z, Li Y, Huang J, Chen R, Xing W. One-step semi-continuous cyclohexanone production via hydrogenation of phenol in a submerged ceramic membrane reactor. Chem Eng J. 2016;284:724–32.

[12] Dosta J, Nieto JM, Vila J, Grifoll M, Mata-Álvarez J. Phenol removal from hypersaline wastewaters in a membrane biological reactor (MBR): operation and microbiological characterisation. Bioresour Technol. 2011;102(5):4013–20.

[13] Andrić P, Meyer AS, Jensen PA, Dam-Johansen K. Reactor design for minimizing product inhibition during enzymatic lignocellulose hydrolysis II. Quantification of inhibition and suitability of membrane reactors. Biotechnol Adv. 2010;28:407–25.

[14] Sengur R, Deveci G, Kaya R, Turken T, Guclu S, Imer DY, et al. CFD modeling of submerged membrane bioreactors (sMBRs): a review. Desalinat Water Treatment. 2015;55(7):1747–61.

[15] Naessens W, Maere T, Ratkovich N, Vedantam S, Nopens I. Critical review of membrane bioreactor models – Part 2: hydrodynamic and integrated models. Bioresour Technol. 2012;122:107–18.

[16] Boyle-Gotla A, Jensen PD, Yap SD, Pidou M, Wang Y, Batstone DJ. Dynamic multidimensional modelling of submerged membrane bioreactor fouling. J Memb Sci. 2014;467:153–61.
[17] Ndinisa NV, Fane AG, Wiley DE, Fletcher DF. Fouling control in a submerged flat sheet membrane system: part II—two-phase flow characterization and CFD simulations. Sep Sci Technol. 2006;41(7):1411–45.
[18] Vargas A, Moreno-Andrade I, Buitrón G. Controlled backwashing in a membrane sequencing batch reactor used for toxic wastewater treatment. J Memb Sci. 2008;320(1):185–90.
[19] Trad Z, Vial C, Fontaine JP, Larroche C. Modeling of hydrodynamics and mixing in a submerged membrane bioreactor. Chem Eng J. 2015;282:77–90.
[20] Yan X, Wu Q, Sun J, Liang P, Zhang X, Xiao K, et al. Hydrodynamic optimization of membrane bioreactor by horizontal geometry modification using computational fluid dynamics. Bioresour Technol. 2016;200:328–34.
[21] Cui ZF, Chang S, Fane AG. The use of gas bubbling to enhance membrane processes. J Memb Sci. 2003;221(1):1–35.
[22] Braak E, Alliet M, Schetrite S, Albasi C. Aeration and hydrodynamics in submerged membrane bioreactors. J Memb Sci. 2011;379(1):1–18.
[23] Wei P, Zhang K, Gao W, Kong L, Field R. CFD modeling of hydrodynamic characteristics of slug bubble flow in a flat sheet membrane bioreactor. J Memb Sci. 2013;445:15–24.
[24] Qi C, Wang J, Lin Y. New insight into influence of mechanical stirring on membrane fouling of membrane bioreactor: mixed liquor properties and hydrodynamic conditions. Bioresour Technol. 2016;211:654–63.
[25] Liu X, Wang Y, Waite TD, Leslie G. Fluid structure interaction analysis of lateral fibre movement in submerged membrane reactors. J Memb Sci. 2016;504:240–50.
[26] Du X, Qu FS, Liang H, Li K, Bai LM, Li GB. Control of submerged hollow fiber membrane fouling caused by fine particles in photocatalytic membrane reactors using bubbly flow: shear stress and particle forces analysis. Separation Purif Technol. 2017;172:130–39.
[27] Meng L, Cheng JC, Jiang H, Yang C, Xing WH, Jin WQ. Design and analysis of a submerged membrane reactor by CFD simulation. Chem Eng Technol. 2013;36(11):1874–82.
[28] Vlaev SD, Tsibranska I. Shear stress generated by radial flow impellers at bioreactor-integrated membranes. Theor Foundations Chem Eng. 2016;50(6):959–68.
[29] Brannock M, Wang Y, Leslie G. Mixing characterisation of full-scale membrane bioreactors: CFD modelling with experimental validation. Water Res. 2010;44(10):3181–91.
[30] Chen C, Bin L, Tang B, Huang S, Fu F, Chen Q, et al. Cultivating granular sludge directly in a continuous-flow membrane bioreactor with internal circulation. Chem Eng J. 2017;309:108–17.
[31] Le-Clech P, Chen V, Fane TA. Fouling in membrane bioreactors used in wastewater treatment. J Memb Sci. 2006;284(1):17–53.
[32] Wicklein E, Batstone DJ, Ducoste J, Laurent J, Griborio A, Wicks J, et al. Good modelling practice in applying computational fluid dynamics for WWTP modelling. Water Sci Technol. 2016;73(5):969–82.
[33] Bentzen TR, Ratkovich N, Rasmussen MR, Madsen S, Jensen JC, Bak SN Numerical modelling of non-Newtonian fluid in a rotational cross-flow MBR. In: 6th IWA Specialist Conference on Membrane Technology for Water and Wastewater Treatment, 2011.
[34] Karpinska AM, Bridgeman J. CFD-aided modelling of activated sludge systems–A critical review. Water Res. 2016;88:861–79.
[35] Liu X, Wang Y, Waite TD, Leslie G. Numerical simulation of bubble induced shear in membrane bioreactors: effects of mixed liquor rheology and membrane configuration. Water Res. 2015;75:131–45.
[36] Judd SJ, Le-Clech P, Taha T, Cui ZF. Theoretical and experimental representation of a submerged membrane bio-reactor system. Membr Technol. 2001;135:4–9.

[37] Yang J, Vedantam S, Spanjers H, Nopens I, Van Lier JB. Analysis of mass transfer characteristics in a tubular membrane using CFD modeling. Water Res. 2012;46:4705–12.

[38] Yang M, Yu D, Liu M, Zheng L, Zheng X, Wei Y, et al. Optimization of MBR hydrodynamics for cake layer fouling control through CFD simulation and RSM design. Bioresour Technol. 2017;227:102–11.

[39] Kurita T, Kimura K, Watanabe Y. Energy saving in the operation of submerged MBRs by the insertion of baffles and the introduction of granular materials. Separation Purif Technol. 2015;141:207–13.

[40] Ding A, Liang H, Li G, Derlon N, Szivak I, Morgenroth E, et al. Impact of aeration shear stress on permeate flux and fouling layer properties in a low pressure membrane bioreactor for the treatment of grey water. J Memb Sci. 2016;510:382–90.

[41] Praneeth K, Moulik S, Vadthya P, Bhargava SK, Tardio J, Sridhar S. Performance assessment and hydrodynamic analysis of a submerged membrane bioreactor for treating dairy industrial effluent. J Hazard Mater. 2014;274:300–13.

[42] Shimizu Y, Uryu K, Okuno YI, Watanabe A. Cross-flow microfiltration of activated sludge using submerged membrane with air bubbling. J Ferment Bioeng. 1996;81(1):55–60.

[43] Böhm L, Drews A, Prieske H, Bérubé PR, Kraume M. The importance of fluid dynamics for MBR fouling mitigation. Bioresour Technol. 2012;122:50–61.

[44] Ghidossi R, Veyret D, Moulin P. Computational fluid dynamics applied to membranes: state of the art and opportunities. Chem Eng Process. 2006;45:437–54.

[45] Germain E, Stephenson T, Pearce P. Biomass characteristics and membrane aeration: toward a better understanding of membrane fouling in submerged membrane bioreactors (MBRs). Biotechnol Bioeng. 2005;90:316–22.

[46] Sofia A, Ng WJ, Ong SL. Engineering design approaches for minimum fouling in submerged MBR. Desalination. 2004;160:67–74.

[47] Hong H, Zhang M, He Y, Chen J, Lin H. Fouling mechanisms of gel layer in a submerged membrane bioreactor. Bioresour Technol. 2014;166:295–302.

[48] Wu J, He C, Jiang X, Zhang M. Modeling of the submerged membrane bioreactor fouling by the combined pore constriction, pore blockage and cake formation mechanisms. Desalination. 2011;279(1):127–34.

[49] Zhang M, Peng W, Chen J, He Y, Ding L, Wang A, et al. A new insight into membrane fouling mechanism in submerged membrane bioreactor: osmotic pressure during cake layer filtration. Water Res. 2013;47(8):2777–86.

[50] Zarragoitia-González A, Schetrite S, Alliet M, Jáuregui-Haza U, Albasi C. Modelling of submerged membrane bioreactor: conceptual study about link between activated slugde biokinetics, aeration and fouling process. J Memb Sci. 2008;325(2):612–24.

[51] Li X, Mo Y, Li J, Guo W, Ngo HH. In-situ monitoring techniques for membrane fouling and local filtration characteristics in hollow fiber membrane processes: a critical review. J Memb Sci. 2017;528:187–200.

[52] Aslam M, Charfi A, Lesage G, Heran M, Kim J. Membrane bioreactors for wastewater treatment: a review of mechanical cleaning by scouring agents to control membrane fouling. Chem Eng J. 2017;307:897–913.

[53] Fane AG. Submerged membranes. In: Li NN, Fane AG, Ho WS, Matsuura T, editors.. Advanced membrane technology and applications. New Jersey: Wiley, 2008: 239–70.

[54] Menniti A, Kang S, Elimelech M, Morgenroth E. Influence of shear on production of extracellular polymeric substances in membrane bioreactors. Water Res. 2009;43(17): 4305–15.

[55] Parvareh A, Rahimi M, Madaeni SS, Alsairafi AA. Experimental and CFD study on the role of fluid flow pattern on membrane permeate flux. Chin J Chem Eng. 2011;19(1):18–25.

[56] Meng F, Zhang S, Oh Y, Zhou Z, Shin HS, Chae SR. Fouling in membrane bioreactors: an updated review. Water Res. 2017;114:151–80.
[57] Ratkovich N, Hunze M, Nopens I. Hydrodynamic study of a hollow fiber membrane system using experimentally and numerically derived surface shear stresses. Multiphase Sci Technol. 2012;24(1):47–66.
[58] Ratkovich N, Bentzen TR. Comparison of four types of membrane bioreactor systems in terms of shear stress over the membrane surface using computational fluid dynamics. Water Sci Technol. 2013;68(12):2534–44.
[59] Duc ENC, Fournier L, Levecq C, Lesjean B, Grelier P, Tazi-Pain A. Local hydrodynamic investigation of the aeration in a submerged hollow fibre membranes cassette. J Memb Sci. 2008;321:264–71.
[60] Khalili A, Mehrnia MR, Mostoufi N, Sarrafzadeh MH. Analyze and control fouling in an airlift membrane bioreactor: CFD simulation and experimental studies. Process Biochem. 2011;46:1138–45.
[61] Jafarkhani M, Moraveji MK, Davarnejad R, Moztarzadeh F, Mozafari M. Three-dimensional simulation of turbulent flow in a membrane tube filled with semi-circular baffles. Desalination. 2012;294:8–16.
[62] Ahmed S, Seraji MT, Jahedi J, Hashib MA. CFD simulation of turbulence promoters in a tubular membrane channel. Desalination. 2011;276:191–98.
[63] Prieske H, Drews A, Kraume M. Prediction of the circulation velocity in a membrane bioreactor. Desalination. 2008;231:219–26.
[64] Marcos B, Moresoli C, Skorepova J, Vaughan B. CFD modeling of a transient hollow fiber ultrafiltration system for protein concentration. J Memb Sci. 2009;337(1–2):136–44.
[65] Liang YY, Fimbres-Weihs G, Setiawan R, Wiley D. CFD modelling of unsteady electro-osmotic permeate flux enhancement in membrane systems. Chem Eng Sci. 2016;146:189–98.
[66] Ameur H, Sahel D. Effect of the baffle design and orientation on the efficiency of a membrane tube. Chem Eng Res Des. 2017;117:500–08.
[67] Ratkovich N, Chan CCV, Berube PR, Nopens I. Experimental study and CFD modelling of a two-phase slug flow for an airlift tubular membrane. Chem Eng Sci. 2009;64:3545–720.
[68] Faridirad F, Zourmand Z, Kasiri N, Moghaddam MK, Mohammadi T. Modeling of suspension fouling in nanofiltration. Desalination. 2014;346:80–90.
[69] Du X, Qu F, Liang H, Li K, Chang H, Li G. Cake properties in ultrafiltration of TiO2 fine particles combined with HA: in situ measurement of cake thickness by fluid dynamic gauging and CFD calculation of imposed shear stress for cake controlling. Environ Sci Pollut Res. 2016;23(9):8806–18.
[70] Ghidossi R, Carretier E, Veyret D, Dhaler D, Moulin P. Optimizing the compacity of ceramic membranes. J Memb Sci. 2010;360:483–92.
[71] Rahimi M, Madaeni SS, Abolhasani M, Alsairafi AA. CFD and experimental studies of fouling of a microfiltration membrane. Chem Eng Processing: Intensification. 2009;48:1405–13.
[72] Alexiadis A, Wiley DE, Vishnoi A, Lee RHK, Fletcher DF, Bao J. CFD modelling of reverse osmosis membrane flow and validation with experimental results. Desalination. 2007; 217(1–3):242–50.
[73] Liang YY, Chapman MB, Weihs GAF GA, Wiley DE. CFD modelling of electro-osmotic permeate flux enhancement on the feed side of a membrane module. J Memb Sci. 2014;470:378–88.
[74] Willems P, Deen NG, Kemperman AJB, Lammertink RGH, Wessling M, Van Sint Annaland M, et al. van der Meer WGJ. Use of particle imaging velocimetry to measure liquid velocity profiles in liquid and liquid/gas flows through spacer filled channels. J Memb Sci. 2010;362:143–53.
[75] Lee K-J, Wu R-M. Simulation of resistance of cross-flow microfiltration and force analysis on membrane surface. Desalination. 2008;233(1–3):239–46.

[76] Karabelas AJ, Kostoglou M, Koutsou CP. Modeling of spiral wound membrane desalination modules and plants – review and research priorities. Desalination. 2015;356:165–86.
[77] Li M, Bui T, Chao S. Three-dimensional CFD analysis of hydrodynamics and concentration polarization in an industrial RO feed channel. Desalination. 2016;397:194–204.
[78] Zhuang L, Guo H, Da G, Xu Z-L. Effect of the inlet manifold on the performance of a hollow fiber membrane module- A CFD study. J Memb Sci. 2017;526:73–93.
[79] Kaya R, Deveci G, Turken T, Sengur R, Guclu S, Koseoglu-Imer DY, et al. Analysis of wall shear stress on the outside-in type hollow fiber membrane modules by CFD simulation. Desalination. 2014;351:109–19.
[80] Fimbres-Weihs GA, Wiley DE. Review of 3D CFD modeling of flow and mass transfer in narrow spacer-filled channels in membrane modules. Chem Eng Processing: Intensification. 2010;49(7): 759–81.
[81] Li Y-L, Tung K-L. CFD simulation of fluid flow through spacer-filled membrane module: selecting suitable cell types for periodic boundary conditions. Desalination. 2008;233(1–3):351–58.
[82] Vinther F, Pinelo M, Brøns M, Jonsson G, Meyer AS. Predicting optimal back-shock times in ultrafiltration hollow fiber modules II: effect of inlet flow and concentration dependent viscosity. J Memb Sci. 2015;493:486–95.
[83] Setiawan R, Ratnayake P, Bao J, Fimbres-Weihs GA, Wiley DE. Reduced-order model for the analysis of mass transfer enhancement in membrane channel using electro-osmosis. Chem Eng Sci. 2015;122:86–96.
[84] Zourmand Z, Faridirad F, Kasiri N, Mohammadi T. Mass transfer modeling of desalination through an electrodialysis cell. Desalination. 2015;359:41–51.
[85] Xie P, Murdoch LC, Ladner DA. Hydrodynamics of sinusoidal spacers for improved reverse osmosis performance. J Memb Sci. 2014;453:92–99.
[86] Kawachale N, Kirpalani DM, Kumar A. A mass transport and hydrodynamic evaluation of membrane separation cell. Chem Eng Processing: Intensification. 2010;49:680–88.
[87] Amokrane M, Sadaoui D, Koutsou CP, Karabelas AJ, Dudeck M. A study of flow field and concentration polarization evolution in membrane channels with two-dimensional spacers during water desalination. J Memb Sci. 2015;477(1):139–50.
[88] Sarkar D, Datta D, Sen D, Bhattacharjee C. Simulation of continuous stirred rotating disk-membrane module: an approach based on surface renewal theory. Chem Eng Sci. 2011;66(12): 2554–67.
[89] Krzeminski P, Leverette L, Malamis S, Katsou E. Membrane bioreactors–a review on recent developments in energy reduction, fouling control, novel configurations, LCA and market prospects. J Memb Sci. 2017;527:207–27.
[90] Pimentel GA, Almeida P, Hantson AL, Rapaport A, Wouwe AV. Experimental validation of a simple dynamic model of a laboratory scale recirculating aquaculture system fitted with a submerged membrane bioreactor. Biochem Eng J. 2017;122:1–12.
[91] Robles A, Ruano MV, Ribes J, Seco A, Ferrer J. A filtration model applied to submerged anaerobic MBRs (SAnMBRs). J Memb Sci. 2013;444:139–47.
[92] Kang C, Hua J, Lou J, Liu W, Jordan E. Bridging the gap between membrane bio-reactor (MBR) pilot and plant studies. J Memb Sci. 2008;325:861–71.
[93] Liu X, Wang Y, Waite TD, Leslie G. Numerical simulations of impact of membrane module design variables on aeration patterns in membrane bioreactors. J Memb Sci. 2016;520:201–13.
[94] Oberkampf WL, Trucano TG, Hirsch C. Verification, validation, and predictive capability in computational engineering and physics. Appl Mechanics Rev. 2004;57(5):345–84.
[95] Stern F, Wilson RV, Coleman HW, Paterson EG. Comprehensive approach to verification and validation of CFD simulations-Part 1: methodology and procedures. Trans-Am Soc Mech Engi J Fluids Eng. 2001;123(4):793–802.

[96] Fimbres Weihs GA, Wiley DE. CFD analysis of tracer response technique under cake-enhanced osmotic pressure. J Memb Sci. 2014;449:38–49.
[97] Monfared MA, Kasiri N, Salahi A, Mohammadi T. CFD simulation of baffles arrangement for gelatin-water ultrafiltration in rectangular channel. Desalination. 2012;284:288–96.
[98] Abbasi Monfared M, Kasiri N, Salahi A, Mohammadi T. Modeling ultrafiltration of gelatin–water suspension by computational fluid dynamics. Chemichal Eng Res Des. 2012;90(8):1098–104.
[99] Brannock MWD, Wang Y, Leslie G. Evaluation of full-scale membrane bioreactor mixing performance and the effect of membrane configuration. J Memb Sci. 2010;350:101–08.
[100] Koutsou CP, Yiantsios SG, Karabelas AJ. Effects of spacer geometrical characteristics and Schmidt number. J Memb Sci. 2009;326:234–51.
[101] Martinelli L, Guigui C, Line A. Characterisation of hydrodynamics induced by air injection related to membrane fouling behavior. Desalination. 2010;250:587–91.
[102] Fortunato L, Jeong S, Wang Y, Behzad AR, Leiknes T. Integrated approach to characterize fouling on a flat sheet membrane gravity driven submerged membrane bioreactor. Bioresour Technol. 2016;222:335–43.
[103] Fortunato L, Leiknes T. In-situ biofouling assessment in spacer filled channels using optical coherence tomography (OCT): 3D biofilm thickness mapping. Bioresour Technol. 2017;229:231–35.
[104] Reiss LP, Hanratty TJ. An experimental study of the unsteady nature of the viscous sublayer. AIChE J. 1963;9:154–60.
[105] Vlaev SD, Nikov I, Martinov M. Shear and skin friction on particles in power-law fluids agitated by flat-blade and fluid foil impellers. Chem Eng Sci. 2006;61:5455–67.
[106] Koutsou CP, Karabelas AJ. Shear stresses and mass transfer at the base of a stirred filtration cell and corresponding conditions in narrow channels with spacers. J Memb Sci. 2012; 399-400:60–72.
[107] Chan CCV, Bérubé PR, Hall ER. Shear profiles inside gas sparged submerged hollow fiber membrane modules. J Memb Sci. 2007;297(1):104–20.
[108] Böhm L, Kraume M. Fluid dynamics of bubble swarms rising in Newtonian and non-Newtonian liquids in flat sheet membrane systems. J Memb Sci. 2015;475:533–44.
[109] Vlaev SD, Tsibranska I, Dzhonova D, Georgiev D. Preconditions of separations in STR with integrated membrane for energy-saving by recovery of value-added materials. Machines, Technologies, Materials, Int Sci J. 2016;9:22–25.
[110] Wang Y, Brannock M, Cox S, Leslie G. CFD simulation of membrane filtration zone in a submerged hollow fibre membrane bioreactor using porous media approach. J Memb Sci. 2010;363(1-2):57–66.

Maciej Staszak

11 Applicability of DFT model in reactive distillation

Abstract: The density functional theory (DFT) applicability to reactive distillation is discussed. Brief modeling techniques description of distillation and rectification with chemical reaction is provided as a background for quantum method usage description. The equilibrium and nonequilibrium distillation models are described for that purpose. The DFT quantum theory is concisely described. The usage of DFT in the modeling of reactive distillation is described in two parts. One of the fundamental and very important component of distillation modeling is vapor-liquid equilibrium description for which the DFT quantum approach can be used. The representative DFT models, namely COSMO-RS (Conductor like Screening Model for Real Solvents), COSMOSPACE (COSMO Surface Pair Activity Coefficient) and COSMO-SAC (SAC – segment activity coefficient) approaches are described. The second part treats the way in which the chemical reaction is described by means of quantum DFT method. The intrinsic reaction coordinate (IRC) method is described which is used to find minimum energy path of substrates to products transition. The DFT is one of the methods which can be used for that purpose. The literature data examples are provided which proves that IRC method is applicable for chemical reaction kinetics description.

Keywords: DFT, reactive distillation

11.1 Introduction

This article is divided into three main parts, the first introducing to the traditional concepts of reactive distillation modeling and clarifying the prospective need for quantum calculations. The second provides more details about density functional theory usage in vapor–liquid equilibria. This element of distillation modeling description is very important for accurate simulation due to the fundamental impact of vapor–liquid equilibria on the components distribution among the phases during the process. The third part presents quantum approach to chemical reaction kinetics description. For reactive distillation, it is another very important aspect which also has large effect on the accuracy and the realism of the results obtained.

This article has previously been published in the journal *Physical Sciences Reviews*. Please cite as: Staszak, M. Applicability of DFT model in reactive distillation. *Physical Sciences Reviews* [Online] **2017**, *2* (12). DOI: 10.1515/psr-2017-0144

https://doi.org/10.1515/9783110482065-011

There exist many highly specialized software tools available for engineers which enable fast and reliable design calculation. ChemCad, Hysys, UniSim are a few examples of them. Up to now no quantum method is adapted or included in commercial packages for distillation or reactive distillation design. But it is only a matter of time when such methods become wider available. Such a progression is visible in the field of vapor–liquid equilibria, where the COSMO methods are used to simulate distillation. Up to now only in the research field such quantum approach can be met, for cases with complex equilibria like three-phase distillation [1], or for the cases when the solvent screening is very difficult by traditional methods [2, 3].

One of the key advantages that a designer might find useful when performing calculations based on quantum approaches is that there should not exist a requirement to perform experiments prior to calculations. Performing *ab initio* (from the beginning) quantum calculations that prepares model for design phase, enables vapor–liquid equilibria and chemical reaction kinetics to be accounted for without additional experimental measurements. The disadvantage might be the high quantum calculation workload which for *ab initio* methods (like, e. g. MP2) is unavoidable. The solution proposed to that problem is the idea of databank creation where all fundamental molecular quantities (charge density surfaces, activation energies, etc.) are already calculated. Developers of one of the leading quantum software tools existing on the market, COSMOlogic, made the efforts to create such database available. Constantly the work of researchers is devoted to develop such databanks for wider use.

11.2 Distillation process

Distillation is well-known industrial process for separating different kind of liquid mixtures. This introductory article will give a concise description of rectification modeling which is the base for further text concerning molecular description of thermodynamic equilibrium.

The different tendency of the mixture components to gather between liquid and vapor phase is the key basis of this physical process. Industrial distillation is often referred as rectification, which is typically conducted in column apparatus where the inner, working elements providing phase mixing are trays or packing. Illustrative sketch of rectifying column is presented in Figure 11.1.

This is also typical that distillation is used to intensify chemical reaction especially in the case of reversible, equilibrium one. In this case the equilibrium shift toward higher conversion is realized by removing the reaction products from the reaction mixture. The most typical design and simulation approach used in many modern software packages is to divide the column height into sections. Depending on its construction the section may refer to trays or package segment. Every section is then balanced based on the approach used. Such a description is referred to as "tray-

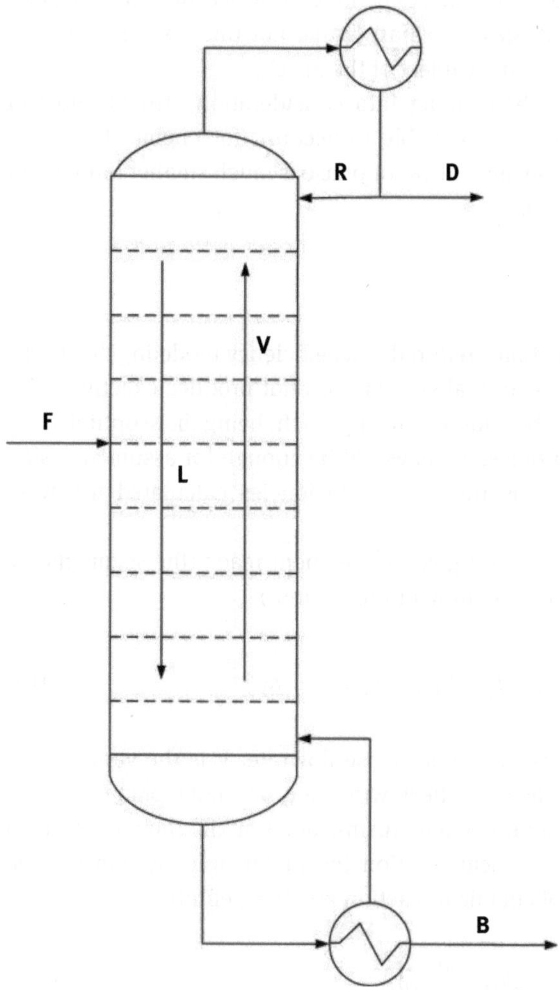

Figure 11.1: Distillation column schematics with designation of flows used in the text.

by-tray" model and consists of several equations in one set. Besides the steady-state or dynamics approach that can be used in such models are divided by the way the interphase mass transfer is treated. The most popular is the so-called equilibrium approach in which the phases are assumed to be perfectly mixed at every column segment and being in equilibrium condition. The second so-called nonequilibrium and more complex approach is to refer for mass transfer at every stage assuming that stage is not at equilibrium conditions.

The model is built for N segments, which can be numbered from top to bottom, but this is not compulsory. At the top, the condenser constitutes unique equation and at the bottom the reboiler constitutes also specific equation, both of them describing the mass and energy balances around them. It depends on the level of complexity and details demanded of how to treat these auxiliary units. For this text, they will be treated as simple as possible due to the subject of the article.

In the most extended case when taking into consideration also the flow of chemical reaction, the model must be capable to account for chemical reaction kinetics. This is also typical that the mass of vapor phase is much smaller than liquid and is neglected during calculations.

11.2.1 Equilibrium approach

The equilibrium approach is sometimes referred to as efficiency modeling due to the fact that the actual mixing process in real system does not produce exactly equilibrium conditions. Consequently, the equilibrium approach, being more optimistic in the way that the smaller number of segments would be enough for assumed design requirements, must be corrected by the use of efficiencies calculated for every segment.

The transient total mass balance for given i^{th} segment reads (the segments are assumed to be numbered from top to bottom of the column):

$$\frac{dm_i}{dt} = L_{i-1} - L_i + V_{i+1} - V_i + F_i + \Delta v \tag{11.1}$$

In the above formulation, the L is the liquid phase flowrate, V is the vapor phase flowrate and F is the feed stream flowrate. The flowrate may use molar or mass base of units' description. The m is the amount of mass (using moles or kilograms) contained at i^{th} stage. When considering chemical reaction the molar units system is most suitable due to the natural way of chemical reaction kinetics definition:

$$\Delta R_{i,j} = \frac{m_i}{\rho_i} v_j r_{i,j} \tag{11.2}$$

The $\Delta R_{i,j}$ is the source of ith specie due to chemical reaction. The variable $r_{i,j}$ is the rate of component j^{th} creation/consumption due to the chemical reaction at i^{th} stage, while r_i is the density of the mixture at i^{th} stage. In the case of nonreactive flow the $\Delta R_{i,j}$ is equal zero. The Δv is the molar change due to the reaction stoichiometry.

The total mass balance is used in the formulation for j^{th} component mass balance at i^{th} stage given below:

$$\frac{d(m_i x_{i,j})}{dt} = L_{i-1} x_{i-1,j} - L_i x_{i,j} + V_{i+1} y_{i+1,j} - V_i y_{i,j} + F_i x_{F,j} + \Delta R_{i,j} \tag{11.3}$$

The variable $x_{i,j}$ refers to molar (mass) fractions of ith component at j^{th} stage. The subscript F refers to feed composition. In the description above the feed is assumed to be liquid but it is not a process limitation but only a simplification applied here. This is typical that column contains one or a few only feed streams or in the case of batch distillation it may contain no feeds at all.

The energy balance is the important part of the balance equations which allows to estimate the temperatures along the column. The energy balance for i^{th} segment can be presented by the relation:

$$\frac{d(m_i H_i)}{dt} = L_{i-1}H_{i-1}^L - L_i H_i^L + V_{i+1}H_{i+1}^V - V_i H_i^V + F_i H_F^L + \Delta Hr \cdot \Delta\Re_i \tag{11.4}$$

The variable H refers to stream enthalpy for given phase at given stage. The ΔHr is the reaction enthalpy and the $\Delta\Re_i$ reads:

$$\Delta\Re_i = \frac{m_i}{\rho_i} r_i \tag{11.5}$$

where r_i is the overall rate of chemical reaction at i^{th} stage.

The above equations form the differential equations set describing whole column at its every stage. The key element of the distillation process description is the vapor–liquid equilibrium (VLE). The usual way of estimating the equilibrium is to use some of the well-known and established methods for equilibrium conditions estimations. In general the equilibrium relation is presented in the form:

$$y_j = K_j x_j \tag{11.6}$$

This formulation relates the concentration of jth component in the vapor to its concentration in the liquid on every stage. The equilibrium constant, being in fact a parameter, K_j is the subject of calculation by several methods, which will be discussed in detail in Section 2.3.

The tray efficiency or overall column efficiency must be applied to account for realism of the distillation process. The simplest definition is the overall efficiency E_o which is defined according to the equation below:

$$E_O = \frac{N_{eq}}{N_{act}} \tag{11.7}$$

In the above N_{eq} is the number of trays used by equilibrium approach, N_{act} is the actual tray number. Such overall efficiency can be applied to some specified column section or to column as a whole. The exact calculation of such efficiency is a complex task and for general cases some estimations are only be proposed [4, 5] but for selected cases the procedures are designed, e. g. for alkane mixtures fractionation [6].

The most typical formulation for tray efficiency is the Murphree efficiency (E^M) which is given by:

$$E_{i,j}^{M} = \frac{y_{i,j} - y_{i+1,j}}{y_{i,j}^{*} - y_{i+1,j}} \tag{11.8}$$

In the above formulation, the $y_{i,j}^{*}$ is the equilibrium vapor concentration of jth specie at i^{th} tray to that of liquid phase. The efficiencies are correlated to the trays hydrodynamics which can be calculated based on well-known tray models of AIChE [7], Chan-Fair [8, 9] or Zuiderweg [10] method.

The model presented is sufficient for calculating basic mass and energy balance which gives as a result the components concentrations and temperature profiles along the column. To be able to describe the hydraulics of the process, additional equations and procedures must be added. By applying the column geometry by defining the tray sizing, the total volume for selected stage can be calculated. Based on mass balance solution from the above equations the amount of liquid at every stage is calculated m_i. In the case when the plates are constructed with weirs the Francis formula for liquid flow can be used. The excess amount of liquid at given plate can be calculated from the geometric plate volume and liquid mass m_i. This gives the possibility to estimate the liquid flow over the weir to the tray below. For the vapor phase flowrates the different approach can be used. The mechanistic calculation for pressure evolution along the column can be calculated by applying the pressure drop calculation for given trays or packing. The reboiler pressure increase (due to heating and liquid vaporization) is then the driving parameter to estimate the vapor flows along the column. Calculation of the actual pressure drop between reboiler and a tray above and then tray-to-tray pressure drops along the whole column enables to perform the estimation of the actual vapor flowrates. The additional data which must be given prior to this algorithm of calculation is top pressure. The detailed correlations for pressure drop for different flow regimes and different types of trays and weirs or packings are given by Kister [11].

11.2.2 Nonequilibrium approach

The nonequilibrium approach is a modification to the equilibrium description in the way that additional mass fluxes are considered, namely component interphase fluxes. The fundamental elements of nonequilbrium distillation approach are:

- material balances,
- energy balances,
- equilibrium relations,
- mass and energy transfer models

The phases in this approach are balanced by independent equations. The mass balance model formulation for j^{th} component for liquid phase then reads:

$$\frac{d(m_i^L x_{i,j})}{dt} = L_{i-1}x_{i-1,j} - L_i x_{i,j} + J_{i,j}^L + F_i^L x_{F,j} + \Delta R_{i,j}^L \tag{11.9}$$

and for the vapor phase:

$$\frac{d(m_i^V y_{i,j})}{dt} = V_{i+1}y_{i+1,j} - V_i y_{i,j} + J_{i,j}^V + F_i^V y_{F,j} + \Delta R_{i,j}^V \tag{11.10}$$

The mass transfer streams for i^{th} component for both phases $J_{i,j}^L$ and $J_{i,j}^V$ are typically calculated for binary mixture by the use of Fick's law. On the other hand for multicomponent mixtures the Maxwell–Stefan theory is used which is more appropriate due to the diffusion coefficients formulation which do not show dependency on the components concentration. These approaches are discussed in detail elsewhere [12–18].

The total mass balance is defined for both phases and for liquid is given by:

$$\frac{dm_i^L}{dt} = L_{i-1} - L_i + J_i^L + F_i^L + \Delta v \tag{11.11}$$

Consequently, for vapor phase it reads:

$$\frac{dm_i^V}{dt} = V_{i+1} - V_i + J_i^V + F_i^V + \Delta v \tag{11.12}$$

The J_i^L and J_i^V are total mass transfer streams between phases.

The energy balance for the liquid phase can be written as follows:

$$\frac{d(m_i^L H_i^L)}{dt} = L_{i-1}H_{i-1}^L - L_i H_i^L + F_i^L H_F^L + \Delta Hr^L \cdot \Delta R_i^L + \xi_i^L \tag{11.13}$$

and for the vapor phase consequently:

$$\frac{d(m_i^V H_i^V)}{dt} = V_{i+1}H_{i+1}^V - V_i H_i^V + F_i^V H_F^V + \Delta Hr^V \cdot \Delta R_i^V + \xi_i^V \tag{11.14}$$

In the above equations, the energy streams ξ_i^L and ξ_i^V represent the energy sink or source due to interphase transfer.

The equilibrium is assumed only to exist at the interface between phases. The value of K_j^{int} is evaluated for the conditions (components concentrations, temperature, pressure) at the interface.

$$y_j^{int} = K_j^{int} x_j^{int} \tag{11.15}$$

11.2.3 Vapor–liquid equilibrium

The equilibrium and mass transfer approaches both require a method of VLE estimation. The choice of specific method is based mainly on the type of distilled mixture components and the secondary on the actual pressure-temperature range.

The traditional methodology is to use one of the following thermodynamic approaches:
- activity methods
- equation of state methods
- special methods
- quantum and molecular methods

A very short description of them is presented below (as they are not based on any quantum or molecular mechanic method) and the latter will be discussed in details later in the text.

11.2.3.1 Activity methods

This approach is developed from the principia relating excess functions with activity coefficients. For the so-called regular liquid solutions (characterized by nonzero heat of mixing while entropy obeys ideal mixing rule) the activity coefficient of ith component is related to the excess Gibbs free energy G_i^E by relation:

$$\gamma_i = \exp\left(\frac{G_i^E}{RT}\right) \tag{11.16}$$

The typical models that are based on that definition are: Margules one and two parameter models [19, 20], van Laar [21], Wilson [22], NRTL (NonRandom TwoLiquids) [23], UNIFAC [24], UNIQUAC [25]. The models mentioned are the subject of extensions which provide additional useful properties like three-phase predictions (modified Wilson [26], UNIFAC LLE [27]) or polymer solutions property estimation (UNIFAC for polymers [28]).

11.2.3.2 Equation of state methods

Describing thermodynamic state functions by single relation is the focus of research which began from the introduction Clapeyron equation of state. Relating the amount of matter along with space occupied for given pressure and temperature is the main objective of this methodology. The inaccuracy of Clapeyron equation led to further study which resulted in Van der Waals introduction of attraction and volumetric term into the equation of state. Further modifications allowed more accurate prediction of phase behavior by cubic equations by Redlich and Kwong [29], which was then later corrected by Soave [30]. Modifications, which aim into more accurate description, were further introduced (e. g. Peng-Robinson [31] with its modifications). Further development leads to quartic equations of state which are reported to be of increased accuracy [32].

At the same time, the statistical associating fluid theory gave rise to so-called SAFT equations of state family. The SAFT equations of state are based on the equilibrium description by residual Helmholtz energy which originates from:

- model of hard spheres effect
- dispersion effects
- molecular chains effects
- association effects

The resulting family of equations are typically presented as the summation of all contributing Helmholtz energies:

$$A_{res} = A_{hardspheres} + A_{dispersion} + A_{chains} + A_{associating} \tag{11.17}$$

Each of the residual part is calculated by models discussed elsewhere [33–37].

11.3 Short outline of quantum modeling by DFT

11.3.1 Density functional theory

In the chemical calculation field, the highest accuracy for the molecular system can be achieved by solving the Schrödinger equation by specified approximation scheme without the need to rely on direct experimental measurements. The wave function of a molecular system can be found by solving the quantum equations directly using so-called *ab initio* methods. On the other hand, the density functional theory (DFT) can be used in cases where knowledge about wave function is not of highest importance. The DFT concept is utilized when the electron charge density is sufficient for describing the required molecular properties. The dominant role of DFT methods in the field of quantum calculations is due to their computational efficiency and high accuracy. The DFT theory was subject of computational research since the seventies of last century, but the fastest development occurred since nineties. From the applicative point of view the requirement to solve electron charge density as a function of position only (three spatial coordinates) is much an advantage over the *ab initio*, or Born–Oppenheimer [38–40] approximation. In the latter, the wave function is not only a function of spatial coordinates position but also of spin coordinate of each electron, which for system containing N electrons results in the $4N$ dimension space to be considered. The specific advantage is also sometimes recognized by the fact that wave function is not measureable while electron density is a property which can be directly measured by, e. g., X ray diffraction experiments. In fact, the discussion still holds if the wave function is a real physical entity or if it is only a mathematical concept which is appropriate for particle and molecular system description. The DFT is developed without using any variable parameters and thus in its origin it is an *ab initio* type method. In the DFT approach any ground state property can be described by the electron charge density [41]. The total energy of the *N*-particle system can be represented by energy components like kinetic energy of the moving electrons, nuclear-electron attraction potential energy, the repulsion energy of electron–electron system, and exchange correlation which describe other interactions between electrons. This is represented by the equation which reads [42]:

$$E[\rho] = T[\rho] + U[\rho] + E_{xc}[\rho] \tag{11.18}$$

The most problematic and computationally challenging term in the above is the last component which represents the exchange correlation $E_{xc}[\rho]$. The mathematical approach for computation of this term is discussed in detail elsewhere [43–50]

One of the widest used numerical approximation to this term is the B3LYP [51] energy functional (Becke, three parameter Lee-Yang-Parr). Such combinative approach to evaluating hybrid functional approximations led to improved description of many properties of molecular systems. The increased quality in atomization energies, vibration frequencies and bond lengths over the description by simple *ab initio* methods are its main advantage. The B3LYP correlation was demonstrated to be of comparable accuracy to preceding *ab initio* quantum methods [52] like e. g. Møller–Plesset perturbation theory [53].

11.3.2 Basis sets

The exact solution to Schrödinger equation is not possible except hydrogen and helium atoms. Numerical approaches rely on specified and defined approximation. The wave function is formulated as a vector span over infinite dimension space, which is one of the reasons why the numerical solution is not able to reproduce it in exact way. The general method of approximating the wave function is to use functional basis sets. They are finite sets of orthonormal functions which form a solution to the quantum problem which is solved. The resultant representation is then an approximation to the actual orbitals by linear combination of such functions. The physically most appropriate basis sets are Slater-type orbitals [54] (STO), which represents the solution to any atoms with one electron (hydrogen-like atoms). The most widespread functional basis sets are formed by linear combination of Gaussian-type functions [55] (GTO). The Pople [56] basis sets are represented by symbols X-YZg in which *X* is the number of Gaussian functions covering each core atomic orbital basis function, the variables *Y* and *Z* represent the valence orbitals by corresponding number of Gaussian primitive functions. The basis sets are subject to intensive studies and are discussed in literature along with their improving and developing [56–60].

This very short outline does not exhaust the topic of quantum modeling and is provided to give background for next chapters.

11.4 Quantum approach to VLE calculations

11.4.1 Continuum solvation models

The dielectric continuum solvation models are the methods of quantum calculations which provide the way to accurately describe fluid properties. The advantage of this family of methods lies in the fact that the solvent is represented by the mean continuous field rather than by system of explicitly positioned molecules. Consideration of solvent-solute molecules interaction by the averaged dielectric field significantly reduces the computational workload. In the case of explicit solvent molecules interactions, the time of calculations is often unaffordable long. Such approach enables the liquid phase calculations with the effectiveness and robustness comparable to gas phase quantum models. The quantum calculations must be conducted prior to the use of the results obtained from continuum solvation models in VLE estimations. Therefore, there is significant preparation stage at which the solvent molecules must be characterized by means of quantum calculated charge density surfaces estimations, which may take quite a long time. However, this preparation must be done one time only and in fact the user may have choose to use some of the existing databases of molecules [61].

Regarding the type of approach that is represented by such models they are classified as activity methods. The advantage of using the continuum solvation models lies in the fact that there is no need to use any group contribution parameters (like in the case of UNIFAC method), nor they require any parameters adjustment (like binary interaction parameters in Wilson or NRTL methods).

11.4.1.1 COSMO-RS

The early works devoted to VLE calculations by the use of quantum approach were works of Klamt [62, 63] and Taylor [64]. The continuum solvation model (COSMO) and its extension to real solvents by COSMO-RS model implementation was included in simulation software ChemSep [65] that was used to predict and simulate the process of distillation unit. COSMO-RS stands for Conductor like Screening Model for Real Solvents and is an equilibrium approach of quantum chemistry which is capable of predicting the chemical potentials in mixtures. This method is aimed at screening charge density σ on the surface of molecules to estimate the chemical potential μ of each mixture component. The implicit solvation approach represents the fluid space by continuous medium.

The COSMO-RS is a two-stage model [66]. At the first stage quantum calculations are done for every component of the mixture analyzed. The COSMO calculations are used to obtain the molecule environment imitated by virtual conductor. In such an environment, the molecules of solute generate a polarization charge density σ at the surface of the molecule which produces a back-charge acting on the molecule itself and generating higher polarized electron density in comparison with vacuum. During this quantum calculation step the energetic optimum is searched for the molecule structure contained in the virtual solvent. For the COSMO-RS model the standard quantum method used is DFT with defined basis set. Such calculation can be performed by the use of many software tools for example Gaussian [67], Turbomole [68] and Jaguar [69].

The second stage involves statistical thermodynamics calculations based on the previously structure and charge density calculated results. The liquid system is considered to be molecules ensemble, span over a lattice. The fluid particles positioned on the lattice can be factorized into three components, which reads:

$$Z = Z^0 Z^C Z^R \tag{11.19}$$

The Z^0 factor is the entropic contribution to the system approximated by:

$$Z^0 = N \sum_i x_i \ln x_i \tag{11.20}$$

which enumerates permutations of identical particles. The factor Z^C is called the combinatorial factor. It represents the partition sum of molecules ensemble which interact only due to steric constraints. The Staverman–Guggenheim [70, 71] equation is used to estimate this factor:

$$\ln Z_{SG}^C = -N \sum_i \left(x_i \ln \frac{\Phi_i}{x_i} + \frac{z}{2} x_i q_i \ln \frac{\Theta_i}{\Phi_i} \right) \tag{11.21}$$

The above formulation is also popular in many equilibrium activity or group contribution models like UNIQUAC or UNIFAC. The variables x_i, Φ_i, Θ_i are mole, volume and surface fractions of ith component, respectively. The variables r_i and q_i are relative volume and relative area. The parameter z is the coordination number of the lattice depending on the fluid spatial model applied, which is commonly assumed to be 10. For example cubic model of molecular packing results in $z = 6$ (Figure 11.2), while for hexagonal model $z = 12$.

The Z^R factor represents the residual contribution and is caused by non-steric molecular, electrostatic and hydrogen bond interactions. It is the most important factor for liquids. The formulation for this part is derived from the assumption of surface–pair interaction models that such residual interactions can be outlined as local pairwise interactions of surface segments. The partition factor Z^R is given by:

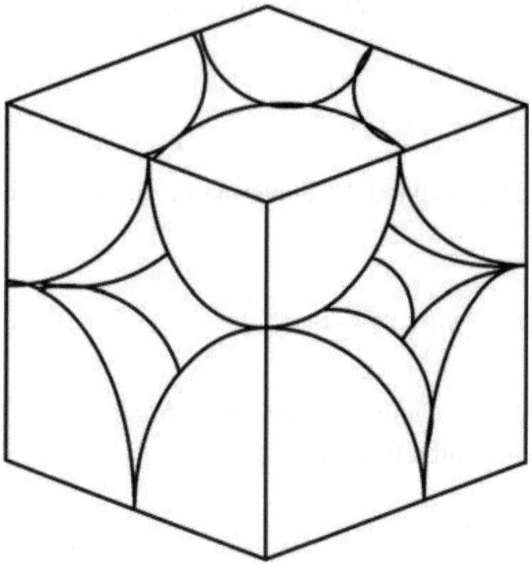

Figure 11.2: Example of cubic model lattice with coordination number z = 6.

$$Z^R = \sum_P \exp\left(-\frac{\sum_{\mu\nu} p_{\mu\nu}(P)\varepsilon_{\mu\nu}}{kT}\right) \tag{11.22}$$

where P counts all possible total pairs of segments. The μ and ν represent different types of surface segments and $\varepsilon_{\mu\nu}$ is the energy of interaction of $\mu\nu$ pair. The function $p_{\mu\nu}(P)$ is the total number of pairs of kind $\mu\nu$. Several models exist to approximate its value since it is quite complex task to evaluate all different configurations considering the number of pairs of type $\mu\nu$.

11.4.1.2 COSMOSPACE

The COSMO Surface Pair Activity Coefficient (COSMOSPACE) model is utilized to estimate the partition sum Z. By assuming that all of the molecules in the ensemble have different surfaces for example due to vibrational effects, and considering M number of segments created by contact pairs it follows that $M/2$ pairs can be formed. Every pair can be placed in different lattice segments, which consequently results in $M!$ different placements. In the light of the above the formulation for Z reads:

$$Z = \sum_P \exp\left(-\frac{\sum_{i=1}^{M/2} \varepsilon_{\nu(2i-1;P)\nu(2i;P)}}{kT}\right) \tag{11.23}$$

It is significant to realize that the expression above is formulated for only one segment type ν, but for very large number of this type segments. The $\nu(i;P)$ indicates the segment of type ν residing on site i in placement P.

Finally, two parts contribute to COSMOSPACE model activity coefficients of ith component in the mixture:

$$\ln \gamma_i = \ln \gamma_i^C + \ln \gamma_i^R \tag{11.24}$$

where the combinatorial part is estimated by modified [72] Stavermann–Guggenheim expression:

$$\ln \gamma_i^C = 1 - \Phi_i' + \ln \Phi_i' - \frac{z}{2} q_i \left(1 - \frac{\Phi_i}{\Theta_i} + \ln \frac{\Phi_i}{\Theta_i}\right) \tag{11.25}$$

in which variables x_i, Φ_i, Φ', Θ_i are mole, two volume and surface fractions of ith component, respectively. These can be calculated from relations:

$$\Phi_i' = \frac{x_i r_i^{c_{comb}}}{\sum_i x_i r_i^{c_{comb}}} \tag{11.26}$$

in the above c_{comb} is adjustable combinatorial parameter.

$$\Phi_i = \frac{x_i r_i}{\sum_i x_i r_i} \tag{11.27}$$

and

$$\Theta_i = \frac{x_i q_i}{\sum_i x_i q_i} \tag{11.28}$$

In the above r_i and q_i are relative volume and surface area respectively. The residual contribution to the eq. (11.24) is given by:

$$\ln \gamma_i^R = \sum_\nu n_i^\nu (\ln \gamma^\nu - \ln \gamma_i^\nu) \tag{11.29}$$

The residual activity coefficient is a function of n_i^ν which is the number of type ν segments on i^{th} molecule, the type ν segment activity coefficient γ^ν in the mixture and the type ν segment activity coefficient γ_i^ν for pure i^{th} coefficient. Finally the segment activity coefficients resulting from Z^R, an ensemble of pairwise interacting segments, are given by relation:

$$1\gamma^\nu = \sum_\mu \tau_{\mu\nu} \Theta^\mu \gamma^\mu \tag{11.30}$$

where the interaction parameter $\tau_{\mu\nu}$ for physical consistency is given by symmetric matrix which elements read:

$$\tau_{\mu\nu} = \exp\left(-\frac{u_{\mu\nu} - \frac{1}{2}(u_{\mu\mu} + u_{\nu\nu})}{RT}\right) \tag{11.31}$$

in which u is the interaction energy between groups $\mu\nu$, $\mu\mu$ and $\nu\nu$. Equation (11.30) is the general equation of COSMOSPACE model and is solved using iterative procedure. The model depends only on a few adjusted parameters for each element to be modeled and dos not rely on any functional group. Consequently, any parameterized variable is totally general and can be used to calculate properties of any compound.

The first step of the calculation procedure is the preparation the molecular model by the use of some specific software tools for modeling three-dimensional chemical structures [73, 74]. The second calculation step is to perform DFT computation with dielectric continuum solvation model, which can be implemented in some software tools [75]. These computations return the screening charge density on surface of the molecule and its total energy. In the next stage, molecules of solvent and solute are taken into account as an ensemble of pairwise interacting surfaces. The particular type of the intermolecular interactions (i. e., electrostatic interactions and hydrogen bonds) is expressed by the screening charge densities of the contacting surface segment types.

Comparison [76] of the COSMO-RS model with more traditional models like UNIFAC, modified UNIFAC and ASOG proved it to be best for systems with alkyl halides or aromatics as solutes in water. The model was less successful for nonaqueous mixtures but good results were obtained for the mixtures that contained alkyl halides, ethanol solutions of alkanes, and ketones in alkanes.

11.4.1.3 COSMO-SAC

Essentially similar in origin to the COSMO-RS , the COSMO-SAC model [77], is theoretically rather different. COSMO-SAC (SAC – segment activity coefficient) resolves some problems that exist in the COSMO-RS model. The model is based on group contribution solvation (GCS) model [78] where the activity coefficients are estimated from the solvation free energy of molecules in a solution. The COSMOS-SAC equation for activity coefficients reads:

$$\ln \gamma_{i/S} = n_i \sum_{\sigma_m} p_i(\sigma_m)(\ln \Gamma_S(\sigma_m) - \ln \Gamma_i(\sigma_m)) + \ln \gamma_{i/S}^{SG} \tag{11.32}$$

In the above formulation the σ_m is the segment with charge density at a fixed position in the solution, $\Gamma_s(\sigma)$ is the segment activity coefficient, $\ln \gamma_{i/S}^{SG}$ is the Staverman –Guggenheim combinatorial term:

$$\ln \gamma_{i/S}^{SG} = \ln\frac{\phi_i}{x_i} + \frac{z}{2} q \ln\frac{\theta_i}{\phi_i} + l_i - \frac{\phi_i}{x_i}\sum_j x_j l_j \tag{11.33}$$

and $p_i(\sigma)$ is the probability of finding a segment with a surface charge density σ, n_i is the number of surface segment in which the molecule contributes.

The segment activity coefficient is defined by the relation:

$$\ln \Gamma_S(\sigma_m) = -\ln\left(\sum_{\sigma_n} p_S(\sigma_n)\Gamma_S(\sigma_n)\exp\left(\frac{-\Delta W(\sigma_m,\sigma_n)}{RT}\right)\right) \tag{11.34}$$

in which the $\Delta W(\sigma_m, \sigma_n)$ is the exchange energy, namely the energy necessary to obtain one (σ_m, σ_n) pair from a neutral pair for which charge densities σ_m and σ_n are equal zero. This energy depends on hydrogen-bonding (*hb*) interactions, electron acceptor (*acc*) and electron donor (*don*) charge densities. The formulation is given by:

$$\Delta W(\sigma_m, \sigma_n) = \left(\frac{\alpha'}{2}\right)(\sigma_m + \sigma_n)^2 + c_{hb}\max(0, \sigma_{acc} - \sigma_{hb})\min(0, \sigma_{don} + \sigma_{hb}) \tag{11.35}$$

The max and min functions in the formulation (35) indicate the larger or smaller values from the arguments in brackets, respectively. The nonelectrostatic contribution is assumed to be constant and cancel out in the formulation because the nonelectrostatic energy is assumed to be constant.

The computational procedure which is performed during COSMO-SAC calculations can be divided into four steps. The first step is to obtain the charge density σ profile for each component in a mixture using a quantum chemistry package which has COSMO model implemented [79]. For example, using the DFT with Becke-Perdew (BP) version of correlation VWN–BP (Vosko-Wilk-Nusair [80]) for many electron system of the spin-polarized homogeneous electron gas functional at the DNP (double numeric with polarization functions) basis set level, the equilibrium geometry of the molecules in the ideal gas phase is established. After that, the COSMO calculation is performed to obtain the ideal screening charges on the molecular surface for each molecule. The screening charge densities σ^* as the result of the COSMO calculations are averaged to give the "apparent" charge density σ on a standard surface segment using the following expression [64]:

$$\sigma_m = \frac{\sum_n \sigma_n^* \frac{r_n^2 r_{eff}^2}{r_n^2 + r_{eff}^2}\exp\left(-\frac{d_{mn}^2}{r_n^2 + r_{eff}^2}\right)}{\sum_n \frac{r_n^2 r_{eff}^2}{r_n^2 + r_{eff}^2}\exp\left(\frac{-d_{mn}^2}{r_n^2 + r_{eff}^2}\right)} \tag{11.36}$$

In the above formulation, the radius of the standard surface segment r_{eff} is given by:

$$r_{eff} = \sqrt{\frac{\alpha_{eff}}{\pi}} \tag{11.37}$$

while the radius of nth segment reads:

$$r_n = \sqrt{\frac{\alpha_n}{\pi}} \tag{11.38}$$

The variable d_{mn} is the distance between m and n segments. The constant α can be derived based on electrostatics [81, 82]. The charge density profile is calculated from:

$$p_S(\sigma) = \frac{\sum_i x_i n_i p_i(\sigma)}{\sum_i x_i n_i} = \frac{\sum_i x_i A_i p_i(\sigma)}{\sum_i x_i A_i} \tag{11.39}$$

where A_i is the surface area of ith molecule and the probability of finding a segment with a surface charge density σ in pure liquid i^{th} specie is:

$$p_i(\sigma) = \frac{n_i(\sigma)}{n_i} = \frac{A_i(\sigma)}{A_i} \tag{11.40}$$

The second step of the calculation process is the calculation of restoring the free energy of the solute in the mixture $\Delta G^{*res}_{i/S}$ and for i^{th} component liquid phase $\Delta G^{*res}_{i/i}$. By applying iterative procedure on eq. (11.34) the free energies can be obtained by the relation:

$$\frac{\Delta G^{*res}_{i/S}}{RT} = \sum_{\sigma_m}\left(n_i(\sigma_m)\frac{\Delta G^{*res}_{\sigma_m/S}}{RT}\right) = n_i \sum_{\sigma_m} p_i(\sigma_m) \ln \Gamma_S(\sigma_m) \tag{11.41}$$

The third step the molecular volume and surface area acquired from the COSMO computation of each compound are normalized to a standard volume and surface area of a functional group to give the r and q parameters, which are then substituted into the Staverman–tavermanrs, which are then eq. (11.33).

The final fourth step consists of calculating the activity coefficient. The calculated free energies and the Staverman–Guggenheim contribution from the previous two steps are used in the COSMO-SAC model (32) to obtain the value of activity coefficient.

The quantum mechanics is required to estimate only the σ profiles for the molecules in the first step. This calculation must be done only once for each molecule regardless of the mixture in which the component is to be used.

The DFT quantum calculations become constantly more popular in the engineering area despite such methods are not widely incorporated in modern software flowsheeting tools. Comparable, and in some areas better than for traditional VLE algorithms, results show that this tool can become a wide used approach in the design area and in a simulation software as well. The requirement to perform lengthy quantum calculations for new compounds need not be an obstacle because in the case of traditional approaches the new components binary interactions parameters also need to be estimated, typically by the means of experimental measurements.

11.5 Quantum approach to chemical reaction kinetics description

The rectification is often used as a multifunctional process together with chemical reaction [83, 84]. Such process intensification is advantageous especially for equilibrium reactions where removing the products from reacting mixture shift the reaction toward higher conversion. During the design phase, the key part of the modeling is to achieve correct and accurate kinetics description of the reaction. This enables the designer to create project which matches the design criteria set.

11.5.1 Intrinsic reaction coordinate

The intrinsic reaction coordinate is a minimum energy path solving method that connects substrates, the transition state and products of the reaction. Any parametrization of reaction path s is called reaction coordinate that can be given by:

$$\mathbf{x}(s) = \left(x^1(s), \ldots, x^n(s)\right)^T \tag{11.42}$$

The minimum energy path or reaction path is a line in coordinate space, which connects two minima by passing the saddle point, the transition structure of a potential energy surface (Figure 11.3). The energy of the saddle point is assumed to be the highest value which is placed along the reaction path.

There exist several models for tracking the reaction path, e. g. Jasien and Shepard [85], Elber and Karplus [86], Gonzalez and Schlegel [87] models. The IRC can be solved by starting at the transition state and following the steepest descent pathway down to the substrate and product minima according to formulation [88]:

$$\frac{dx(s)}{ds} = -\frac{g(\mathbf{x})}{|g(\mathbf{x})|} \tag{11.43}$$

where s is the arc length along the path, $\mathbf{x}$ is the coordinate vector, and g is the gradient of the potential energy surface (PES). Due to usually very high stiffness (large value of

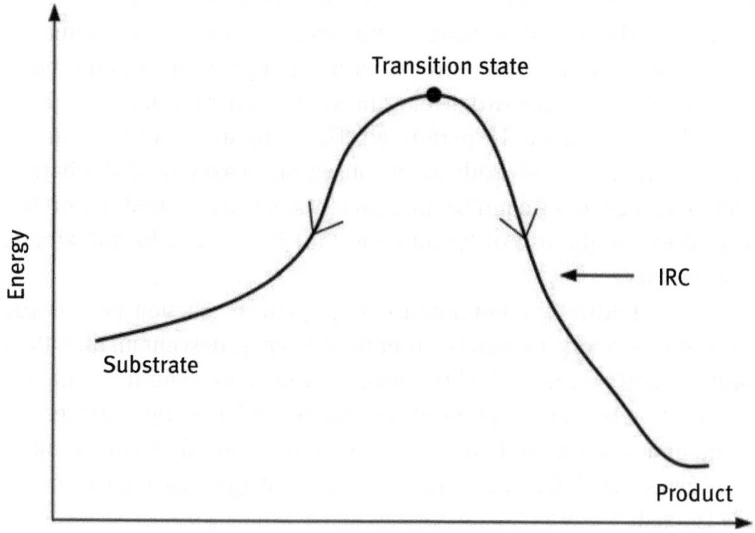

Figure 11.3: Intrinsic reaction coordinate presenting energy states of substrate, transition state and reaction product.

maximal to minimal eigenvalues ratio of the matrix representing the equations) of the above equation some special numerical techniques must be used to solve it.

The kinetic reaction constants which are of most importance can be estimated using Eyirng equation that resembles Arrhenius equation. The Eyring equation reads:

$$k = \frac{k_B T}{h} e^{-\frac{\Delta G^{Act}}{RT}} \tag{11.44}$$

where k_B is the Boltzmann's constant, h is the Planck's constant and ΔG^{Act} is the Gibbs activation energy.

The activation energy is the difference of energies found for different states of the molecular system. In the case of chemical reaction this is the difference between substrate and product energetic states at equilibrium and energy state of transition structure (high energy state). The starting structure for vibrational analysis is the high energy state, when performing frequency analysis the zero point energies of the molecular system are found. The zero point energy corresponds to a state of minimum possible energy. For example the zero point energy for hydrogen atom after interpolation to absolute zero temperature is given by $h\nu/2$. The gradient search is performed both ways to find substrate and product assuming correct transition structure is provided. Several methods are capable to calculate the optimized structures and among them is DFT approach.

The typical approach to IRC is firstly to find the equilibrium and transition states of the reaction system. The transition state in the simplest case can be found by manipulating the equilibrium resulting structures by changing for example bond angles, to obtain a structure of about two to four times higher energy. This is the point where actual IRC calculations are to be performed. The computation of IRC paths is done by two-way direction method. Both ways consist on relaxation of the barrier (transition) structure energy to the equilibrium states. Many software tools report the zero-state energy which by the use of formulation (44) is the base for the kinetic reaction rate description.

Another method of finding the minimum energy path or reaction path is the application of Newton trajectory approach rather than steepest descent method [89]. The starting point is a pathway described by means of geometric definition which is considered as an reaction path. Only properties of the potential energy surfaces are taken into account, and no transient behavior of the molecule is taken into account. This idea can be generalized that any gradient direction $\mathbf{g}(\mathbf{x})$ selected over the potential surface is fixed:

$$\frac{\mathbf{g}(\mathbf{x})}{|\mathbf{g}(\mathbf{x})|} = \mathbf{r} \tag{11.45}$$

where $\mathbf{r}$ is selected unit vector of the direction of the search and the corresponding curve is Newton trajectory. A curve fits the search direction $\mathbf{r}$ when the gradient of potential energy surface stays parallel to the gradient $\mathbf{r}$ at every point along the curve $\mathbf{x}(\mathrm{t})$.

Generally, the IRC can be defined by a variational integral, which in general form reads:

$$I(a,b) = \int_a^b F(x_1(t),\ldots,x_n(t),x'_1(t),\ldots,x'_n(t))dt \tag{11.46}$$

which depends on n continuously differentiable functions $\mathbf{x}(t)$ in an n-dimensional configuration space. The IRC is frequently named as the minimum energy path. Moreover there are other reaction path models. The term minimum energy path is used for the whole category of these pathways.

$$I_{RP} = \int_a^b E(\mathbf{x}(t))l(\mathbf{x}'(t))dt = \min! \tag{11.47}$$

where min! stands for minimization of the functional [90], a and b are the parameters of substrate and product [91]. The function $E(\mathbf{x})$ is n-dimensional potential energy surface. Different approach to reaction path is proposed by introduction the path length L in the denominator as [92]:

$$I_{ORP} = \frac{1}{L(a,b)} \int_a^b E(\mathbf{x}(t))l(\mathbf{x}'(t))dt \tag{11.48}$$

In the formulations (47) and (48) for the non-local variational integral the function $l(\mathbf{x}'(t))$ is given by formula:

$$l(\mathbf{x}'(t)) = \sqrt{\mathbf{x}'(t)^T \mathbf{x}'(t)} = \sqrt{\sum_{k=1}^{n} \mathbf{x}'_k(t)^2} \tag{11.49}$$

Several methods, that are proposed to calculate minimum of certain integral, as a result give energy path which can be different. Such that the results satisfy the relations provided, but the question still exist which path is the true reaction path between substrates and products.

There are literature reports illustrating the usefulness of the IRC method to describe the reaction mechanism and kinetics. The calculations of chemical reaction rate of F^- + CH_3OOH reported by literature [93] are promising and the accuracy of calculated value with comparison with the experimental value is reported to be excellent. Value determined from the simulation is k = (1.70 ± 0.07) × 10^{-9} cm^3/molecules and the experimental k = 1.23 × 10^{-9} cm^3/molecules.). As an example of reaction which could be conducted in distillation column and is analyzed by quantum DFT calculations is the esterification reaction [94]. The methanol and acetic acid (and also its halides) are taken into account and corresponding kinetic properties are calculated. The values are of reasonable accuracy when comparing with available experimental data. The reaction of gas phase decarboxylations of the β-keto carboxylic acids $XCOCH_2COOH$ (X = H, OH, and CH_3) [95] is another example. The optimization of structures was applied at the MP2/6-31G* level of theory. The predicted values energy barrier are reported to be with good agreement with experimental data obtained for various solvents. Another example were authors present good agreement of calculated data with experimentally measured data is a OH radical reaction with hydrofluorocarbon [96]. They used typical quantum approach MP2 level theory together with B2LYP functional using Pople basis set 6-31G* and 6-311++G**.

11.6 Summary

The presented quantum approaches which may utilize the DFT as a quantum tool are intended to predict vapor–liquid equilibria and chemical reactions rates. Both areas are of fundamental importance for reactive distillation simulations. Although the presented approaches are popular in computational chemistry rather than in industrial design in chemical engineering field, they can be useful for a mixture systems which are poorly examined experimentally. The ability of COSMO methods family to predict highly complex, non-ideal behavior of mixtures, including three-phase systems is large advantage. The traditional group contribution method like UNIFAC allows to create user component from predefined functional group. But for the case when no functional group is presented in the UNIFAC database, one cannot consider the component to be non-ideal, which may generate undesirable loss in calculation accuracy. On the other side the COSMO methods family do not rely on any functional groups and needs to use some quantum approach (MP2, DFT etc.) to estimate the charge density surfaces of the molecules of the mixture components.

The design of multifunctional reactors like in the presented case the reactive distillation columns requires not only vapor–liquid equilibria knowledge but also chemical reactions rates. The typical approach relies on experimental results. Such results, depending on the complexity of the reactive system, contain typically the values of activation energy, the frequency factors, etc. To obtain pilot design of reactive distillation system which can be required to estimate the approximate costs of the equipment and the process, there is no need to perform experiments. Such experiments besides the costs may also require quite complex equipment and analytic hardware. For the initial estimation of the design it can be quite useful to rely first on the result from quantum chemical reaction rate estimation, which in fact proved to be quite accurate. Incorporating such a method is a question of future designers' tool development although.

Funding: The work was supported by grant 03/32/DSPB/0707.

References

[1] Gutiérrez JP, Meindersma GW, De Haan AB. COSMO-RS-based ionic-liquid selection for extractive distillation processes. Ind Eng Chem Res 2012;51(35):11518–11529.
[2] Scheffczyk J, Redepenning C, Jens CM, Winter B, Leonhard K, Marquardt W, et al. Massive, automated solvent screening for minimum energy demand in hybrid extraction–distillation using COSMO-RS . Chem Eng Res Des Part B 2016;115:433–442.
[3] Jongmans MT, Schuur B, De Haan AB. Ionic liquid screening for ethylbenzene/styrene separation by extractive distillation . Ind Eng Chem Res 2011;50(18):10800–10810.
[4] Biddulph MW, Kalbassi MA. New column for measurement of multicomponent distillation design efficiencies. Chem Eng Res Des 1990;68(5):453–456.
[5] Lockett MJ. Distillation tray fundamentals. Cambridge, England and New York: Cambridge University Press, 1986.
[6] Klemola KT, Ilme JK. Distillation efficiencies of an industrial-scale i-butane/n-butane fractionator. Ind Eng Chem Res 1996;35(12):4579–4586.
[7] AIChE. Bubble tray design manual: prediction of fractionation efficiency. New York: AIChE, 1958.

[8] Chan H, Fair JR. Prediction of point efficiencies on sieve trays. 1. Binary systems. Ind Eng Chem Proc Des Deu 1984;23:814–819.
[9] Chan H, Fair JR. Prediction of point efficiencies on sieve trays. 2. Multicomponent systems. Ind Eng Chem Proc Des Deu 1984;23:820–827.
[10] Zuiderweg FJ. Sieve trays – A view of the state of the art. Chem Eng Sci 1982;37:1441–1464.
[11] Kister H. Distillation design. New York: McGraw-Hill, 1992
[12] Sundmacher K, Hoffmann U. Development of a new catalytic distillation process for fuel ethers via a detailed nonequilibrium model. Chem Eng Sci 1996;51(10):2359–2368.
[13] Peng J, Edgar TF, Eldridge RB. Dynamic rate-based and equilibrium models for a packed reactive distillation column. Chem Eng Sci 2003;58(12):2671–2680.
[14] Higler AP, Taylor R, Krishna R. Nonequilibrium modelling of reactive distillation: multiple steady states in MTBE synthesis. Chem Eng Sci 1999;54(10):1389–1395.
[15] Baur R, Higler AP, Taylor R, Krishna R. Comparison of equilibrium stage and nonequilibrium stage models for reactive distillation . Chem Eng J 2000;76(1):33–47.
[16] Taylor R, Krishna R. Modelling reactive distillation . Chem Eng Sci 2000;55(22):5183–5229.
[17] Higler A, Taylor R, Krishna R. Modeling of a reactive separation process using a nonequilibrium stage model. Comp Chem Eng 1998;22:S111–S118.
[18] Katariya AM, Kamath RS, Moudgalya KM, Mahajani SM. Non-equilibrium stage modeling and non-linear dynamic effects in the synthesis of TAME by reactive distillation. Comp Chem Eng 2008;32(10):2243–2255.
[19] Prausnitz JM. Molecular thermodynamics of fluid-phase equilibria. Englewood Cliffs, New York: Prentice-Hall, 1969.
[20] Pitzer KS, Simonson JM. Thermodynamics of multicomponent, miscible, ionic systems: theory and equations. J Phys Chem 1986;90(13):3005–3009.
[21] Nickmand Z, Aghamiri SF. A New modified van Laar model for associating mixtures. J Dispersion Sci Technol 2010;31(12):1638–1647.
[22] Hanks RW, Tan RL, Christensen JJ. Limits on the simultaneous correlation of gE and hE data by the NRTL, LEMF and Wilson's equations. Thermochimica Acta 1978;23(1):41–55.
[23] Arlt W, Onken U. Liquid-liquid equilibria of organic compounds: measurement, correlation and prediction. Chem Eng Commun 1982;15(1–4):207–213.
[24] Eng R, Sandler SI. Vapor-liquid equilibria for three aldehyde/hydrocarbon mixtures. J Chem Eng Data 1984;29(2):156–161.
[25] Malanowzki S, Skjold-Jørgensen S, Rasmussen P, Fredenslund A. Simultaneous representation of binary VLE, LLE and HE data using the UNIQUAC model. Chem Eng Sci 1981;36(10):1727–1730.
[26] Nagata I, Tamura K, Yamada T. Correlation of liquid-liquid equilibria in aqueous and organic systems using a modified Wilson model. J Solution Chem 1996;25:567–587.
[27] Yee D, Simonetty J, Tassios D. Prediction of vapor-liquid equilibrium from ternary liquid-liquid equilibrium data. Ind Eng Chem Process Des Dev 1983;22(1):123–129.
[28] Surana RK, Danner RP, De Haan AB, Beckers N. New technique to measure high-pressure and high-temperature polymer-solvent vapor-liquid equilibrium. Fluid Phase Equilib 1997; 139(1):361–370.
[29] Redlich O, Kwong JN. On the thermodynamics of solutions. V. An equation of state fugacities of gaseous solutions. Chem Rev 1949;44(1):233–244.
[30] Soave G. Equilibrium constants from a modified Redlich-Kwong equation of state. Chem Eng Sci 1972;27(6):1197–1203. ISSN 0009-2509.
[31] Peng DY, Robinson DB. A new two-constant equation of state. Ind Eng Chem Fundam 1976;15:59–64.
[32] Zhi Y, Meiren S, Jun S, Lee H. A new quartic equation of state. Fluid Phase Equilib 2001; 187–188:275–298.

[33] Huang SH, Radosz M. Equation of state for small, large, polydisperse, and associating molecules. Ind Eng Chem Res 1990;29:2284–2294.
[34] Chapman WG, Gubbins KE, Jackson G, Radosz M. New reference equation of state for associating liquids. Ind Eng Chem Res 1990;29:1709–1721.
[35] Fu YH, Sandler S. A simplified SAFT equation of state for associating compounds and mixtures. Ind Eng Chem Res 1995;34:1897–1909.
[36] Huang SH, Radosz M. Equation of state for small, large, polydisperse, and associating molecules: extension to fluid mixtures. Ind Eng Chem Res 1991;30:1994–2005.
[37] Mansoori GA, Carnahan NF, Starling KE, Leland TW. Equilibrium thermodynamic properties of the mixture of hard spheres. J Chem Phys 1971;4:1523–1525.
[38] Born M, Oppenheimer R. Zur quantentheorie der molekeln. Annalen der Physik 1927; 389(20):457–484.
[39] Sutcliffe BT, Woolley RG. On the quantum theory of molecules. J Chem Phys 2012;137:22–31.
[40] Sutcliffe BT, Woolley RG. Comment on "On the quantum theory of molecules" [J. Chem. Phys. 137, 22A544 (2012)]. J Chem Phys 2014;140:3–4.
[41] Hohenberg P, Kohn W. Inhomogeneous electron gas. Phys Rev 1964;136 :B864–B870.
[42] Kohn W, Sham LJ. Self-consistent equations including exchange and correlation effects. Phys Rev 1965;140:A1133–A1140.
[43] Perdew JP, Chevary JA, Vosko SH, Jackson KA, Pederson MR, Singh DJ, et al. Atoms, molecules, solids, and surfaces: applications of the generalized gradient approximation for exchange and correlation. Phys Rev B 1992;46:6671–6676. Erratum Phys Rev B 1993, 48, 4978.
[44] Perdew JP, Burke K, Ernzerhof M. Generalized gradient approximation made simple [Phys. Rev. Lett. 77, 3865 (1996)]. Phys Rev Lett 1997;78:1396–1402.
[45] Becke AD. Density-functional exchange-energy approximation with correct asymptotic behavior. Phys Rev A 1988;38:3098–3100.
[46] Śmiga S, Buksztel A, Grabowski I. Chapter 7 – density-dependent exchange–correlation potentials derived from highly accurate ab initio calculations. In: Hoggan P, editor(s). Advances in quantum chemistry. Cambridge, Massachusetts: Academic Press, 2014
[47] Cohen AJ, Mori-Sánchez P, Yang W. Challenges for density functional theory. Chem Rev 2012;112(1):289–320.
[48] Bilc DI, Orlando R, Shaltaf R, Rignanese GM, Íñiguez J, Ghosez P. Hybrid exchange-correlation functional for accurate prediction of the electronic and structural properties of ferroelectric oxides. Phys Rev B 2008;77:165107–165129.
[49] Baer R, Neuhauser D. Density functional theory with correct long-range asymptotic behavior. Phys Rev Lett 2005;94:043002–5.
[50] Sen KD, Luque FJ. First-order correlation-kinetic contribution to Kohn–Sham exchange charge density function in atoms, using quantal density functional theory approach. Int J Quantum Chem 2005;101:231–238.
[51] Becke AD. Density-functional thermochemistry. III. The role of exact exchange. J Chem Phys 1993;98:5648–5652.
[52] Kim K, Jordan KD. Comparison of density functional and MP2 calculations on the water monomer and dimer. J Phys Chem 1994;98(40):10089–10094.
[53] Møller C, Plesset MS. Note on an approximation treatment for many-electron systems. Phys Rev 1934;46:618–622.
[54] Davidson E, Feller D. Basis set selection for molecular calculations. Chem Rev 1986;86(4): 681–696.
[55] Gill PM. Molecular integrals over Gaussian basis functions". Adv Quantum Chem 1994;25: 141–205.

[56] Ditchfield R, Hehre WJ, Pople JA. Self-consistent molecular-orbital methods. IX. An extended Gaussian-type basis for molecular-orbital studies of organic molecules. J Chem Phys 1971; 54(2):724–728.
[57] Manninen P, Vaara J. Systematic Gaussian basis-set limit using completeness-optimized primitive sets. A case for magnetic properties. J Comput Chem 2006;27(4):434–445.
[58] Kresse G, Furthmüller J. Efficiency of ab-initio total energy calculations for metals and semiconductors using a plane-wave basis set. Comput Materials Sci 1996;6(1):15–50.
[59] Zakharov M. Performance of numerical atom-centered basis sets in the ground-state correlated calculations of noncovalent interactions: water and methane dimer cases. Int J Quantum Chem 2013;113(15):1899–1918.
[60] Ema I, García De La Vega JM, Ramírez G, López R, Fernández Rico J, Meissner H, et al. Polarized basis sets of Slater-type orbitals: H to Ne atoms. J Comput Chem 2003;24:859–868.
[61] Klamt A, Eckert F. Prediction of vapor liquid equilibria using COSMOtherm. Fluid Phase Equilib 2004;217(1):53–57.
[62] Klamt A. Conductor-like screening model for real solvents: a new approach to the quantitative calculation of solvation phenomena. J Phys Chem 1995;99(7):2224–2235.
[63] Klamt A, Jonas V, Bürger T, Lohrenz JC. Refinement and Parametrization of COSMO-RS . J Phys Chem A 1998;102(26):5074–5085.
[64] Taylor R, Kooijman HA, Klamt A, Eckert F. Distillation simulation with COSMO–RS. Presented at the Distillation & Absorption Conference, Baden-Baden, Germany, 2002.
[65] Kooijman HA, Taylor R. ChemSep – another software system for the simulation of separation processes. CACHE News 1992;35:1–9.
[66] Klamt A, Krooshof GJ, Taylor R. COSMOSPACE : alternative to conventional activity-coefficient models. AIChE J 2002;48:2332–2349.
[67] Hehre WJ, Lathan WA, Ditchfield R, Newton MD, Pople JA. Gaussian 70. Quantum chemistry program exchange, Program No. 237, 1970.
[68] Ahlrichs R, Bär M, Häser M, Horn H, Kölmel C. Electronic structure calculations on workstation computers: the program system turbomole. Chem Phys Lett 1989;162(3):165–169.
[69] Bochevarov AD, Harder E, Hughes TF, Greenwood JR, Braden DA, Philipp DM, et al. Jaguar: a high-performance quantum chemistry software program with strengths in life and materials sciences. Int J Quantum Chem 2013;113(18):2110–2142.
[70] Staverman A. The entropy of high polymer solutions. Generalization of formulae. Recueil des Travaux Chimiques des Pays-Bas 1950;69(2):163–174.
[71] Guggenheim EA. Mixtures: the theory of the equilibrium properties of some simple classes of mixtures, solutions and alloys. Wotton-under-Edge: Clarendon Press, 1952.
[72] Bosse D, Bart HJ. Viscosity calculations on the basis of Eyring's absolute reaction rate theory and COSMOSPACE . Ind Eng Chem Res 2005;44(22):8428–8435.
[73] ACD/Labs Toronto. ACD/Structure Elucidator, version 15.01. Toronto, ON, Canada: Advanced Chemistry Development, Inc., 2015.
[74] Evans David A. History of the Harvard ChemDraw Project. Angewandte Chemie International Edition. 2014 8 11;53(42):11140–11145. DOI: 10.1002/anie.201405820
[75] TURBOMOLE GmbH. TURBOMOLE: program package for ab initio electronic structure calculations. Leverkusen, Germany: COSMOlogic, 2017.
[76] Putnam R, Taylor R, Klamt A, Eckert F, Schiller M. Prediction of infinite dilution activity coefficients using COSMO-RS . Ind Eng Chem Res 2003;42(15):3635–3641.
[77] Lin ST, Sandler SI. A priori phase equilibrium prediction from a segment contribution solvation model. Ind Eng Chem Res 2002;41(5):899–913.
[78] Lin ST, Sandler SI. Infinite dilution activity coefficients from Ab Initio solvation calculations. AIChE J 1999;45:2606–2618.

[79] Accelrys Software Inc. Cerius2, Dmol3, version 4.0. San Diego, CA: Molecular Simulations Inc., 1999.
[80] Vosko SH, Wilk L, Nusair M. Accurate spin-dependent electron liquid correlation energies for local spin density calculations: a critical analysis. Can J Phys 1980;58(8):1200–1212.
[81] Klamt A, Schüürmann GJ. COSMO: a new approach to dielectric screening in solvents with explicit expressions for the screening energy and its gradient. Chem Soc Perkin Trans 1993;2:799–805.
[82] Klamt A. Conductor-like screening model for real solvents: a new approach to the quantitative calculation of solvation phenomena. J Phys Chem 1995;99:2224–2230.
[83] Mansouri SS, Sales-Cruz M, Huusom JK, Woodley JM, Gani R. Integrated process design and control of reactive distillation processes. IFAC-PapersOnLine 2015;48(8):1120–1125.
[84] Pérez-Cisneros ES, Mena-Espino X, Rodríguez-López V, Sales-Cruz M, Viveros-García T, Lobo-Oehmichen R. An integrated reactive distillation process for biodiesel production. Comp Chem Eng 2016;91:233–246.
[85] Jasien PG, Shepard R. A general polyatomic potential energy surface fitting method. Int J Quantum Chem 1988;34:183–198.
[86] Elber R, Karplus M. A method for determining reaction paths in large molecules: application to myoglobin. Chem Phys Lett 1987;139(5):375–380.
[87] Gonzalez C, Schlegel HB. Reaction path following in mass-weighted internal coordinates. J Phys Chem 1990;94(14):5523–5527.
[88] Hratchian HP, Schlegel HB. Using Hessian updating to increase the efficiency of a Hessian based predictor-corrector reaction path following method. J Chem Theory Comput 2005; 1(1):61–69.
[89] Hirsch M, Quapp W. Reaction pathways and convexity of the potential energy surface: application of newton trajectories. J Math Chem 2004;36(4):307–340.
[90] Zeidler E. Nonlinear functional analysis and its applications, II/B: nonlinear monotone operators. New York: Springer-Verlag, 1990.
[91] Sevick EM, Bell AT, Theodorou DN. A chain of states method for investigating infrequent event processes occurring in multistate, multidimensional systems. J Chem Phys 1993;98(4): 3196–3212.
[92] Pratt LR. A statistical method for identifying transition states in high dimensional problems. J Chem Phys 1986;85(9):5045–5048.
[93] López JG, Vayner G, Lourderaj U, Addepalli SV, Kato S, DeJong WA, et al. A direct dynamics trajectory study of $F^- + CH_3OOH$ reactive collisions reveals a major non-IRC reaction path. J Am Chem Soc 2007;129(32):9976–9985.
[94] Lawal MM, Govender T, Maguire GE, Honarparvar B, Kruger HG. Mechanistic investigation of the uncatalyzed esterification reaction of acetic acid and acid halides with methanol: a DFT study. J Mol Model 2016;22:235–246.
[95] Huang CL, Wu CC, Lien MH. Ab initio studies of decarboxylations of the β-Keto Carboxylic Acids $XCOCH_2COOH$ (X = H, OH, and CH_3). J Phys Chem A 1997;101(42):7867–7873.
[96] Ali MA, Rajakumar B. Kinetics of OH radical reaction with CF_3CHFCH_2F (HFC-245eb) between 200 and 400 K: G3MP2, G3B3 and transition state theory calculations. J Mol Struct 2010; 949(1–3):73–81.

Index

https://doi.org/10.1515/9783110482065-012

Bei Fragen zur Produktsicherheit wenden Sie sich bitte an:
If you have any questions regarding product safety,
please contact:

Walter de Gruyter GmbH
Genthiner Straße 13
10785 Berlin
productsafety@degruyterbrill.com